Advocacy after Bhopal

2 pg on
the question of
fortuns vision of ethics
resonating with me
Do I subscribe to her
vision.

when he is naked
YEAH

we just drew
a younger
stu

Advocacy after Bhopal

Environmentalism, Disaster, New Global Orders

KIM FORTUN

THE UNIVERSITY OF CHICAGO PRESS
CHICAGO AND LONDON

KIM FORTUN is associate professor in the Science and Technology Studies Department at Rensselaer Polytechnic Institute.

The University of Chicago Press, Chicago 60637
The University of Chicago Press, Ltd., London
© 2001 by The University of Chicago
All rights reserved. Published 2001
Printed in the United States of America

10 09 08 07 06 05 04 03 02 01   1 2 3 4 5

ISBN: 0-226-25719-3 (cloth)
ISBN: 0-226-25720-7 (paper)

Library of Congress Cataloging-in-Publication Data
Fortun, Kim.
    Advocacy after Bhopal : environmentalism, disaster, new global orders / Kim Fortun.
        p.   cm.
    Includes bibliographical references (p.   ) and index.
    ISBN 0-226-25719-3 (cloth : alk. paper)—ISBN 0-226-25720-7 (pbk.: alk. paper)
    1. Disaster victims—Services for.   2. Environmental policy—Citizen participation.
3. Social responsibility of business—Environmental aspects.   4. Bhopal Union Carbide Plant
Disaster, Bhopal, India, 1984.   5. Disaster relief—India—Case studies.   I. Title.

HV553 .F65   2001
363.7′058′0954—dc21
                                                                                          00-046716

CONTENTS

FIGURES

Acknowledgments

The Bhopal disaster has brought many different people together, for many different kinds of collaborative work. This book is one example. I have received so much help writing it and putting it together that it is impossible to express my appreciation. A few people can nonetheless be named, and offered special thanks. The Bhopal Gas Affected Working Women's Union must be thanked for all their teaching and good care while I was in Bhopal—Rehana Begum, Hamida Bi, and Mohini and Jabbar Khan especially. T. R. Chouhan also deserves special thanks for teaching me the technical details, as well as for his friendship. Chouhan's family also must be thanked, as must the Bhopal Group for Information and Action—Rajiv Lochan, Surajit Sarkar and Vani Subramaniam, T. J. Birdi, and Sathyu Sarangi especially. Ward Morehouse is also a special figure. He has been a great example, while helping me make connections between India and the United States. It was through Ward that I was introduced to the people in Communities Concerned About Corporations. Nan Hardin, Diane Wilson, and Karen Lynne offered particular inspiration and help, as did members of the Oil, Chemical and Atomic Workers Union in Texas City, Texas. In West Virginia, Steve Midkiff, Pam Nixon, and Wendy Radcliffe deserve special thanks.

The help I have received in various university settings has also been invaluable. My Ph.D. advisor, Michael Fischer, must be thanked for his patience and his fine-grained readings (of more versions of this book than can be counted), and, most important, for helping me believe in the importance of critical scholarship and in the experimentation needed to link scholarship to a changing world. George Marcus has also been an important influence and an invaluable source of moral support. George's commitment to the work of anthropology, as well as to reinventing anthropology, has been a great motivation. Sharon Traweek, Steve Tyler, and Julie Taylor were also important influences in the Anthropology Department at Rice. Sharon taught me how feminist critiques could be threaded through my scholarship, as well as through everyday life. Steve Tyler gave me my first glimpse into the powerful workings of language. Julie Taylor introduced me to the ways our constructions of history shape what we make of the world. My friends in the graduate program at Rice were also great teachers—Melissa Cefkin, Jeff Petry, Laurel George, and Bruce Grant especially. Bruce read the first complete draft of this book—an even more unwieldy version than the one here—so he gets special accolades for his patience, advice, and particularly sharp eye for purple prose. Graduate students at Rice in 1991 also read the first draft of the book and shared their commentary with me. This was a special privilege. Jae Chung, Timothy Dylan Wood, Erica Schreiber, and Lisa Breglia must be thanked.

People in the Science and Technology Studies Department at Rensselaer, where I have worked since completing my degree, have also been a great help. Art Fricke and David Levinger read the first draft of the manuscript and gave me great comments. Todd Cherkasky co-taught many courses with me as I was writing, read many things as I got them completed, and helped me juggle teaching, writing, and the many other things I was trying to do. The juggling act was also greatly supported by Marge McLeod and Kathie Vumbacco. Belief that it was all worth it was sustained by the many graduate students I have had the privilege to work with.

Dan Price, Joe Dumit, and Kaushik Sunderrajan also read the first draft of the book and gave me detailed comments. It's a different book because of these comments. All the people who read and commented on the first draft of the book—including the reviewers for the University of Chicago Press—really do deserve special thanks. It was invaluable to have specific comments to direct my rewriting and reorganization.

David Brent, my editor at University of Chicago Press, has also been great to work with and deserves special thanks not only for helping with this book, but also for encouraging new kinds of ethnographic writing more generally. Nadine Kavanaugh, Jenni Fry, and Sherry Goldbecker have helped me through the publication process and are also greatly appreciated.

My parents, Joe and Jeane Laughlin, and my two sisters, Misty and Merida, also deserve special thanks. They've had to live with me—despite my distraction and anxiety—as well as without me, during the many years I was in India, as well as over the many years when I kept writing, rather than returning phone calls. My appreciation for my family is truly beyond words. As is my appreciation for Mike Fortun—husband, editor, writer of paragraphs I could not write. Mike does have to share responsibility for this book; I'm sure he'll also share my relief that it's finally leaving the house and going out in the world on its own. So that we can continue other work together.

PROLOGUE: THE TIMES

1984 Global agricultural output slows, reversing steady gains in yields sustained since 1950. The drop from a 3 to a 1 percent annual increase in food production does not keep up with population growth, leading to significant decreases in per capita availability, reliance on surplus stocks, and steady rises in food costs. Declining output is attributed to the ecological side effects of Green Revolution techniques, initiated after World War II by public-sector research centers and then marketed by giant chemical companies like Monsanto, DuPont, and Union Carbide. These corporations respond to the shrinking markets of the 1980s through intensified investment in biotechnology, hoping to "collaborate" with environmental problems by de-

In the early morning hours of December 3, 1984, Union Carbide's pesticide plant in Bhopal, India, released over forty tons of toxic gas into the atmosphere. The number killed in the immediate aftermath remains debated, ranging from two to fifteen thousand. Reliable estimates—based on reports of missing persons, the numbers of shrouds sold, and the amount of cremation wood consumed—indicate that 10,000 people died within the first few days. Since then, the Bhopal disaster has been widely recognized as "the world's worst industrial disaster." Some argue that Bhopal has provided an "innovative model of law" for dealing with global distributions of technological risk. Others say that Bhopal is what globalization looks like on the ground.

Union Carbide, a multinational chemical company headquartered in

veloping crops that prosper in regions that had become infertile.

1984   Union Carbide's plant in Bhopal manufactures only one-quarter of its licensed capacity for 5,250 tons per year of methyl isocyanate–based pesticides. Huge stocks are stored in tanks designed in anticipation of direct sales in excess of the plant's own formulation requirements for the production of the pesticide Sevin. Famine in India during the 1980s had indebted farmers and squashed expected growth in demand. By 1984, local managers were directed to close the plant and prepare it for sale. When no buyer was evident in India, plans were made to dismantle key units for shipment to another Third World country.

1984   Bruce Springsteen sings "Born in the USA" to fans around the world. Ronald Reagan sweeps his election, just before headlines announce that America the Beautiful is in hock, with each citizen owing part of an escalating national debt. Junk bonds fuel "merger mania," and the possibility of hostile takeover haunts every CEO's dreams. Federal researchers announce that cholesterol is a cause of coronary heart disease, prompting food processors to seek chemical substitutes for

the United States, was manufacturing pesticides intended to bolster agricultural production as part of the Green Revolution. In 1984, Bhopal was a city in the heartland of India with a population of approximately 900,000. One reason Union Carbide built a plant in Bhopal was because the region was considered "backward" and was thus targeted for industrial development by the Indian government. But another reason was because Bhopal was already well connected by road and rail to all major ports connecting India to the rest of the world. Bhopal was supposed to be easy to get out of. On the night of the disaster, it was not. The city had become a node in a complex global system, but the transactions were skewed. The world came into Bhopal, but residents had no way out.

Most of those killed were poor. Many lived in slum colonies adjacent to the Carbide plant, in flimsy houses that offered little protection from the weather—or from airborne toxics. The railway station and surrounding areas were also hard hit. It is known that H. S. Bhurve, the station superintendent, died as he worked to divert incoming trains. Meanwhile, people working, waiting, and sleeping on the platform were asphyxiated. The courtyards of Bhopal's hospitals became lined with bodies. Rumors circulated about the dead being dumped in surrounding forests or many meat and dairy products. Meanwhile President Reagan, in the words of Sierra Club Conservation Director Carl Pope, "reinvented the environmental movement by his contempt for it."

1984   George Orwell's reference point for a time when neither love nor possession is permissible and advocacy is against the law. Winston Smith, Orwell's antihero, becomes an icon on all sides of the political spectrum. Apple borrows the image to announce the release of its Macintosh personal computer, fueling speculation on possibilities for mass democratization and the role of "knowledge workers" and "symbolic analysts" in a postcapitalist economy.

1984   Prime Minister Indira Gandhi sends troops into the Golden Temple, leading to her assassination by Sikh bodyguards. Rioting breaks out throughout India. The police stand by, as do most political parties. The voluntary sector emerges as the most reliable avenue of response, particularly in Delhi.

1984   Rajiv Gandhi comes to power, securing the legacy of India's Congress Party by pass-

in the nearby Narmada River. The foundations of systemic distrust had already been laid.

Within a few weeks, a new world began to emerge in Bhopal, referred to by many as "the second disaster." Many victims remained in hospitals, but many more lined up to receive free rations of wheat, sugar, and oil, distributed within a government program that would evolve into the official, long-term rehabilitation scheme. American lawyers had descended on Bhopal "like vultures," tracking down corpses and force fitting their meaning onto retainer agreements, often signed with a thumbprint and taken away by the trunk load. Doctors at Bhopal's hospitals claimed that the worst was over, but acknowledged that they knew nothing about possible long-term effects.

Over 600,000 people received free rations during the first few months of 1985, despite arguments that there were only 250,000 people living in areas affected by the gas and that an insidious precedent was being set by such arbitrary numeration. Work sheds were established to provide some women with jobs; no rehabilitation centers were created for men, as bureaucrats argued that any income they generated would be frittered away on drink— laying the ground for a reorganization of gender roles in Bhopal, with contradictory effects. Victims began to be registered as official

ing the first legislation to liberal-
ize India's economy, breaking
with planning orientations that
had shaped Indian development
since independence.

1984   The renowned Urdu poet
Faiz dies.

1984   Indian public interest
attorney M. C. Mehta visits the
Taj Mahal for the first time. The
monument's marble is yellow
and pitted by pollutants from
nearby industries. Mehta files
his first environmental case in
the Supreme Court of India,
launching an environmentalism
framed by a key question about
industrial facilities: temples or
tombs? Later Mehta initiates litigation to clean up the Ganges
River after it caught fire due
to the amount of industrial ef-
fluents in the water. In 1996,
Mehta won the Goldman Environmental Prize, credited with
over forty landmark judgments
and with bringing environmental
protection into India's constitutional framework. Lois Gibbs
received the Goldman Prize
in 1990. In 1992, the Goldman
Prize was awarded to Mehta
Patkar, a lead activist in the Indian movement to halt a World
Bank–funded hydroelectric
project on the Narmada River.

1984   The Resource Conservation and Recovery Act (RCRA)

claimants for final compensation,
sparking intense controversy over
diagnostic taxonomies and geographical indicators of exposure.

With time, as paper proof of victimization became necessary, identification of deaths in Bhopal as
officially "gas related" became increasingly contested. Middlemen
emerged to help secure this paper
proof, bringing further corruption to
Bhopal and complicating the role of
voluntary-sector political activists,
who also occupied a social space
somewhere between the powerful
and the powerless—at crossroads
unmarked by signs indicating exactly where expertise was needed
and how it should be deployed.

The desperate need for paper
proof of victimization has become
part of the legacy of Bhopal, as
has a need for new idioms through
which disaster can be represented,
both textually and socially. What
counts as expertise has been complicated. Scientific inputs have been
crucial to direct diagnostic and
therapeutic agendas, but also to
validate legal arguments. Science
has also been unsettled—by the
rigorously nonlinear, unpredictable,
cumulative effects of toxic chemicals. The science needed to underwrite fair distributions of compensation in Bhopal is a different
science than that which can remedy
the suffering of individuals. Meanwhile universalist claims remain important to counter suggestions that
life in Bhopal is worth less than

is amended, emphasizing human health as a goal of U.S. environmental policy. At the same time, human health is made subject to cost-benefit analyses. The new RCRA also grants the Environmental Protection Agency (EPA) authority to certify sites appropriate for treatment, storage, and disposal facilities, positioning the agency in opposition to local communities—who initiate a phase of environmentalism shaped by a key slogan: NIMBY: Not In My Backyard. The infamous Cerrel Report— which encourages companies to site disposal facilities in rural areas without much education— becomes a key reference for the environmental justice movement. Opposition to the siting of treatment and disposal facilities within the United States leads to increases in the export of hazardous waste.

1984   Rakesh Sharma becomes India's first man in space.

1984   India's national broadcasting company, Doordarshan, commissions "Hum Long" (We People), inspired by Mexican "prodevelopment soaps." Only a few years before, for the occasion of the Asian Games in 1982, the Indian government had initiated the expansion of television beyond a few metropolitan areas and introduced

that in Connecticut or that double standards are legitimate grounds on which to build a new world order.

Bhopal's second disaster has emerged from a tangle of crosscutting complex systems, where law, science, and economics collide. It is a disaster that has persisted, and operated cumulatively, drawing in a spectrum of issues that can't be contained by old blueprints for social change. The "people" cannot represent themselves in this rehabilitation effort. Nor can the state stand in as guardian. Technoscience must be condemned, while it is relied on. Legality must be pursued, while acknowledged as an insufficient remedy. Corporate negligence must be disclosed, even while recognizing that the knowledge on which charges of negligence hinge has become increasingly difficult to locate. Knowledge in and about the contemporary global system has become increasingly distributed. Charging the CEO of Union Carbide with homicide has been one attempt to draw accountability out of a complex system. Other ways to justice implode the law's usual demand for clear designations of human agency, of cause and effect.

It was within the implosions of Bhopal's second disaster that I worked as an advocate for gas victims and became an ethnographer haunted by the uneven distributions of risk and reward that characterize contemporary global order. The timing of my work was significant. I color broadcasting. "Hum Long" was supposed to teach family planning in a series of 156 episodes about the life of a lower-middle-class urban family. The first few shows were mercilessly didactic; no one watched. Show fourteen brought the shift: "Hum Long" became a breathless soap opera shorn of a schoolmarm's edge. The show became a great success. By 1985, Doordarshan was permitting private sponsorship of programs, leading to Ramanand Sagar's renowned reproduction of the Hindu epic *Ramayana.* Sixty million watched, many of whom burned incense or took purificatory baths before gathering for what became a Sunday morning ritual throughout India. Doordarshan earned $20 million in advertising revenues. Doordarshan's monopoly remained relatively unthreatened until the Gulf War, when CNN began transmitting into India.

1984   R. E. Freeman's *Strategic Management: A Stakeholder Model* is published. A stakeholder model of management is supposed to harmonize the interests of owner and nonowner groups, including employees, community members, lenders, suppliers, and customers. Freeman argues that the stakeholder model requires "a radical rethinking of our model of the

arrived in Bhopal in early 1990, at high noon on the so-called second disaster, a year after the out-of-court settlement of the Bhopal case by order of the Indian Supreme Court. The settlement provoked mass protest rallies. Critics argued that the settlement confirmed that life in the Third World is cheap and without the protection of law. In later proceedings contesting the constitutionality of the role played in the litigation by the government of India, which had represented gas victims in the capacity of *parens patriae,* the Supreme Court did acknowledge that the case had departed from principles of "natural justice" by failing to give victims the right to "opt out." The settlement of the Bhopal case was nonetheless upheld, in October 1991— a few months after the rupee was devalued 30 percent and made partially convertible, ushering India even more fully into the new world of globalization (Dehejia 1993, 92).

Meanwhile a grassroots environmental movement was emerging in India to challenge what would count as national development. Bhopal became a symbol. Of the side effects of the Green Revolution. Of a corrupt state. Of the ways science can be used to legitimate uneven distributions of risk and reward. Of the elitism of environmental politics concerned more about tigers than toxics. Of globalization and multinational corporations out of control. By 1990, when I arrived in Bhopal, firm" and revision of a pivotal assumption, codified by law: that managers have a fiduciary responsibility to serve a single organizational purpose—profit maximization.

1984 India amends the Foreign Contribution Regulation Act of 1976 to mandate a "Council on Voluntary Agencies" to formulate and implement a code of conduct for nongovernmental organizations. Drafts of India's Seventh Five Year Plan begin to circulate, "elevating" voluntary agencies to the role of "key actors" in the implementation of antipoverty programs. Most interpret these moves as an attempt to both legitimate and supervise the voluntary sector. Some argue that these changes could provide important mechanisms of accountability. Others warn of the dangers of co-optation by the state.

1984 Lester Brown's first annual *State of the World* is published. The concept of sustainable development also receives attention at the October meeting of the World Commission on Environment and Development. The United Nations Environment Program is also on board, emphasizing "nontraditional economic criteria" in evaluating cost effectiveness, community self-reliance, and the need for

progressive outrage over the Bhopal case was directed as much at the government of India as at Union Carbide. The "selling out of justice" was blamed on a system that linked the two together, with increasing force as pressure from the International Monetary Fund escalated into initiatives for overall liberalization of India's economy.

In the United States, an already vibrant grassroots movement focused on toxic wastes was being extended to address hazardous chemical production—provoked, in part, by a series of plant disasters and what people refer to as "almost Bhopals." The repercussions of free trade agreements were felt there as well. The possibility of relocating industrial facilities abroad provided leverage in corporate negotiations with labor unions and local communities. Increasing reliance on untrained contract labor in chemical plants was an important issue, as was the unreliability of the U.S. EPA. Reagan's appointees had been ousted, and the U.S. Congress had committed to rebuilding the EPA as a viable regulatory authority. Citizens and workers in high-risk communities weren't convinced, allying instead with environmental justice activists working to demonstrate how distributions of environmental risk synergize with the risks of race and class.

But the Berlin Wall had fallen, setting the stage for international trade talks that promised to harmo-

"people-centered initiatives."
Sustainable development becomes associated with reconcili-
ation: between the development
community and conservationists, between economic and environmental priorities, between
the First and Third Worlds.

1984 "Ecological modernization" is promoted at the
Organization for Economic Cooperation and Development's
Conference on Environment and
Economics, to encourage a restructuring of capitalist political
economy without the need for
complete transformation.

1984 *In Search of Answers:
Indian Women's Voices from
Manushi* is published.

1984 Famine ravages Ethiopia.
The concert Live Aid, organized
in response, becomes a model
for a series of "socially responsible" benefit concerts. Rock
stars become social activists, advocating live via satellite. "We
Are the World" becomes the
new anthem.

1984 Union Carbide ends
forty years of operations at federal facilities in Oak Ridge, Tennessee, established as part of the
Manhattan Project to separate
bomb-grade uranium.

---

nize global order. Lenin and Marx
were turned into "little more than
icons" of obsolete political ideas.[1]
New communications technologies
were linking dispersed movements
for democratization. The Gulf War
was fought.[2] And multilateral institutions like the World Bank and the
United Nations were promising to
pluralize participation in decision-
making. Plans for the Earth Summit to be held in Rio de Janeiro in
1992 suggested that environmental
politics would provide a platform
for major political realignments—
between countries, and between
governments, corporations, and increasingly powerful nongovernmental organizations.

It was a time rife with contradiction. The demise of an overtly
bipolar world order complicated
straightforward implementation of
honored political programs. So did
shifting class configurations and the
creation of subject positions on the
boundaries between different interests. Struggles to articulate transnational allegiances intensified, as
did hostility to the contamination
caused by transnational flows of
people, products, and ideas. As India debated the virtues of access to
Hollywood films, Americans turned
against new immigrants with increasing vengeance. As religious
intolerance escalated in India, riots
rocked Los Angeles.[3] But many still
described the emerging world order in terms of harmonization. Dip-

1984 Union Carbide India Limited begins celebration of its "Fiftieth Anniversary Jubillee."

1984 December 3. A press conference is scheduled at Union Carbide headquarters in the United States. Journalists expect to hear about how the company has been restructured to meet the demands of a globalizing economy. Instead, Carbide must announce the Bhopal disaster.

lomats involved in the Uruguay Round of the General Agreement on Tariffs and Trade were particularly eloquent, as were many environmentalists. The signing of the Montreal Protocol in 1987 was an important point of reference, as was the 1989 ban on ivory products, instituted by the Convention on Trade in Endangered Species after declaring the African elephant an endangered species.[4] For some, the settlement of the Bhopal case was yet another sign that global cooperation was indeed possible. For others, the settlement of the Bhopal case signaled denial of the past, while portending how power would operate in the future.

People around the world were asking difficult questions. Was it the best of times or the worst of times? Should the forces catalyzing greater global interconnection be supported or countered? How could the scope and complexity of emerging world orders be represented? What could an individual know? What could an individual do?

Could environmentalism provide idioms and visions for collective action?

What was the legacy of Bhopal?

INTRODUCTION

Advocacy, Ethnography, and Complex Systems

The Fieldwork

The timing of my work in Bhopal was out of joint in more ways than one. I arrived in Bhopal in February 1990, one year after the out-of-court settlement of the Bhopal case by the Indian Supreme Court. I did not plan to stay. I had come to India to do anthropological research on environmental politics in Madras (now Chennai). I traveled to Bhopal to collect material illustrative of the background from which concern about chemical pollution had emerged. Immediately, it was clear that Bhopal could not be conceived as a "case study," a bounded unit of analysis easily organized for comparative ends. To the contrary, Bhopal showed no evidence of boundaries of time, space, or concept. Bhopal, as I encountered it, was a disaster that entangled the local and the global, the historic and the future, continuity and dramatic change. Only later would I begin to understand the deeply normative implications of how Bhopal is encased—in writing by management experts in particular. In 1990, newly arrived in Bhopal, I knew only that I was, indeed, at the scene of disaster, where injustice was complicated by grossly inadequate modes of conception and description, where everyday life screamed for rectitude, without prescriptions for anything more than symptomatic relief.

Far from seeming a stable point of reference, Bhopal seemed more like a whirlwind—a maelstrom produced by opposing currents, sucking everything into an upward spiral—with gas victims at storm center. There was the immediate and obvious need to disrupt official claims that the Bhopal case was settled by promise of cash compensation to the few able to prove their status as victims. So I stayed in Bhopal, where my Tamil language skills were largely useless, knowing only that there was a call to respond to blatant injustice. English language writing skills were the only resource I had to share, and even these were not primed for the task at hand. My task as advocate was to translate disaster into the languages of law, science, and bureaucracy. And I was far from fluent in these languages. I had not prepared for working in Hindi. Nor had I prepared for working in the languages of press releases and affidavits, especially within the time frame of politics.

I arrived in Bhopal "after the fact," after the "full and final" settlement of the Bhopal case. Nonetheless, there was immediate work to be done, writing responses to the process and terms of the out-of-court settlement in press releases, political pamphlets, and legal petitions. I began this work through affiliation with the Bhopal Group for Information and Action (BGIA), a small group of political activists who served as interlocutors and translators for the Bhopal Gas Affected Working Women's Union, the largest organization of victims and the only one with sustained participation in the legal proceedings.

While I was in Bhopal, the size of BGIA ranged from two to seven people. I was the only one who wasn't Indian and, for much of the time, was the only woman. Other members of the group came from different regions of India and from different ideological backgrounds, but all were college-educated and English-speaking, though some insisted on "thinking in Hindi" as a way of refusal and critique. During my last year in India, we were joined by a working-class British-Indian woman, trained as a lawyer. With time, she developed an admirable role "between" the rigid social classifications with which the group continually struggled. Neither foreigner nor native, neither elite nor subaltern, she became a model for translating between social and cultural differences without denying their force.

In the main, BGIA structured its relations with gas victims through involvement with the Bhopal Gas Affected Working Women's Union, despite continual controversy over how the collaboration should be configured. The Women's Union was formed in 1986 in response to urgent needs for medical and economic relief. By the early 1990s, the Women's Union had a membership of approximately twenty thousand gas victims. Leadership was provided by Abdul Jabbar Khan and a steering committee of twenty women. Union funds were raised through monthly dues of Rs. 5 (15¢), which supported Women's Union commitments to avoid dependence on outsiders, but not much else.

Acute shortages of resources plagued all Union efforts. One way this shortage was offset was through persistent collaboration. The Women's Union built itself into many different social networks, linking it to other grassroots organizations, to national organizations, and to international organizations. There were links with tribal activists in the mining districts of Chattisgarh, with villagers resisting the Narmada hydroelectric project, and with villagers organized to challenge conservation officials, hired to protect forests on their ancestral lands. There were links to various middle-class groups, including the Medico Friends Circle, a coalition of physicians and public health professionals, and the Bhopal Gas Victims' Support Group, a Delhi-based coalition of lawyers, journalists, and others who have provided support throughout the legal proceedings of the Bhopal case. There also were links to the International Coalition for Justice in Bhopal, organized out of New York to connect environmental and consumer groups from different countries. In all, the Women's Union was "the grassroots," but networked nationally and globally. A product of the Bhopal disaster, the Women's Union reiterated its intricate morphology.

With time, the social networks in which BGIA and the Women's Union circulated became my own. My "informants" came to include political activists spread across India, working within an emerging grassroots environmental movement. Some of these activists were based in metropolitan areas, at documentation centers, or at other sites where their expertise could be used as critique; other activists lived in rural locales, working to organize community participation and economic development. Some had been drawn into activism because of commitment to issues of environmental justice; most had been involved in social justice campaigns for a long time and were just beginning to shape the issues they engaged in accord with environmentalist vocabulary.

I met these activists at their home sites, as well as at meetings organized to bring activists together to build an environmental justice movement with national scope. Many activists also came to Bhopal, en route to work elsewhere, to participate in public demonstrations addressing the needs of gas victims or to help with health surveys or other short-term projects. Each year's anniversary rally was a particularly important opportunity for building collaborative relationships. Through work and conversation with these activists, I learned about India, about how globalization was working, and about how environmentalism could be understood as a political strategy for bringing together people, issues, and ideologies that once seemed unrelated—without reducing diversity to benign pluralism. The eloquence of these activists regarding the many ways violence operates within and across national boundaries was extraordinary. It is through them that I learned to understand Bhopal in ways that drew out systemic critiques, without forgetting the harsh particularity of every day in the queues of ration shops, dispensaries, and claims courts.

The activists with whom I engaged most directly were those who occupied social roles similar to those within BGIA. In part, this was due to language, on at least two registers. Most of these activists spoke English and often admitted being most comfortable thinking and talking politics in English because they had learned to think and talk politics amidst the English language dynamics of university life. English also operated as a default language, to accommodate activists from South India who either didn't know Hindi or resented it at least as much. The politics of language choice were always contested, signaling the contradictions of a postcolonial context still operating by logics of divide and rule. My own position was predictably ironic. Acknowledging the way language inflected politics, I should have taken the time to work in Hindi. But there was always a press release or affidavit needing to be written, in English.

Throughout my time in India, advocacy directed my practice and outlook. When I asked about the successes of literacy programs in Kerala, I wanted to understand how we could structure literacy campaigns among gas victims. When I visited participatory health projects, I worked to imagine viable alternatives to the government hospitals of Bhopal. When I learned about the network of groups opposing nuclear energy, I worried about the isolation that so often undermines local-level initiative. Throughout, I wrote—a field journal, but also political statements of many kinds. Letters to newspapers, or to the prime minister. Reports on the health status of victims, based on surveys BGIA organized. Petitions to the courts and proposals for rehabilitation initiatives for the Ministry of Gas Relief. Pamphlets for activists at other sites wanting to participate in remembrance of Bhopal.[1]

The political statements I helped produce in Bhopal were directly responsive to particular turns of events and to the times in which they were produced. Most often we worked parallel to developments in the legal proceedings aimed at overturning the settlement of the Bhopal case. Local turns of events also dictated our focus, with heightened urgency after the BJP, a Hindu fundamentalist political party, came to power in Madhya Pradesh in November 1989. Meanwhile, communal tensions were brewing throughout India, Rajiv Gandhi was assassinated, and the stability of the central government wavered, though not enough to interrupt negotiations over the General Agreement on Tariffs and Trade—which, for many environmental justice activists, represented the demise of official commitments to social as well as economic democracy.

My fieldwork in India ended in June 1992, during a particularly desperate period in Bhopal, when inaction had settled over the city with a dreadful force. The out-of-court settlement had been upheld by the Indian Supreme Court in October 1991; gas victims were concerned that interim relief payments would be discontinued any day; faith that the claims courts to be established in Bhopal would ever deliver justice was minimal. The work of BGIA and the Women's Union lost the focus of court proceedings. Local needs for health and economic rehabilitation persisted, but demanded intensely collaborative approaches—of which we were no longer capable. Acrimony between the Women's Union and BGIA escalated. BGIA split apart. Frustration corroded everything.

So I left Bhopal, extraordinarily aware that the disaster continued. The challenge was to find a way to keep Bhopal proximate, in my line of sight, centering my understanding of how the world works. More specifically, I needed to respond to Union Carbide's claim that "it [Bhopal] can't happen here"—in chemical communities spread across Texas, Louisiana, West Virginia, and elsewhere in the United States.

This stage of my research began in 1993, through work with community groups, labor unions, environmentalists, government regulators, and corporations. The networks I moved through extended directly from the networks I worked within in India. The most direct connection was the Council on International and Public Affairs, which runs the International Coalition for Justice in Bhopal and helped jump-start a national alliance of grassroots groups called Communities Concerned about Corporations (CCC). My association with CCC allowed me to participate in the collaborative efforts of labor unions, residents, and environmentalists responding to political processes that heightened the threat of job flight, plant shutdowns, and regulatory failure in the United States. As part of a specific commitment to acknowledge the legacy of the Bhopal disaster, a primary focus of this alliance was on the sites and issues affected by Union Carbide. CCC also played a lead role in keeping the Bhopal disaster in public memory, waging a campaign to displace Earth Day with "Corporate Clean-Up Day"—on December 3.

My work within environmental justice activism in the United States was not as concentrated as in India. I participated in many organizational meetings and in brief campaigns, but didn't directly negotiate with the Environmental Protection Agency (EPA), Monsanto, or doctors ill-prepared for my concerns. To learn how these negotiations worked, I visited many activists in their home communities, relying much more on directed interviews than was necessary in Bhopal. Easy access to telephones was also important.

My goal was to avoid isolating Bhopal in space or time by continually seeking new connections—connections that drew out the complex systems that continue to bind me to Bhopal. There have been many hooks, and learning to prioritize them has been a formidable challenge. I haven't been able to keep up with all the articulations, much less order their significance. Pamphlets produced at the grassroots. Articles from legal journals. Annual reports. Updates on developments in toxicology. Case studies in management textbooks. All of these came to seem significant. As did articles on environmental issues in *Forbes* magazine and in the seemingly endless number of newsletters produced by environmentalists. I learned about Geographical Information Systems and how they could be used by labor unions to map health effects in and around industrial facilities. I read about how sociologists have studied corporations, mapping the social networks of directors and the structure of firm decision-making. I waded through Toxic Release Inventories. And I've continually returned to the binder of political statements I helped produce in Bhopal, recalling themes and events of disaster with everyday detail.

Oscillation between different sources of data became an important research strategy. It seemed to be a way to keep Bhopal from being ghettoized and forgotten. Constant movement between different orderings of Bhopal has, however, been unsettling. The confidence said to come with knowing your material well has been forever forestalled. Instead of cohering with time, my expertise has been increasingly dispersed. Working within such dispersion has often been frustrating and has always been overwhelming. But it has been a way to learn about how things come to count as significant or not—through the tactics of participation-observation. In process, simple critiques based on what dominant interpretations of the Bhopal case exclude have lost their force. The question is not *whether* interpretations exclude, but how and to what effect.

The Analysis

How has the Bhopal disaster been produced, suffered, and remembered? What has been the shape and tone of advocacy in response to the disaster? How have the dynamics of late-twentieth-century world order affected what is possible and believable? How and for whom has environmentalism been a resource?

These questions have oriented my analysis of the Bhopal disaster, helping me understand how environmentalism has emerged as a powerful cultural and political-economic response to globalization—and how the advocacy of people responding to the Bhopal disaster has shaped what environmentalism has become. By focusing on different ways the Bhopal disaster has been remembered, I have tried to understand how historical perspective is built into law, policy, bureaucratic initiative, civic action, and commercial endeavor—and how environmental catastrophe repeats patterns of victimization associated with race, class, gender, and colonialism. I wanted to understand how knowledge of the past becomes ethical knowledge that shapes and legitimates certain constructions of the future. I also wanted to understand the systems within which the Bhopal disaster has operated, so that I could describe specific mechanisms that connect geographically dispersed actors and specific mechanisms that grant these actors differential possibilities for working well within these systems. I wanted to know if power was operating in new ways.

My broadest commitment was to understanding how advocates responding to the Bhopal disaster have anteriorized the future—through legal precedents and the structure of rehabilitation schemes, but also more subtly. By establishing what counts as adequate description, explanation, and social response in the wake of disaster. By establishing how the past should be encountered.[2]

Thus, the importance of the settlement of the Bhopal case as a reference point. When the settlement was announced, on February 14, 1989, Chief Justice Pathak argued that "the enormity of human suffering occasioned by the Bhopal disaster and the pressing urgency to provide immediate and substantive relief to victims of the disaster" made the case "preeminently fit for overall settlement" (Cassels 1993, 222). Pathak's message was that the appropriate response to disaster is to manage it—to simplify it sufficiently to provide a final solution. Continuing disaster at the local level had to be systematically discounted. The development of forms of responsibility appropriate for a globalized world order was foreclosed. Law became a means of exorcism.

Critical response to the settlement of the Bhopal case involved gas victims, plant workers, health professionals, legal experts, and others. Addressing the problems of Bhopal became a way to chart a range of systemic failures, exacerbated by their tight coupling. Poverty, coupled with toxics. Failures of science, exacerbated by corrupt bureaucracies and politicized legal regimes. Trade liberalization, alongside the decline of labor unions. Technology transfer, without building in social and cultural diversity. The transnationalization of both corporations and nongovernmental organizations, without modes of accountability that crossed territorial borders.

Particularly recurrent themes revolved around the knowledge practices and institutions of modernity—denounced for atomizing social problems in order to control them—that were said to have domesticated disaster, veiled it, and removed it from public purview. One activist explained the problem, and the necessary response: "the main stream press, Union Carbide and even the Government of India have represented the Bhopal case as an isolated event, without source and finalized by distribution of cash compensation. There has been an overall attempt to encapsulate and exorcise, all within the logic of the market. Our task is to subvert this encapsulation."

The settlement of the Bhopal case was also called a failure of historiography. According to environmental justice activists in India, the settlement veiled continuing disaster at the local level, shrouded the need to reassess how we chart progress and shut down productive transactions between past and future. Sathyu Sarangi, a lead activist with the BGIA, articulated the conundrum: "History as we know it can not account for disaster—disasters are externalities, written off official balance sheets, lost in a logic that assumes that today is better than yesterday, even if the same percentage of the population lives below the poverty line and government hospitals only offer symptomatic relief." Sarangi's message was ironic: we must remember, recognizing that habitual ways of recollection fail.

For activists like Sarangi, the challenge was to unsettle all settlement of the Bhopal disaster. During the years I was in Bhopal, I worked within this challenge in very direct ways. Writing ethnography, the settlement has remained important, both literally and figuratively. Writing about the settlement is a way to show how globalization is "justified"—through accounting processes that foreclose certain lines of inquiry, disable certain forms of knowledge, and legitimate discriminatory social categorizations. It is a way to show how the Third World poor are sucked into destructive crosscurrents, without means to challenge them. It is a way to show how rhetorics promising harmonization of world order depend on systematic exclusion of the particularities of globalization on the ground, in everyday experience.

The settlement of the Bhopal case invokes a need for accounts of the disaster that show how it continues. Across time. Across space. At the intersection of crosscutting forces. Practical strategies for putting together such an account had to be designed. I knew that I needed to track both micro and macro processes. To triangulate interpretive and political-economic perspectives. To account for paradox—the contradictory ways science, technology, law, nationalism, and other phenomena operate as crucial social resources in some registers and locales, as mechanisms that legitimate uneven distributions of risk and reward in others. To account for the ways political authority is both catalyzed and corroded—by monetary might, but also by myriad other operations of power. To account for complex systems without losing ethnographic detail.

But these imperatives are abstract. Putting words together on a page required something more specific. I needed structure to help me write around the way remembrance of Bhopal had been structured before.

One key move was to turn the practice of advocacy into an empirical focus. My own advocacy would become an object of analysis alongside the advocacy of others, subject to similar questions. Geographical contiguity could not, however, delimit relevance. Advocacy had to become an axis around which I could spin an account of the Bhopal disaster in global terms.

Two methodological decisions became important in the backdrop design. The first was about parameters—about how I would delineate the "site" of my research—about what it meant, in practice, to produce an ethnographic account of a global disaster. The second decision regarded a way of understanding the social formations of disaster—recognizing that I should not assume that the social formations I accounted for were either stable or homogenous.

The decision about how to delineate the "site" of my research was partly motivated by the teachings of George Marcus and Michael Fischer.[3] In the mid-1980s, Marcus and Fischer argued that "the task lies ahead of reshaping our dominant macro frameworks for the understanding of historic political-economy, such as capitalism, so that they can represent the actual diversity and complexity of local situations for which they try to account in general terms" (1986, 88). The challenge for anthropologists was to find ways to embed rich descriptions of local cultural worlds in larger, impersonal systems of political economy:

> What we have in mind is a text that takes as its subject not a concentrated group of people in a community affected in one way or another by political-economic forces, but "the system" itself—the political and economic processes spanning different locales, or even different continents. Ethnographically, these processes are registered in the activities of dispersed groups or individuals whose actions have mutual, often unintended consequences for each other, as they are connected by markets and other major institutions that make the world a system. Pushed by the holism goal of ethnography beyond the conventional community setting of research, these ideal experiments would try to devise texts that combine ethnography and other analytic techniques to grasp whole systems, usually represented in impersonal terms, and the quality of lives caught up in them. (Marcus and Fischer 1986, 91)

Marcus and Fischer gave two reasons for this experimentation. The first reason was because established systems perspectives were descriptively out of joint with the reality to which they were meant to refer. In other words, experimentation was necessary to better understand and represent the world. The second reason also resonates with the concerns of advocacy. Marcus and Fischer argued that the purpose of anthropology—and particularly its experimental ventures at the end of the twentieth century—is "cultural critique"—a continually comparative effort that uses understanding of different experiences to destabilize and revitalize our efforts toward a more humane society. Ethnography becomes a means to imagine social alternatives, disrupting conventions that have become naturalized and immune to change (Marcus and Fischer 1986).[4]

Marcus and Fischer acknowledged, however, that the tension between strong cultural and strong political-economic analysis "is primarily a problem of representation or textual construction, rather than a difference of good intention or political conviction" (1986, 86). In other words, textual strategies for bridging micro and macro levels of analysis remained to be worked out. There were no models for writing the disaster.[5] Bhopal nonetheless had to be accounted for—and accounted for as a global disaster.

Writing about the Bhopal disaster across locales, rather than in terms of a spatially circumscribed site of research, made things complicated. The Bhopal disaster continues in time and crosscuts spatial boundaries. It resides at the intersection of different legal regimes and where bureaucracy collides with everyday life. It is located in competing definitions of health, fairness, and progress. It inhabits the discourse and social style of corporate environmentalism. The meaning of disaster cannot be stabilized. But, written as disaster, Bhopal becomes a prism for drawing a shifting world order into visibility. A world order beset by environmental problems and politics. A world order rife with calls to advocacy.

Who inhabits this world? How, once social and territorial boundaries have been destabilized, can one discern what a community is and who is part of it? How does one account for the way disaster *creates* community?

Within environmental politics, these questions are usually responded to in terms of "stakeholders"—groups of people who have a stake in decisions to be made by corporations, government agencies, or other organizational bodies within which decisions by a few people can affect many. A stakeholder model recognizes different social positions and different ways of perceiving both problems and solutions. The model has merits. It has pushed corporations to include nonshareholders in their calculations of stakes. Workers, people living near production facilities, vendors, and others have been offered a place at the table. It has also helped draw once marginalized players into policy formulation and evaluation.

But there are problems with the stakeholder model that the Bhopal disaster makes visible. The value of the model is its recognition of difference.

In use, however, the goal is usually to manage difference by forcing diversity into consensus. And each stakeholder community is usually considered to be epistemologically homogeneous and epistemologically consistent. Members of any given stakeholder community are assumed to think alike. And they can't seem to learn, remaining unchanged over time: within a stakeholder model, government agencies are expected to think objectively, procedurally, and politically. Corporations are expected to think objectively, technically, financially, and legally. Rationality is demonstrated by both. Citizens are not expected to be rational. Their "frame of reference" is subjective, intuitive, and experiential (Shrivastava 1987, 90).

Like most pluralist models, the stakeholder model can't seem to tolerate much complexity—or much dissent. I needed a different model. Yet I also needed to better understand how a stakeholder model *could* work. I needed to turn the stakeholder model into a resource, rather than insisting only on its insufficiencies. So I made the decision to rely on a stakeholder model a little differently than usual—miming the model so that I could see its limits—organizing my own thinking around "indigenous categories" to better understand the merits and problems of those categories.

In my account, stakeholder communities become "enunciatory communities." Some, like gas victims, are relatively tied to one locale; others are more dispersed and include corporate and government officials, medical and legal professionals, and environmental activists working at various tiers of regional, national, and transnational organizations. Like stakeholder communities, enunciatory communities often share certain interests. But they do not necessarily think in the same way about what those interests mean or about how those interests can be protected. Sometimes shared interest is not even what holds an enunciatory community together.[6]

Instead of locating enunciatory communities via indicators of interest or epistemological habit, I have focused on fields of force and contradiction—on the double binds that position enunciatory communities within new world orders. These double binds are more than "context" conventionally conceived. Enunciatory communities do not exist prior to the double bind, with which they deal according to an already coherent identity. Enunciatory communities are produced by double binds, which call into play both new and entrenched ways of engaging the world. The "identity" of enunciatory communities is strategically configured; collectivity is not a matter of shared values, interests, or even culture, but a response to temporally specific paradox.

My conception of enunciatory communities draws on the communicational theories developed by Gregory Bateson and colleagues in the 1950s and 1960s. Originally trained as an anthropologist, Bateson had some prior interest in "problems of classification in the natural sciences" and took as a new "point of departure" (for a proposal to the Rockefeller Foundation) Bertrand Russell and Alfred North Whitehead's *Principia Mathematica* (1910).[7] Where Russell and Whitehead analyzed the classification of abstract logical statements and the paradoxes that they often entailed, Bateson and his colleagues were more interested in the actual operation of such statements and their paradoxes in "real-world" contexts. A classic example is the statement that a therapist might make to a patient, or a parent to a child: "I want you to disobey me." To obey the statement is to disobey it; to disobey it requires obeyance. The contradiction is produced by the condensation of messages of different logical types in one experiential field, from which there is no escape. Bateson was particularly interested in double binds produced by family interaction.[8] I want to understand the double binds produced by environmental crisis within globalization.

The contradictions faced by Indian activists working on the litigation of the Bhopal case illustrate how double binds are structured and produce enunciatory communities. Gas leaked in Bhopal on December 3, 1984. Within a few days, middle-class progressives from across India mobilized to help—both in Bhopal and in major cities where governmental and press representatives could be lobbied. These activists came from different professional and ideological backgrounds. They came together in response to the Bhopal disaster, and within its contradictions. One key contradiction revolved around choice of a forum for the litigation of the Bhopal case. On the one hand, the trial *should* have been held in the United States so that the role of Union Carbide Corporation—the parent company—would be built into the way the case was presented—making an important statement about the responsibility of multinational corporations for the activities of their foreign subsidiaries. On the other hand, the trial *should* have been held in India to demonstrate that Indian courts could successfully oversee adjudication of disputes involving Indian citizens—making an important statement about national sovereignty and the capabilities of the postcolonial state.

The contradictions faced by grassroots organizations, whether in India or in the United States, are also illustrative: Grassroots organizations emerge because decisions affecting the local level are being made by outsiders. Grassroots organizations *should* therefore prioritize initiatives to build institutional structures for effective local-level decision-making. But the resources (including cultural and political authority) necessary to institutionalize local-level decision-making are often available only through the law. Grassroots organizations *should* therefore engage in legal battles, even though the leadership required often works against efforts to sustain broad local-level participation in decision-making. The law can create a space for grassroots organizations to work, while undermining the very modes of sociality such space was to protect.

A double bind situation is not simply a situation of difficult choice, re-solvable through reference to available explanatory narratives. "Double bind" denotes situations in which individuals are confronted with dual or multiple obligations that are related and equally valued, but incongruent. As John Weakland explains, understanding of double bind "is only ham-pered by most attempts at analytic simplification (e.g. 'What is the *real* message?'). The multiplicity and complexity of messages, their interrela-tions and reciprocal qualification, must be attended to and taken into ac-count simultaneously" (Weakland 1976, 312).

Double-bind situations create a persistent mismatch between explana-tion and everyday life, forcing ethical agents to "dream up" new ways of understanding and engaging the world. They provided a lens for observing experiences produced by established rules and systems, yet not adequately described in standard explanations of how these systems function and change. They are a way to locate enunciatory communities within demands for language for which there are no indigenous idioms.

Thinking in terms of enunciatory communities is a way to account for the emergence of new subject positions, as entrenched signifying systems are being challenged and displaced. Subjects are drawn into new realities and fields of reference. Traditional constructs of society and culture no longer seem adequate. Enunciatory communities emerge in response, as a social register of profound change.

Enunciatory communities are different from stakeholder communities in a number of ways.[9] First, their identity cannot be divorced from context. If citizens are irrational, it is because they are responding to an irrational context. Second, the identity of enunciatory communities is assumed to be fissured within, even when members themselves insist otherwise. Enun-ciatory communities are, by definition, morphologically complex. Third, enunciatory communities are not expected to devolve into benign plural-ism. They cannot be aggregated into a harmonious whole. Their differences are considered a resource, rather than a problem.

Most important are the discontinuities. Enunciatory communities do not remain the same across time or across space. They are chameleonlike, morphing in response to the interplays in which they find themselves, learning as they go—developing new strategies at every turn.

Enunciatory communities are both subjects of and subject to change.[10] They change as they change the systems in which they are embedded—

redirecting and reconfiguring the very force fields that constitute them as enunciatory communities.

An enunciatory community, like a global disaster, is not a "unit" of analysis. Like a global disaster, an enunciatory community is an emergent effect of crosscutting forces. The architecture of an enunciatory community—like that of a global disaster—is open. Meaning and power distributed within and by enunciatory communities—as by global disaster—operate through complex systemic interaction. Both therefore resist macrotheorization.

Enunciatory communities, like a global disaster, are generalizable. They recur in different times and places. But every example of disaster, or of an enunciatory community, disrupts generalization. There is always discontinuity between the general and the particular—and it is highlighted, rather than effaced.

Enunciatory communities, like disaster, provide a way to work within one of the most stubborn dilemmas of advocacy: the need to render one's object of concern in all its particularity so that justice is not lost to grand schemes and glossy claims—while, at the same time, showing how "the problem" one is concerned about crosscuts time and space, demanding a systemic response. Oscillation between the horns of this dilemma is not impossible. But it does require new analytic strategies and new ways of understandings advocacy itself.

The Focus

Diverse individuals and organizations have entered the contest over how the Bhopal disaster should be understood. Continuing debates at the local level address the adequacy of the centralized, hospital-based system of government health care, the justice distributed in compensation courts, and efforts to sustain victims' organizations. At the national level, the Bhopal disaster has provoked intense debate over Nehruvian dreams to make big industrial projects the temples of modern India. Some cite the Bhopal case to demonstrate the failure of Leninism, insisting that it proves that the primary enemy is not private property, but the state, as primary manager of surplus value and lead mimic of the Western fetish for "development," whatever the costs. Others say the Bhopal disaster is a key index of efforts to dismantle Indian commitments to "scientific socialism." Government officials reminded the press that the out-of-court settlement of the Bhopal case proves India's receptivity to foreign investment.

United States courts denied jurisdiction in the Bhopal case, arguing that asymmetry between the First and Third Worlds had been overcome, allowing India to "stand tall before the world and determine her own destiny."

The U.S. Congress responded with new regulatory requirements, often referred to as "Bhopal's Babies." Some argue that the new regulations threaten the viability of the U.S. economy, undermine corporate rights to control information, and institutionalize unscientific modes of risk assessment. Union Carbide has accepted "moral responsibility but no liability" for the Bhopal disaster and has told people in the United States that "it can't happen here." Union Carbide also claims to have "gone green," leading industrywide commitments to sustainable development, citizen participation in risk assessment, and "voluntary compliance."

Activists in the U.S. environmental justice movement responded by refocusing protest away from the state and directly against the commercial corporation—reworking strategies inherited from the civil rights movement. New social coalitions have been formed, linking labor, community, and environmental activists to fundamentally challenge how the risks and rewards of industrial society are distributed.

Academic theorists also entered the fray. Legal theorists argued that the consolidation of claims in the Bhopal case provides an innovative model for dealing with the uneven distribution of risks and rewards in industrial society. Theorists of "strategic management" argued that Union Carbide's response to the disaster initiated an "avalanche of change" that harmonizes opposition between economic growth and environmental safety and between corporations and their neighbors. One management account draws us into the enduring controversy, foregrounding a chart that illustrates seven different calculations of the death rate caused by the Bhopal disaster as a way to suggest how competing perspectives complicate efforts to understand multinational corporate activity. The differences are sobering: The Indian government counted 1,754 dead and 200,000 injured. Indian newspapers counted 2,500 dead and 200,000–300,000 injured. United States newspapers counted the dead as 2000+ and the injured as 200,000. Voluntary organizations counted 3,000–10,000 dead and 300,000 injured. The Delhi Science Forum, one of the groups I worked with, listed 5,000 dead and 250,000 injured. Eyewitness interviews claim 6,000–15,000 dead and 300,000 injured. "Circumstantial evidence"—based on shrouds sold, cremation wood used, and missing persons estimates—finds 10,000 dead (Shrivastava 1983, 65).

I learned of these conflicting responses to the Bhopal disaster through different modes of ethnographic engagement. At the outset, only one thing was persistently clear: work on the Bhopal case would be tangled in vehement struggle over how contemporary political economy should be interpreted and institutionalized. All debate surrounding the Bhopal case compounded with emergent discourses attempting to position people, nations, and corporations within the New World Order. Environmentalists from across the political spectrum participated in this synthetic process, insisting that the "situational particularities" of the Bhopal case operated as evidence of certain understandings of the past and of certain directions for the future. Details of the disaster were cited, or effaced, to substantiate widely disparate interests, values, and political strategies.

With time, I would recognize not only different attributions of meaning to Bhopal, but also different rhetorical styles, different fields of reference, and different ways of constructing legitimacy. Bhopal would become a site from which India would be articulated with a globalizing economy, where citizens would develop roles at odds with national identity, where environmentalism would become a politics of fissure, rather than harmony. Bhopal would come to mean many things, indexing discord and malfunction on many registers.[11]

To explicate this discord, I have focused my inquiry on and through the social practice of advocacy. Advocacy brings together people from different places, perspectives, and interests, throwing into visibility the harsh divides and asymmetries upon which global order operates. And advocacy shows the limits of many received ideas—about change, consciousness, and responsibility. Environmental advocacy is a particularly rich focus, entwined as it is with science (as both culture and practice), with technology, and with manifest concern for the future.

Advocacy is usually thought of as a specific, easily recognizable practice involving argument for a cause or on behalf of another. It is associated with direct pleading in courtrooms, political pamphlets, or other fora where disputants struggle to assert their particular view of what is and what must be. My conception of advocacy is broader and less straightforward. The advocacy described here is not limited to overt argument or intentional action. Nor is it limited to those instances in which people assert what they believe to be true in the expectation that others will concede. Desire for rational consensus is not always the goal here. Advocacy without the guarantees of teleology is of particular interest. Intentionality is examined, but not assumed. The challenge here is to uncouple advocacy and modernist ideals.

I define advocacy as a performance of ethics in anticipation of the future. The stress on performance is a way to put ethics in the world, where the mismatch between ideal and reality demands creative response—where there is never only one way to do things. Culture matters, as does the structure of language, discourse, and social relations. Time is also important. The historical context of advocacy is given significance, as is the experience of interpretation complicated by intersubjectivity. Advocates are situated within historically specific force fields constituted by broad political-economic, technical, and sociocultural change. But they are also situated in reciprocal relation to other advocates, even if geographically distant, whose intended as well as unintended actions influence what is perceived as good and possible.

The advocacy described here is situated within particular worlds of practice, amidst competing demands and the forces of social, technological, and political-economic change. I describe different styles and strategies of advocacy, and very different conceptions of justice, nationalism, science, and the role of law in society. And I show how advocacy is at work even when not conceived as such, embodied in the structure of sentences, narratives, personhood, and social relations. The form of environmentalism is considered as important as its content. Configuration of "the problem" is recognized as an important ethical move.[12]

This book reassesses what advocacy is, and where and how ethics are enacted. It is a book about the rhetorics and practices of contemporary environmentalism. And it is a book about ethnography, in globalizing, politicized contexts. But it is first and foremost a story about advocates responding to the Bhopal disaster amidst the dramatic change brought about by the end of the Cold War, globalization, and other processes that have fundamentally challenged established political programs. These are the people who make environmentalism happen, from many different political positions, but all confronted with sharply conflicting demands—whether between environmentalism and economic growth, between community autonomy and governmental support, or between ideological commitments and everyday life.

The advocates I describe here are of all political persuasions. Union Carbide is described as an advocate, as are gas victims themselves, and various groups working alongside them or on their behalf. They advocate different things, but all are critically concerned with how advocacy works. They all want their advocacy to be effective, though criteria for evaluating effectiveness differ markedly. Most consider advocacy a problematic enterprise, if only because they acknowledge the strength of those who oppose or compete with them. Focusing on the work of advocates is thus a way to study crosscutting social and political economic systems as they are interpreted by the individuals involved in them. It is also a way to study how systems change and how advocacy makes a difference.[13]

The Text

In February 1989, the Bhopal case was settled out of court for $470 million. At that time, 600,000 people had registered as claimants. The Manville asbestos case was settled for $2.5 billion to cover 60,000 claimants. The Dalkon Shield settlement for $2.9 billion covered 195,000 claimants. In

August 1994, as people around the world prepared to commemorate the tenth anniversary of the disaster, a total of 82,523 cases had been cleared by claims courts in Bhopal. The award for each death ranged from $4,000 to $12,000.[14]

These quantifications and comparisons resist memory, intellectual engagement, and political change. Abstracted from context, they become mere blips in the litany of death figures that structure twentieth-century history, lost in other numbers citing the threat of population growth, deforestation, and declining terms of trade. I have tried to configure Bhopal differently, exploring while challenging conventional modes of remembrance, reaching for ways to describe a forcefully interconnected world system that is too complex to fully understand, yet calls for a response.

I wanted to provide readers a glimpse into the unrepresentably complex world system in which we live—pulling the "whole world" into an ethnographic field of vision to mark the limits of all ways of representing "the whole," but also to show how ethical moves in one locale resonate with those in other locales, with intended as well as unintended consequences—which transform the entire system, repositioning everyone within it.

The challenge was to offset "settlement" of the Bhopal case, demonstrating how the disaster continues to operate—in Bhopal itself, but also elsewhere, in the myriad ways disaster is continually reproduced, distributed, and legitimated, asymmetrically. I wanted to show how globalization operates in ways pluralism cannot describe. Pluralism denotes a reality made up of two or more independent elements, which are often hierarchically organized. What I wanted to describe was a reality made at the intersections of these elements, which are mutually constituted. My focus needed to be on moving objects, continually transformed by the system that their actions create.[15]

This meant that I needed to conceptualize ethnography as a verb.[16] I needed to make ethnography move, to catch up and keep pace with the globalizing orders in which we live. A static account of provincial India, cut off from transnational flows of capital and culture, would not suffice—because a static account would reify difference, rather than engage it (Fabian 1983).

The Bhopal disaster throws the problem of reification into high relief. Difference persists. But it cannot be perceived through tactics of distinction, insisting that a thing is all that and only that which it is not. Idealized versions of scientific method have had to be rethought. Representation of "the actual" remains the goal, but the actual is to be found in processes and intersections, rather than in objects or locales.

Difference within the Bhopal disaster has been produced, and continually reproduced, through transnational transfers of technology, risk, and rights; through the interaction of different legal systems; through the collision of biomedical and indigenous approaches to health care; through rehabilitation regimes that turn women into sole wage earners; through prophylactic measures that promise to secure the future by granting citizens the right to know.[17] Late-twentieth-century world order operates through these interactions. Environmentalism provides explanations, justifications, and agendas for change. Advocacy is one way individuals and communities play their cards, stake positions, and win or lose.

How can one represent the cumulative effect of actions and attitudes distributed throughout a complex system? Toxicologists refer to cumulative effect to name phenomena that exceed what is made visible when an individual chemical is identified, classified, and listed on a toxic registry. Studies of cumulative effect must account for additive effects, antagonistic effects, and synergistic effects—while acknowledging that established protocols for measuring chemical exposure are not sufficient. Measuring the signs of exposed rats against the signs of unexposed rats is not enough. Toxicologists must continue to carefully study the structure and effects of individual chemicals—most of which have never been studied at all. But they also must account for the ways individual chemicals operate in conjunction with other chemicals, with sociohistorical factors, and with specific physiologies.

Accounting for cumulative effect means accounting for the production of meaning through forceful interaction—without ignoring the specific structure and mechanisms of individual components. What was once thought to be separate now needs to be thought of in terms of interpolation. India, interpolated with the United States. Governments, infiltrated by corporations. Everyday life, inflected by legal and scientific categorization schemes. Every narration, interlaced with its narrator. But specificity shouldn't be glossed. This meant that I needed a textual structure that preserved needlepoint analysis of specific locales, while rendering them in dynamic connection. The structure I rely on here only partially accomplishes this double task.

The book spirals outward from the story of gas victims, countering tendencies to think of the disaster as isolated in time and space—as an "incident" (in the words of the chemical industry) that "can't happen here." It tries to represent pluralism in motion. Stories from Bhopal are run alongside stories told by people living near chemical plants in the United States, by professional environmentalists of different political persuasions, and by industry and government representatives. Reciprocal determination is an important theme, as is asymmetry.

Chapters of the book are shaped around enunciatory communities—collectivities formed by the particular demands of disaster. Some emerged in direct response to disaster. Organizations of victims in both India and the United States are examples. So is the collectivity of risk communication experts I call "green consultants." Other enunciatory communities are reconfigurations of already established social formations—Union Carbide, the government of India, middle-class Leftist groups in India. These enunciatory communities encountered the Bhopal disaster with entrenched ways of working together, interpreting the world, and staging their legitimacy. For most, disaster made these things much more difficult. It is such difficulty that my use of enunciatory communities as a textual organizing device is meant to highlight. My goal is to describe not only what advocates have advocated, but also how advocacy has been produced—in different cultural and political-economic contexts, within social formations that often cohere uneasily, if at all.

Comparison between chapters is fostered through continual return to a set of questions I have formulated to query how and where advocacy happens.[18] Return to these questions throughout the book allows one to read horizontally, through each chapter, or vertically, across chapters, each of which is organized around these questions.[19] Within each chapter, the questions provide a structure for different modes and levels of analysis. Tracked across the text, the questions facilitate comparison and scope, rather than depth of description of any one enunciatory community.

The first of the questions works to position advocates within the competing messages that call for their response and provoke the formation of "enunciatory communities." Succinctly, this first question asks, "What double binds call an enunciatory community to speak?" The task of the first question is to establish the field in which an advocate is called to speak and become part of an enunciatory community.

The second question provokes a description of how advocates have acted out their response to the field's contradictions. It asks, "How have advocates strategized and configured their own role in the disaster?"

The third question asks, "What, in sum, has been said about Bhopal? What is the Bhopal disaster said to mean? What does 'Bhopal' stand for or evoke?"

Question four asks about the focusing devices advocates have used in their encounters with the disaster. In the idiom of social scientific method, this question prompts description of an enunciatory community's "core categories"—the mechanisms through which they draw otherwise disparate processes and properties into relation, the mechanisms through which things are granted significance or discounted.

The fifth question asks, "How did advocates find venues for their enunciations? What organizational forums provided a stage? What performance strategies worked? What textual technologies were relied on?"

The sixth question asks, "How is the work of advocacy corroded or undercut?" In some cases, corrosions come from afar, from the brute power of money or the more mundane ordering of bureaucracies. In other cases, the corrosions operate from within, as an effect of how advocacy is imagined and thus practiced.

The seventh question reverses the sixth, asking "What catalyzes advocacy? What gives advocacy momentum? How have the various moves and contexts of advocacy in response to the Bhopal disaster worked synergistically, somehow producing more than the sum of their parts?" [20]

The questions

1. What double binds call advocates to speak?
2. How have advocates strategized and configured their own role in disaster?
3. What is the Bhopal disaster said to mean?
4. What focusing devices have advocates used to help them understand the disaster and connect it to other things?
5. Through what organizational, rhetorical, and textual forms has advocacy been disseminated?
6. What corrodes or undercuts this advocacy?
7. What catalyzes this advocacy?

These questions have been used to prompt ethnographic writing—helping me show where and how advocacy works, accelerating my focus on the interactions that constitute disaster. The questions provide a structure for narrativization of material that resisted insertion into comparative inquiry. They were a means of creating significance for seemingly random data points—conventionally left off the record, externalized as noise.[21]

The questions have not, however, been engaged like a quiz, demanding a direct and complete response. In most cases, the questions are responded to with descriptions of the enunciatory community that is the subject of that particular chapter. In some cases, I write of my own work as an activist working as part of the enunciatory community. In other cases, however, I use the space of the question to describe how, as an ethnographer, I have been able to produce these descriptions. In the chapter about gas victims themselves, for example, the catalyst I describe is not what catalyzes victims' own advocacy, but what catalyzes my description of gas victims. This folding of methodological accounting into what I describe is purposeful. It is a way to highlight—and embody textually—a basic argument about the constructedness of all description and thus the political implications of methodology. Interpolating accounts of my own methods of producing the text into the body of the text admits how method and truth are intertwined—and provides an opportunity to experiment with ways that advocacy—and environmental advocacy, in particular—can deal in truth claims without effacing where they come from.

Interrupting my own narration with the narrations of others also has methodological—and political—purpose. Excerpts from writing done by my informants pepper the text. I helped produce some of the excerpts in my work as advocate. Inclusion of the excerpts is thus a way to illustrate how the articulations of direct advocacy blur into, while remaining distinct from, ethnographic modes of description. The excerpts also help tell the story of Bhopal, providing both detail and exemplification of the tones and tropes deployed. They thus do many kinds of work. They demonstrate the vastly different rhetorical styles and fields of reference relied on to articulate a response to disaster. They show how the techniques of ensuring that certain things don't count operate in practice. They suggest the limits of many of the ideological frames in which enunciations have been force fit. And they allow people often left off the record to participate in the historicization of Bhopal.

But the seeming emphasis on "in their own words" is skewed. Most of the excerpts—and those with gas victims, in particular—were collected for various purposes and publications directly related to the adjudication of the Bhopal case. They are mediated by the demands of advocacy. So I do not include them to move closer to the final truth of Bhopal. Instead, I include them because they harbor potential for ever further elaboration. Sometimes these excerpts substantiate the interpretation I offer. Other times they challenge it. Always, they exceed it. The excerpts are crucial indicators of what we speak of when we refer to New World Order. They are like warning lights in a storm, signaling where history cannot be narrativized into logics of harmonization and benign pluralism.[22]

My overall goal has been to develop ways to understand the Bhopal disaster—and the complex global system that focus on the disaster brings into view—in a way that does not require that we be able to stand outside the system in order to critique it. In a way that takes responsibility for questioning key concepts without losing sight of the importance of work done through and in the name of such concepts. In a way that catalyzes progressive advocacy, while recognizing faults in the system within which progressive advocacy has been built.

The rationale could be thought of in terms of reflexivity. I prefer to think in terms of recursion. Recursion produces meaning through iteration, running back over preceding operations again and again to better understand how they've already determined what the next operation can be. Dialectical confirmation does not have to be the goal. Iterations constantly cross epistemological boundaries, narrative levels, and social landscapes.[23] The differences encountered can become a way to understand the recurrences that make any system a system. Encounters with difference can also be a way to remind oneself that systems, and systemic analyses, always leave some things out.

Reflexivity calls for the ethnographer to position herself. Recursivity positions her within processes she affects without controlling, within competing calls for response. Reflexivity asks what constitutes the ethnographer as a speaking subject. Recursivity asks what interrupts her and demands a reply.

Thinking in terms of recursivity is a way to hold ethnography responsible for advocacy. Attention to recursivity foregrounds how every articulation—whether ethnographic or in direct advocacy—operates on previous articulations, nesting every move and every word within multiple discourses and worlds. These nested worlds may be more or less contiguous with the world one considers primary, whether that is the world of law, literature, community organizing, or anthropology. But they implicate each other in significant ways.[24] What is said in domains of law implicates what it is possible and necessary to say in community organizing—and vice versa. What is said in direct advocacy implicates what it is possible and necessary to say in ethnography.

ONE

Plaintive Response

(1.1) Standing Tall before the World

The Bhopal case was dismissed from U.S. courts by Judge John Keenan on May 12, 1986, on the grounds of *forum non conveniens,* a legal doctrine that posits that significant decisions leading to the case were made elsewhere, making it inconvenient to secure witnesses and evidence in the proposed forum. The doctrine also posits that there is an adequate alternative forum wherein justice can be adjudicated. In Judge Keenan's synthesis, it would have been "sadly paternalistic, if not misguided" for his court to evaluate the operation of a foreign country's laws. He did acknowledge that double standards of industrial safety are not to be encouraged, but also noted that "the failure to acknowledge inherent differences in the aims and concerns of India, as compared to American citizens, would be naive, and unfair to the defendant." His judgment concluded with the argument that retention of the case in U.S. courts "would be yet another act of imperialism, another situation in which an established sovereign inflicted its rules, its standards and values on a developing nation. . . . To deprive the Indian judiciary this opportunity to stand tall before the world and to pass judgment on behalf of its own people would be to revive a history of subservience and subjugation from which India has emerged" (quoted in Cassels 1993, 134).[1]

Plaintiff lawyers, journalists, and environmental activists read the Bhopal case differently, challenging the obviousness of national boundaries and recognizing asymmetry between the First and Third Worlds. They articulated connections between Union Carbide India Limited (UCIL) and

Union Carbide Corporation (UCC), including stock ownership, shared executives and directors, and decisions made at Union Carbide offices in the
United States. Among these were the decision to site the Bhopal plant
barely two kilometers from the main railway station; the decision to switch
to a production process for synthesizing the pesticide Sevin that relied on
methyl isocyanate (MIC), the gas released in Bhopal; and the decision to
use an "open circuit process" requiring bulk storage of MIC. In addition,
when India did "stand tall" and litigate the Bhopal case, UCC was the
defendant and thus the party that negotiated, accepted, and paid 90 percent
of the February 1989 settlement of $470 million. Perhaps the most telling
connections were much more indirect: news of the settlement caused UCC
stock to rise $2 a share, or 7 percent. *really*

But this story cannot be told in oppositional terms. To disagree with
Judge Keenan would be to concede that Indian courts could not perform
their duties. To disagree with Judge Keenan was also an important means
of demonstrating the systemic problems from which disaster emerged. Indian attorney Indira Jaising explains one of the competing messages:

> If the suit had been tried in the United States, its courts would have been
> compelled to confront the question of double-standards not only in threshold
> levels of safety but also in the matter of compensation levels. For an Indian
> court to award lower levels of compensation might look reasonable, for an
> American court to do so, based on the lower standard of living in the Third
> World, would have come across as being blatantly discriminatory. . . . But
> the most important reason that obviously weighed with him was his under-
> standing that the United States did not have a strong interest in the case.
> Thus, he declined to accept the argument that keeping the case in the U.S.
> would compel U.S. multinationals to accept responsibility for the wrongs of
> their subsidiaries overseas, accept responsibility for double-standards and be
> a deterrent against future accidents. . . . The refusal to accept this argument
> is based on a policy decision that it is "not my problem." (1994, 189)

Debate over the jurisdiction issue raged throughout 1985. It reemerged
during the years I was in Bhopal as a result of the possibility of refiling the
case in Texas following a ruling by the Texas Supreme Court that disallowed dismissal of cases on grounds of *forum non conveniens.* The double
binds did not go away.

One constraint was legislative. In March 1985, the government of India
passed the Bhopal Gas Leak Disaster (Processing of Claims) Act. The Bhopal Act granted the government of India exclusive rights to represent victims of the disaster, both in India and abroad. So-called voluntary social
action groups therefore had no legal standing. Applications for intervenor
status had been submitted, but both Union Carbide and the government
of India had opposed the applications, and they were not granted. Lawyers

in the United States argued that the case should be pursued nonetheless because it could be argued that the Bhopal Act disallowed due process because of conflicts of interest.

Lawyers in the United States also told us that there was no time to wait. If we didn't refile in Texas right away, proceedings would be barred due to a statute of limitations that disallowed filing more than two years past the last action on the case in that court. So we had to make a decision fast about whether we would serve as liaisons between the victims and the lawyers, facilitating the certification of plaintiffs required in class action suits. Basically, we had to decide whether we should encourage gas victims to participate in the suit and what our role would be if they decided in favor of participation. Figuring out which gas victims should be approached as representative of all gas victims was also a problem.

Our role in the work to revive the Bhopal case in the United States was always ambiguous and controversial. Some of the controversy stemmed from disagreement over legal strategy among people who had worked on the Bhopal case for many years. Some argued that reinitiating proceedings in the United States would lessen the possibility of overturning the settlement of the case in India. The lawyers argued that simultaneous proceedings in two different forums were standard. The lawyers also argued that any award granted in India would be difficult to collect, since UCIL—the subsidiary that owned the Bhopal plant—did not have sufficient assets and because any collection from the parent company would depend on verification by U.S. courts that the judgment had been obtained through due process—thus leaving collection dependent on judicial interpretation by the same court that had denied jurisdiction of the case.

Another argument for pursing the litigation in the United States regarded the potential amount of damages. Compensatory damages in the United States were, of course, routinely higher than in India. Further, it was possible to sue for punitive damages in a U.S. court, while punitive damages were not an option in India. Suing for punitive damages was attractive to many of the activists I worked with because of the promise of proving liability, rather than merely collecting an award based on a no-fault decision. Punitive damages necessarily hinge on the establishment of liability. They also are much harder to recover through insurance. The prospect of turning the Bhopal disaster into a financial burden for Union Carbide rather than only for victims was indeed attractive.

One problem, however, was that this would entail entanglement with the contingency fee arrangements through which U.S. lawyers were paid for their services. The fee expected by the lawyers we worked with was the standard 33.3 percent of the final award. Many activists working on behalf of gas victims found this grotesque. Contingency fee arrangements for legal services were not permitted in India. So Indian activists were particularly offended by the suggestion that justice had to be paid for, at exorbitant rates.

Previous experience with U.S. lawyers involved in the Bhopal case did not help. The lawyers had begun filing suits in the United States within a few days after the disaster. They also arrived in Bhopal itself, en masse. Their presence no doubt helped galvanize relief work by the Indian government. The magnitude of the disaster was harder to deny once U.S. lawyers began providing expert commentary to the international press. And the U.S. lawyers helped victims understand what the law promised and how their suffering entitled them to a response from Union Carbide and from the United States.

But the conduct of U.S. lawyers in Bhopal was widely perceived as despicable. They worked fast, acquiring trunk loads of retainer agreements in record time. Advertisements promising great rewards were run in local newspapers and posted on walls. Local agents were hired to help and, when necessary, to distribute small fees in exchange for signatures. Gas victims were told little about what it meant to sign a retainer agreement, much less about how they should choose a lawyer. The lawyers tried to work this out among themselves, in slinging matches in which all accused all others of being ambulance chasers.

Some commentators argued that, at the very least, the U.S. lawyers had "broken the pattern of legal resignation" in India (Marc Galenter, quoted by Cassels 1993, 117). Others provided a frontal defense of their actions and their motivations. Legal ethicist Monroe Freedman, for example, asks an evocative question: "What else but the profit motive could have brought to the doorsteps of the impoverished people of India some of the finest legal talent in America?" Melvin Belli, one of the most well known of the U.S. lawyers, argued that "capitalist lawyers are needed in a capitalist society" and that he was a "good capitalist" (quoted in Cassels 1993, 115).

Whether or not U.S. lawyers should receive accolades for "pioneering new forms of accountability" still deserved consideration in 1990, as did arguments that contingency fees are a good way to "bring marginal constituencies into the marketplace of remedies" (Marc Galenter, quoted in Cassels 1993, 115–16). But we had a decision to make, and we had to make it fast. And we had to make it knowing that Benton Musslewhite, the lead lawyer offering to reinitiate the proceedings in the United States, had had his professional license suspended by the Texas Bar based on allegations of misrepresentation and solicitation. Musslewhite argued that he was the object of a smear campaign by the Texas Bar, which he described as a

corporatist enclave made uncomfortable by his antiestablishment challenges. For the appeal of his suspension, Musslewhite received supporting
amicus briefs from the Veterans Association, which he had worked with on
the Agent Orange litigation, and from numerous grassroots groups he had
worked with on environmental problems.

Finally, those of us working through the Bhopal Group for Information
and Action (BGIA) decided that we would encourage and support the reini-
tiation of legal proceedings in the United States. Our apprehension about
this decision did not ever go away. Basic questions continued to haunt us:
Could the law ever be anything other than another technology that repro-
duces and extends capitalist economies? Was it possible to remake the law,
by working within the law?

To: The Editor, *Indian Express—Delhi*

Dear Sir:

We read with interest the report on the legal initiatives against Union Carbide in the United States ("U.S. Lawyers Await Indian Support," *Indian Express* 13-10-90).

The Texas Supreme Court held on 28 March 1990 that *forum non conveniens* can no longer bar personal injury or wrongful death lawsuits filed in
Texas by citizens of countries that maintain equal treaty rights with the
United States. This means that citizens of countries who are members of the
United Nations can seek legal redress for personal injuries caused by American multinationals in Texas State Courts. The 28 March ruling was given in
a case in which Domingo Castro Alfaro and 80 other Costa Ricans filed
damage suits against Dow Chemical Co. and Shell Oil Co. for causing irreversible sterility from repeated exposure to Dibromochloropropane manufactured by the two multinationals. In Texas, there also exists a 1913 statute
that gives foreigners the right to sue in Texas courts. Thus, there is a reasonable basis for taking on Carbide for the genocide in Bhopal in the Texas
courts, particularly after the Alfaro ruling.

The Indian government, as *parens-patriae* of the gas victims had, as a
matter of priority, chosen to sue Union Carbide in the U.S. and was forced to
move the courts in India only after Judge Keenan dismissed the case from a
New York Federal Court on grounds on *forum non conveniens.* Given that
forum non conveniens does not prevent litigation from being pursued in
Texas courts, there are strong grounds for the Indian government to support
legal action in Texas.

We would like to correct some of the mistaken notions of Mr. D. S. Sastri,
the Indian government's Washington based lawyer. Mr. Sastri contends that
"all the organizations that claimed to represent the victims in court had
come into being after the settlement with UCC had been reached." This
is far from the truth; there are countless newspaper reports prior to February 1989 (when the settlement was reached) that have documented the

relentless struggle of victims' organizations for relief and justice. Mr. Sastri would do well to familiarize himself with the extra-legal initiatives of the organizations of gas affected people of Bhopal that date back to the immediate aftermath of the disaster. Mr. Sastri bemoans the fact that no Bhopal organization had come up with an alternate amount of settlement payable by Union Carbide. It needs to be pointed out that several organizations have, after careful analysis, computed the amount of compensation (NOT settlement!) that should be extracted from Union Carbide. Putting forward an alternate settlement amount would virtually mean giving sanction to the illegal, unconstitutional and immoral settlement accepted by the government but never accepted by gas victims. The organizations of gas victims look forward to establishing the liability of Union Carbide and hold the perpetrators of the greatest industrial accident of the century accountable.

It is not money that the Bhopal victims are after, Mr. Sastri, it is justice. And while you may think differently, the two are not the same.

Yours Sincerely,

16 October 1990 /Dr. Rajiv Lochan Sharma/ /Satinath Sarangi/
Bhopal Group for Information and Action-Bhopal
cc Mr. D. S. Sastri

Broad questions about the role of law in society remained unanswered after we decided to pursue a second round of U.S. litigation. Our decision was based on another level of analysis. If U.S. courts delivered a decision on the Bhopal case, it was possible that the court would appoint a body other than the government of India as executor of the award. As we saw it, this opened up the possibility of building a rehabilitation scheme that would not merely duplicate the social and cultural formations that had produced the disaster. The responsibility for gas victims could shift to the voluntary sector.

BGIA was, of course, a key component of the voluntary sector in Bhopal. And it was all but impossible to imagine the scale-up it would take to become better administrators of rehabilitation than those we criticized. But calls to shift social responsibility from the state to the civil sector were being heard throughout India, and from grassroots environmental groups, in particular. Daily life in Bhopal validated many of the critiques upon which the new political vision was being built. But that didn't mean that there were blueprints to guide our everyday moves.

(1.2) Oscillating Roles

The government of India claimed to represent the victims of the Bhopal disaster before the law. Progressive activists were legally barred from assuming this role. So their role has been cast otherwise—in ways that have taught me many things about how the advocacy of middle-class progressives is configured. Advocacy in the name of people less fortunate than oneself can rarely claim the status of representation. To represent is to speak and act by delegated authority, as a substitute or proxy. Most often progressive advocacy works without this authority. A vote is not taken; there is no social contract whereby the advocate is invested with the right to speak on another's behalf, as her higher equivalence.

Middle-class progressive advocacy, as I have observed and practiced it, can't hope to legitimate itself through reference to conventional criteria of legitimacy. Middle-class progressives can't claim the right to speak. They are not authorized by their social position or by possession of the truth. Their authority comes from a conscientious watchfulness that draws out concern and compels a response.

Even the confirmations of peer review are hard to come by. Agreement among progressives themselves on what should count as solid truth or appropriate method is uncommon. The work of constructing truth and method is usually collective. But confirmation of one sure way to do things is rarely the result.

Middle-class advocacy is thus careful—in two ways. It is driven by concern. It proceeds cautiously—responding even without sure footing. Reaching beyond one's expertise becomes a habit. The challenge is to figure out how and when to respond, using whatever resources are available, however imperfect or insufficient.

**AFFIDAVIT SUBMITTED TO THE DISTRICT COURT OF
JEFFERSON COUNTY TEXAS OCTOBER 19, 1990
in BANO BI, ET AL.,
Plaintiffs
v.
UNION CARBIDE CORPORATION, ET AL.,
Defendants**

Before me, the undersigned authority appeared Kim Laughlin [now Fortun], known to be, and after first being sworn, dead upon her oath, state, swear and affirm as follows.

My name is Kim Laughlin; I am over the age of twenty-one years; of sound mind, and make this statement on my own free will and state that it is based on personal knowledge.

I am a doctoral candidate at Rice University in the Department of Anthropology. My educational background has prepared me to assess the role of science and technology in society. The subject of my Ph.D. research is the long-term social and economic impact of the Bhopal gas disaster of 1984.

I resided in Bhopal from September 1990 through June 1991. During this

time I worked directly with the leaders and members of Bhopal Gas Peedit
Mahila Udhyog Sangathan (Bhopal Gas Affected Working Women's Union)
(BGPMUS), the largest victims' organization in Bhopal. I have also worked
with leaders of Bhopal Group for Information and Action (BGIA), an organization of educated volunteers who serve as English language representatives of BGPMUS. I also have had numerous discussions with Indian legal
experts, political leaders, health professionals, and bureaucrats who have
been involved in the Bhopal case. I also have done archival research to analyze documents and other materials generated in the six years since the disaster. The following statements are derived from my research efforts and
material:

I. REGARDING THE MOTION TO REMAND BASED ON THE ISSUE OF IN PERSONAM JURISDICTION

A. Union Carbide Corporation (UCC) is the alter ego of Union Carbide India Limited (UCIL). . . . Specifics connecting UCC to the Bhopal subsidiary
are as follows:

1. 50.9% of UCIL stock was held by UCC.
2. UCC approved a site choice for the Bhopal plant that situated hazardous production facilities adjacent to already existing residential communities and barely two
   kilometers from the main railway station. The Bhopal plant fell out of compliance
   with city ordinances when, in 1974, it began manufacturing rather than simply
   formulating pesticides. As the twenty-first-largest company in India and employer
   of over 10,000 people, UCIL was sufficiently influential to override the city's ob-
   jections to their location through approval from central and state government
   authorities.
3. The decision to use and process methyl isocyanate (MIC) in Bhopal was made by
   UCC. From 1958 to 1973, the end product Sevin was manufactured without using
   MIC. Carbide switched to a MIC dependent method when it became the cheaper
   alternative. During the first years the Bhopal plant was operating, MIC was imported from the parent company. The decision to build a MIC processing unit in
   Bhopal was based on the need to "backward integrate" so that raw materials
   were produced on site, thus saving transportation costs and exploiting economies
   of scale. The context of the decision to manufacture MIC domestically was one
   of increasing competitiveness and industry decline due to local agricultural
   conditions.
4. The process procedure used to manufacture MIC in Bhopal was an open circuit
   process chosen, for economic reasons, over the closed circuit process used by
   Bayer Corporation. Carbide's open circuit process required bulk storage of MIC
   and thus precipitated the storage management problem that led to the disaster.
5. Design of the UCIL plant and particularly the MIC unit was carried out and approved by UCC. Significant plant design features approved by UCC include the
   following:
   a. Regulatory and alarm mechanisms at the Bhopal plant were manual and de-
      pendent on human detection. At a sister plant in Institute, West Virginia, con-
      trol systems were automatic and computer monitored.
   b. Union Carbide made the decision to rely on bulk storage knowing of the dan-
      gers. An affidavit on the decision to bulk store was filed in the Federal District
      Court in Manhattan by Edward Munoz, a retired Vice President of UCC and
      Managing Director of UCIL during the design of the Bhopal MIC unit. Munoz
      states that UCIL personnel preferred a design plan for nominal storage based

solely on downstream process requirements. UCC insisted on large-scale storage.

c. The refrigeration unit was too small (30tn) to help control a runaway reaction. Further, a water based brine solution was used as a coolant whereas in Institute a more expensive chloroform cooling system was used. The brine solution could have been a source of water contamination of storage tank contents.

d. In Bhopal, there was no means for continuous check of the purity of MIC. In case of failure in the final refining process, off-grade MIC would be mixed with previously stored material, introducing large-scale contamination and danger. In Institute, interim tanks were provided between the refining system and the storage tank so that newly produced MIC could be checked for purity.

e. The storage tank was pressurized through copper tubing, making it possible for copper filings to be mixed with MIC. Union Carbide information on MIC itself states that copper can cause a dangerously rapid trimerisation, generating sufficient heat to cause a reaction of explosive violence.

f. Vent gas headers were made of carbon steel. In the absence of check valves, back pressure inside the header would push material back into the storage tank, possibly carrying with it rusted coating from the vent header walls. Carbon causes a catalytic reaction when in contact with MIC.

g. A jumper line modification was made in May 1984 to provide a standby in the event that either the relief valve vent header or the process vent header needed to be shut down for repair. According to the Indian Central Bureau of Investigation inquiry into the disaster, approval for this design modification was given by UCC. The jumper line connected the relief-valve vent header to the process-vent header, allowing water ingress into the MIC storage tank. MIC in reaction with water set off the exothermic process that led to the pressure build up and release of gas into the atmosphere of Bhopal.

h. The vent gas scrubber, intended to neutralize leaks with a caustic soda solution, was built to handle a maximum pressure of 15psi. The rupture disk channeling gas into the scrubber was set to release at 40psi. Thus, even in circumstances of controlled flow, the scrubber could only accommodate 38% of the gas moving through it.

i. The flare tower, though purportedly part of the MIC safety system, was only designed for slow, steady burn off of carbon monoxide during phosgene (a component of MIC) production.

j. The water sprinkling system could not spray high enough to reach the gas. When the company fire truck arrived on the scene, it, too, was unable to spray water the 120 feet to the top of the vent gas scrubber, where the gas was gushing out.

6. Maintenance of the UCIL facility was overseen by UCC. This was demonstrated by periodic safety checks by UCC personnel. On at least three occasions, UCC safety auditors recommended the formulation of an evacuation plan for the residential communities adjacent to the plant. UCC was aware these recommendations were never carried out.

7. Quality and quantity of plant personnel was overseen by UCC. Worker manuals were produced in the United States and distributed in English. Senior plant personnel were given training at a sister plant in Institute, West Virginia; 80% of these workers left UCIL in the four years preceding the disaster due to low morale partially caused by awareness that the Bhopal plant tolerated negligence and lack of safety consciousness. The plant's 1982 operational safety survey documented regular breaches of basic safety rules and warned of the problems that could accompany staff reductions. Please see Appendix 1 for graphs showing staff reductions.

8. UCC financed and managed public relations after the disaster, first by shifting crisis management activities to corporate headquarters in Danbury, Connecticut and
then by hiring Burson Marstellar, one of the largest public relations firms in the
world.

9. UCC managed "Operation Faith," the effort to safely neutralize the MIC remaining in storage tanks after the major leak on 2/3 December.

10. UCC was the defendant in the Indian litigation of the Bhopal case; UCC was the
party that negotiated and accepted the February '89 settlement; UCC provided
nearly 90% of the $470,000,000 settlement amount to the Indian Reserve Bank
(UCIL provided 45,000,000); news of the settlement caused UCC stock to rise
$2 a share.

B. Union Carbide India Limited (UCIL) is present in Texas through their
relationship with Humphreys and Glascow, a Bombay engineering firm
domiciled in Texas under the name of Ensearch. From the period 1972 to
1980, UCIL supervised 55 to 60 engineers employed by Humphreys and
Glascow in detailing summary construction plans provided by UCC.

C. Ensearch, a defendant in this case domiciled in Texas, is responsible for
many of the design decisions that led to the 1984 gas leak. Ensearch was
hired to adapt UCC's "process design package" to the context of the Bhopal
facility. Ensearch was involved in the decisions that led to design and construction much less safe than that in the Union Carbide plant in Institute
West Virginia.

II. REGARDING THE MOTION TO DISMISS BASED ON ADEQUATE
ALTERNATIVE FORUM AND THE MOTION FOR SUMMARY JUDG-
MENT BASED ON THE ALLEGED EFFECT OF THE SETTLEMENT
NEGOTIATED IN THE INDIAN PROCEEDINGS

A. Since 1985, it has become clear that India cannot provide an alternative
adequate forum. The February '89 settlement pre-empted a trial on the merits of the case and came about through numerous failures of due process:

1. On March 29, 1985, the Indian Parliament enacted the Bhopal Gas Leak Disaster
(Processing of Claims) Act, granting Union of India (UOI) the exclusive right to
represent the victims in India or elsewhere. Thereafter, UOI has represented the
Bhopal victims in the capacity of parens patriae. The Bhopal Act itself is a breach
of due process because it involves conflicts of interest: while representing the victims, UOI also owns a substantial portion (22%) of UCIL stock and controls the
courts, since judges are appointed and there are no juries. Thus, in the Indian proceedings, UOI has been plaintiff, defendant and the court.

The overlapping responsibilities of UOI have precluded the possibility of an im-
partial tribunal. This structural problem is magnified by the current political-
economic environment in India. Since the early eighties, the orientation of Indian
development has changed. In the decades following Indian independence in 1947,
there was an overt attempt to maintain self-sufficiency through minimalized dependence on foreign investment and technology transfer. This agenda did not provide
adequate growth to keep up with the demands of a growing welfare state and increasing national indebtedness. Hence, policy changes in the eighties redirected
the economy toward greater utilization of outside resources.

The current dependence of the Indian economy on foreign investment has under-
mined the capacity of the Indian State to fairly evaluate the distribution of risks

and benefits that accrue through industrial development. Because it is obliged to attract foreign investment, the Indian government is forced to let market considerations override all other concerns. Thus, it is impossible for the Indian government to fairly adjudicate the claims of those victimized by corporations fulfilling investment demands. It is publicly recognized in India that a harsh decision against Carbide would be a deterrent to economic goals. The problem of Third World governments being caught in a double-bind wherein they promote their economies at the cost of safety for their citizens is well recognized in the social science literature.

2. There has been a failure of representation on the part of UOI as *parens patriae* of the gas victims. This failure is indicated by the paucity of research carried out before arriving at a settlement amount. At the time of the settlement, fewer than 50,000 victims had been assessed for personal injury. The results of this initial assessment were not made public until April 1989, two months after the settlement was finalized.

The great majority of victims contest the adequacy of the settlement because it is insufficient to cover compensation and rehabilitation costs. Approximately 600,000 persons have filed for damages based on physical proximity to the plant. Medical monitoring of the health status of these claimants alone would cost in excess of $6,000,000. Conservative estimates for full compensation range between one and two billion dollars. Please see Appendix 2 for details. Further, according to the terms of the settlement, the Indian government has first access to reimburse itself for litigation costs, interim relief payments and general economic loss. Government reimbursement alone could easily consume much of the settlement, leaving little for the present and future claims of victims.

Within India, it is not possible to correct the settlement's inadequacies because of UOI's involvement in denial of the magnitude of the disaster. Medical categorization data produced by the Madhya Pradesh State Government and released in 1991 indicates that out of a gas affected population of over half a million people, only forty individuals are permanently disabled. The release of this data generated disbelief and outrage within the medical community. Doctors affiliated with the organization Socially Active Medicos (SAM) have denounced the data as corruption of scientific means for political ends. They critique the testing protocol for being insufficiently thorough and for blatantly ignoring long term and multisystemic ailments. Please see Appendix 3 for elaboration on problems with medical categorization data now available in India.

Attempts to produce alternative medical documentation through the voluntary sector have led to confiscations and arrests. Officially produced data is not available to interested health professionals. Hence, the research basis for representation of gas victims in India is unavailable.

3. Victims were given no notice of the settlement prior to it being finalized and publicly announced. Such notice was clearly possible through public news releases or through mailings to victims included on a computerized list used for registering claims. Notice also could have been given through any of the victims' organizations in Bhopal. The Supreme Court of India itself recognized this failure of due process during its review of the Bhopal Act. While the Supreme Court upheld the validity of the Bhopal Act, it conceded that victims were not given notice.

Former Chief Justice of the Indian Supreme Court P. N. Bhagwati has critiqued the settlement decision for the following reasons:

a. The settlement order emerged from appeal arguments and not from proceedings directed toward final judgment. In disposing of the compensation case, Bhagwati argues that the Indian judiciary overstepped accepted principles of jurisprudence.

b. The settlement amount of $470,000,000 was not based on knowledge of re-

> habilitation costs and "places the value of Indian life at a ridiculously low figure."
>
>   c. Settlement offers and counter-offers were not made in open court but in the Chamber of the Chief Justice. In failing to consult or even notify victims, Bhagwati argues, "the Government forgot that it was not prosecuting a case of its own but was acting as a trustee." (*India Today* March 15, 1989)
>
> 4. The 1989 settlement order did not contain a distribution plan. Such a plan still has not been publicized. Thus, it is not known how much will be allocated to each individual, particularly since the Government has first option to reimburse itself for expenses. Further, the settlement order did not give attention to future claims.
>
> 5. Victims have never been given the opportunity to opt out of the February '89 settlement or of the litigation all together. However, sustained protest by victims has clearly indicated opposition. In a major demonstration protesting the settlement in August 1989, police brutally attacked demonstration participants. This attack is publicly recognized in India as a show of the government's refusal to attend to victims' demands. See Appendix 4 for documentation on victim protest against the settlement. . . .
>
> ## CONCLUSION
>
> Evaluation of the Bhopal phenomena has indicated three major points of particular relevance to this litigation:
>
> 1) UCC played a controlling role in all UCIL functions. Plant design features that precipitated the disaster were put forth and finally approved by the parent company. Other significant factors, including site location and personnel reductions, were also approved by UCC.
>
> 2) The medical categorization data generated by the Madhya Pradesh State Government is grossly unscientific and unacceptable as a basis for rehabilitation efforts. Controversies over this data have revealed UOI's role in denying the magnitude of compensation requirements both to veil their own culpability and to protect broad economic development agendas requiring foreign investment. It has thus become clear that Indian courts, as an arm of UOI, cannot provide an adequate forum to adjudicate the Bhopal case. Conflicts of interest are further complicated by UOI's role as parens patriae representative of the victims.
>
> 3) The settlement amount of $470,000,000 determined in the Indian proceedings is clearly insufficient to compensate and rehabilitate over half a million Bhopal victims. Conservative estimates have costed rehabilitation needs in the range of 2 billion dollars. The "full and final" clause of the settlement is of particular concern due to continued health deterioration and the likelihood of long-term effects.
>
> SIGNED THIS, the 14th day of August, 1991
>   Kim Laughlin
>
> SWORN TO AND SUBSCRIBED before me, the undersigned authority, on this, the 14th day of August 1991.
>   Notary Public, State of Texas

No one language or style of advocacy is sufficient.[2] Movement across different institutional domains demands change in both the style and the substance of one's rhetoric, adherence to different protocols of conduct, and reimagination of what expertise one has and how it should be shared.

Movement across class, race, and gender divides is particularly difficult, demanding continual explanation of what one is doing, and why.

Middle-class progressives have to morph, changing to fit the times and places where there is work to be done. Their advocacy takes many forms, often operating in many places at once—requiring constant movement between different orderings of the problems and recognition that the organizing principles at work in one domain are incongruent with those at work in other domains, and never congruent with "the whole truth." Different kinds of evidence, modes of argumentation, and narrative forms signal different contexts. If shaped for intervention in Supreme Court proceedings, arguments often rely on syllogistic logics that directly engage official statistics. Other narrations highlight the treason in translating continuing disaster into linear logics of cause, effect, and proof. The challenge is not to stand outside established systems, but to find places to work within them—finding repose in none, recognizing that neither poetry nor law is sufficient in itself. In some instances, categorization is necessary, despite the reductions involved. In other instances, less formalistic modes of representation work best. At times, it is important to write in a way that questions the efficacy of all description. The challenge is to move between different places of articulation, recognizing that no one form can tell the whole story.[3]

Places of articulation are not, however, stable or discrete. One is pulled from one articulation to another, into often unfamiliar territory. Talk about health care in Bhopal leads one to talk about the role of multinational pharmaceutical corporations in the Third World, about Gandhian alternatives, about the etiology of toxics, and about the limits of scientific knowledge. Talk about environmentalism as a matter of justice crosses into claims that trees, too, have rights, into feminist critiques of the discourse of rights, into questions about the role of nations and nationalism in contemporary world order, into spaces where one must (definitively) state that the law has failed its promise. Issues and discourses crosscut and reciprocally qualify each other. The "right thing to say" is gauged by different, often incongruent, techniques of measurement.[4]

Collisions between worlds of discourse bring the "whole world" together. It is within these collisions that middle-class progressives work—without a model, figuring out what to do next *within* a system of differences.[5]

Middle-class progressives are catalysts—making small moves within complex systems that can provoke ripples and realignments throughout the system. Effectiveness is often in the details. Style is important, as is the ability to work with many different kinds of knowledge to understand where and when inventive intervention is called for. Catalysts, however, usually operate without affecting the initiating agent. Within disaster, their operation is not so clean. Disaster implodes clear distinction between the active and the passive. One can no longer tell whether subjects are acting or being acted upon. All move within processes that they affect without controlling. Subjects accomplish things that are accomplished in them. Progressive advocates end up in the middle voice.

The middle voice is the modality of disaster. It is a way of speaking that builds in awareness that the system that one critiques also operates within the critique. It is a way of understanding how actors within complex systems are distribution points—sending out messages, but also being encoded by them—making change happen, but also being changed.[6]

(1.3) Justice Before, and After, the Law

The Bhopal case was dismissed from U.S. courts in May of 1986. Three years later—on February 14, 1989—the case was settled out of court for $470 million by an Indian Supreme Court order adjudicating the claims of Union Carbide and the government of India, which had represented victims under the guardianship principle of *parens patriae.* The settlement announcement included no specifics on how the final award was calculated, no plans for compensation distribution, and no explanation for why the Indian government's previous proposal for seeking approximately $3 billion in damages was abandoned. Subsequent explanations by the court stated that counsel for both parties were willing to accept whatever amount the court deemed appropriate, within the range of $426 million, the maximum offered by Union Carbide, and $500 million, the minimum demanded by the Indian attorney general.

Three months later the rationale for the final settlement amount was disclosed—as a medical categorization scheme provided by the Madhya Pradesh state government. Based on this scheme, the court had allotted approximately $14,500 to compensate each of 3000 families in which a member had died. The 30,000 people identified as "permanently disabled" were allotted approximately $5,200 each. The 20,000 people identified as "temporarily disabled" were allotted approximately $3,215 each. A final medical category—for people who had suffered injuries of "the utmost severity"—would be the basis for up to $25,000 in compensation. Another portion of the settlement amount—approximately $15.6 million—was to be used for general medical treatment and rehabilitation. The remainder— approximately $140.6 million—was a general category to cover an estimated 150,000 claims for minor injuries, property damage, business losses, the costs of dislocation, and future injury (Cassels 1993, 229).

The rationale for the settlement amount was contested for many reasons. The most obvious was the adequacy of the amount allocated for people in each category. Even by "Indian standards," the amounts could hardly be called liberal. Further, the validity of the numbers associated with each category—and even the categories themselves—was certainly debatable. And it was unclear how people—particularly those to be accommodated in the largest and most general category—would be able to prove their right to any compensation at all. The vague anticipation of future injury was also a problem, particularly given the paucity of data available on the long-term effects of exposure to MIC or on the other toxins released as by-products.

With time, the Indian Supreme Court did acknowledge a "denial of natural justice" in not allowing victims the opportunity to opt out of the settlement of the Bhopal case. The court nonetheless upheld the structure and process of the case by insisting that the "situational particularities" of Bhopal legitimated an abrogation of rights in order to secure justice. The motivation was bureaucratic, as well as political.

A hasty, definitive settlement was said to be necessary in order to move forward with rehabilitation, which gas victims desperately needed. The medical categorization scheme provided by the government of Madhya Pradesh both justified the settlement amount and provided a structure on which rehabilitation programs could be built. The court's argument thus implied that the delivering of justice to the poor masses of India required subordination of complexity into categories with which bureaucracy could work. In other words, the court forced a fit between social problems and existing institutions and modes of analysis.

The political motivation for the settlement was less overt—and more complicated. The Bhopal Act literally reinstated the government of India as guardian of the Indian people, a role played since independence through the programs of scientific socialism. The question, however, was who counted as "the Indian people." The Bhopal Act established the government of India as the guardian of particular people—the victims of the Bhopal disaster. The government of India, however, was also the guardian of the Indian people in general. The government of India was thus caught within competing obligations. In the view of government officials, the future of India depended on foreign investment. The Bhopal case was an opportunity to demonstrate how India dealt with multinational companies, even within disaster. The particularities of gas victims themselves had to fit within this general frame.

By the court's own admission, the Bhopal Act provided victims access to the law, but not to rights. It denied gas victims the right to represent themselves. And, in effect, it also denied them the chance to "opt out" of representations made by others in their name. In defense of this move, the attorney general argued that while rights are indispensably valuable possessions, they might be theoretically upheld, while the ends of justice are sacrificed. The appropriate response is a curtailment of rights, such that the largest good of the largest number is served. Justice, then, was a utilitarian quantification.

Progressive activists understood the purpose and promise of law otherwise, arguing that the Bhopal case was both a failure of law *and* evidence that the law in itself could never be a means to justice—and thus could never be a final solution.

One critique addressed the dismissal of the Bhopal case from U.S. courts. When Judge Keenan dismissed the case from U.S. federal courts, he insisted that India be "allowed to stand tall before the world and determine her own destiny." Keenan failed to acknowledge systemic constraints on Indian determination of "her own destiny." Most obvious was his failure to recognize the forces set in motion by trade liberalization trends, coupled with the lending requirements of the International Monetary Fund (IMF)—which, in the 1980s, escalated into demands for "structural adjustments" that would encourage foreign private investment in countries holding IMF loans, and possibly seeking further loans, credit, or approval of payment deferral plans. Liberalization of India's economy began as early as 1980, fueled by a massive IMF loan in 1981 (Omvedt 1993, 177). As the settlement was being reevaluated in the early 1990s, the effects were becoming clear. The U.S. corporate commitment to new investment in India, for example, continued to grow, from approximately $20 million in 1990, to $70 million in 1991, to $387 million in 1992. John Macomber, the chairman of the U.S. Export-Import Bank, even described India as the most important new market for the United States in the 1990s (Kreisberg 1993, 59).

Keenan also failed to acknowledge systemic effects in his decision to discount the role of Union Carbide Corporation, the parent company, in the production of the disaster. In dismissing the Bhopal case from U.S. courts, Keenan implied that Union Carbide India Limited, a subsidiary of Union Carbide Corporation, could determine its own destiny—and was thus responsible for the disaster. It was not considered significant that the plant was designed in the United States and operated according to procedural and cultural logics distributed by the parent company.

Like in his judgment of India's autonomy, Keenan's judgment on parent company liability failed to recognize difference as asymmetry. The courts of India were said to be equal to those of the United States in their capacity to provide an autonomous forum in which justice could be determined through due process. Likewise, Union Carbide Corporation and Union Carbide India Limited were considered equally capable of controlling factors that contributed to the gas leak. Once these equivalences were established, place became the basis for sending the case back to India. The gas leak occurred in India and affected people there, so it was more convenient to litigate there. Simple contiguity provided justification, despite the complexity of the case. In denying jurisdiction, Keenan also denied the hierarchical order of relations between the United States and India, as between Union Carbide Corporation and Union Carbide India Limited.

In India, a related logic worked. The Bhopal case was litigated as a class action tort in which mass numbers of claims were consolidated and represented by the government of India, operating as *parens patriae,* or legal guardian. Conceptually, *parens patriae* invokes the authority of a legislature to provide protection to persons who are *non sui juris,* or judicially incompetent. Gas victims were assimilated into this category on the grounds of their lack of resources for pursuing litigation on their own behalf. The problem here was also a matter of difference. The structure of *parens patriae* suggests that the interests of the government of India are the same as those of gas victims. Obvious conflicts of interests—including the fact that government financial institutions owned 22 percent of Union Carbide India Limited—were not considered significant. Nor were the conflicts of interests that arose as the government of India tried to reposition itself on the global stage.

Homogenization of difference was also a problem in the process of the litigation in India. The medical categorization scheme on which the settlement was based was treated as though it matched the reality on the ground in Bhopal. Distributing compensation in accord with the categories included in the scheme was considered the same thing as distributing justice. Systematic mismatch between the categories and the reality they were supposed to represent was effaced.

Progressive critiques of the Bhopal case can be understood in structural terms. They also can be understood in terms of time out of joint. Judge Keenan's decision suggested that world order in the 1980s was *already* sufficiently symmetrical to justify denial of jurisdiction by U.S. courts. Thrown back to India, the Bhopal case ran on paternalistic and scientistic logics not unlike those of colonialism. Gas victims, like colonial subjects, were sucked into the law, but not granted competence. Nor could they opt out. Thus, while the United States pushed India ahead of the times, the government of India reinstated the past.

A different enactment of the Bhopal case would have required a shift in sign systems. Instead of operating by territorial logics, U.S. courts would have had to be as global as the case they encountered, admitting that the Third World poor are *already inside* their jurisdiction and thus could not (logically speaking) be required to beg access. Once named juridically incompetent, gas victims would have had to become legal subjects, with rights and resources to represent themselves.

But turning gas victims into subjects *before* the law would have remained insufficient. Justice would require that gas victims become subjects before *and after* the law.

Progressives responding to the Bhopal disaster have critiqued the law for being insufficiently lawful. The courts handled the task of adjudicating difference badly. But matters are worse than this. Even if the law had been lawful, something would have been left out. A residual would have remained, unencompassable with the law in its most perfect instantiation. Disaster demanded a legal response. But it was also beyond the scope of legal remedy—not only because the injuries suffered by gas victims could never be adequately categorized, much less compensated, but also because of the law's way of being lawful (through adjudication and thus the reduction of dissenting arguments into a single judgment). The law demands generalization. The Bhopal disaster calls for such generalization, but it also calls for something more.

Progressives I worked with expressed their understanding of how the Bhopal disaster "spilled over the law" in different ways. What they expressed provoked chronic awareness of the need to rethink what "counts," as knowledge, as adequate representation, as a justifiable plan of action. Bhopal becomes *Bhopal*—untranslatable, signaling more than can ever be narrativized into languages of law or bureaucracy.

(1.4) Material Effects

Responding to the Bhopal disaster through legal initiative easily collapsed into cynicism about the potential of law to do any good whatsoever. The tensions can be traced throughout my field notes. In one excerpt, my notes trail off with reference to missed deadlines, detail on the legal documents that I needed to retrieve and mail, and a description of how we argued over whether it was worth it at all.

Later that same week I came back to my field journal, writing in excruciating detail of the broken arms and bruised bodies of villagers who had come to Bhopal to prepare a report to be submitted to the governor, pleading for protection from the police. It was a new strategy. The activists living in the affected area called themselves Lohians (Lohia 1955, 1985, 1987). They worked to foster self-sufficiency through small-scale, local initiative that didn't rely on any input from the state. In principle, they scorned legal initiative, arguing against any dependence on abstract, bureaucratized ways of solving social problems. But violence carried out by official representatives of the law could be countered only through the law itself.

Our household helped with press and medical arrangements, cooking, and looking after children. I typed the police report, and two letters to the Supreme Court Legal Aid Committee. One letter was from the villagers, and the other was from a middle-class activist named Sunil. The police report was signed with the imprint of a thumb.

DATE: DECEMBER 1991
FIRST INFORMATION REPORT (COPY)
POLICE STATION: KESLA
CRIME NO.: 48/91

NAME: BUDHRAM
FATHER'S NAME: BHAJNA
CASTE: KORKU

I live in Rajamarihar and I am engaged in agriculture. Today morning my brother Sejram went out between 6 and 7:00 A.M. Around 10 to 12 Sardarji [Sikhs] started beating him. When I went to protect Sejram, I was also beaten. I received injuries on the wrist and thumb. Sejram received injuries on his back, thigh, etc. These people beat other villagers also. Umrao s/o Kanak Singh, my mother Suddo Bai, sustained injuries on the head, left shoulder and left hip. People were beaten here and there. My mother was sitting near the fire and her ten-rupee note was burnt. Baran s/o Teji, Antram, Rameshwar, my wife Sukhwati, received injuries on her left fingers. Phoolwati's matka [water pot] was broken. Sukhdev and Lalsaheb were also injured. Sejram s/o Bhajna, Atarsingh s/o Babloo and Atarsingh's wife also received injuries. The Party was accompanied by Excise Sub Inspector Jain. The attackers had gone to seize liquor. Sejram and Atarsingh are serious. These attackers surrounded and beat villagers. I report for proper action. The report was read to me. It is written correctly.

SD/-

FILE: letter to Legal Aid Committee, Supreme Court, Delhi
DATE: JANUARY 1992

Sir,

We are residents of village Rajamarihar, block Kesla, Dist Hoshangabad, Madhya Pradesh. We are all tribals. We were displaced by the establishment of a proof range around 20 years back. This literally destroyed us. We do not have any proper employment opportunities. By tradition, we brew our own liquor for our own consumption. We are harassed by the local Excise Department and the goondas of the liquor contractor on this ground.

In a recent incident on December 25, 1991, Excise Sub Inspector Mr. K. C. Jain, accompanied by 15–20 hired goondas of the liquor contractor,

attacked our village at around 6:00 A.M. The first villagers to be attacked
were the ones returning after nature call. After this, the "raid" party entered
people's houses and attacked them. They even beat up children, women and
the old. The attackers were armed with lathis and swords. Around 20 people
were injured, out of which 4–5 seriously. In addition, the goonda party also
threatened to rape women. They also took away a lota and a shovel. They
ransacked the houses and damaged things.

The whole incident was reported to the Kesla police station the same day.
Nine of the injured persons were medically examined. The local MLA, Dr.
Sitasharon Sharma, also arrived from Itarsi and demanded strict action. Yet,
the police let the criminals go away. Two days after the incident, on December 27th, villagers held a demonstration at Kesla Police Station, after which
8 goondas were arrested.

1. However, they were charged with very ordinary crimes (such as IPC 147, 148,
   149, 323, 452), which were extremely insufficient, looking at the seriousness of the
   crime and injuries therein. When questioned on this issue, the police officials said
   that new charges would be instituted after medical reports were received. However,
   even 15 days after the incident, medical reports have not been received from the
   hospital, according to Kesla P.S. In the meanwhile, the goondas were released on
   bail just after two days.
2. Excise Sub Inspector K. C. Jain, under whose leadership the attack was planned
   and executed, has neither been arrested nor suspended nor transferred. In other
   words, no action has been taken against him as yet. Jain is known to have committed such atrocities before and always uses private goondas.
3. Although swords were also used by the goonda party during the attack, police have
   only shown confiscation of lathis.

It is quite clear that police are trying to shield and protect the criminals. We
are afraid that this will act as a morale booster for criminals and the incident
might recur. In the whole of the tribal area where we live, excise, police,
forest and other government officials harass people and commit atrocities
day in and day out. Yet, no action is taken against them nor any preventive
measure adopted. Therefore, we seek help in this incident to secure justice
and ensure non-recurrence of such incidents in the future.

Residents of Village Rajamanihar,
Block/ Police Station Kesla, Dist Hoshangabad, (MP)

The letter to the Supreme Court Legal Aid Committee from Sunil, zonal
convenor of Samata Sangathan, began by referencing enclosed documents,
in English translation, which described "an incident of atrocity on tribals by
the Excise Department of Madhya Pradesh and private goondas [thugs]." It
then noted that such atrocities were quite common and requested support
in "fighting the pattern of abuse and to get justice in this particular incident." The Legal Aid Committee was also reminded that one of their own
members had personally visited the village, met and talked to the victims,
and seen their wounds and marks of beating.

Sunil's agenda for action was straightforward: the culprits, including the
excise subinspector, should get proper punishment so that such incidents

didn't happen again. The victims should get proper compensation, since they couldn't afford hospital treatment, and had immediate food shortages, since they were unable to work. And the excise policy and laws of the government should be implemented in ways that didn't turn into atrocities on tribals.

The victimization described in these documents was recurrent and systemic. It couldn't be fully represented in the language of law. Nor could legal language express the irony of asking the law to oversee itself. The law nonetheless had to be called upon to respond.

Terrible cynicism about the law, coupled with awareness of how much law means, has shaped the work of middle-class progressives responding to the Bhopal disaster, as at other sites of recurrent victimization throughout India. Strategizing within and around the law's insufficiency has demanded humility, creativity, and a stubborn commitment to the pursuit of justice—even when imperfect, constrained, and beset by contradiction. The importance of this work is often hard to track, particularly through formal legal decisions. The daily proceedings of the claims courts in Bhopal are a better index:

> *Judge:* You have been categorized as "C" which entitles you to 25,000 rupees in compensation. Do you accept?
> *Plaintiff:* Sir, I have already spent that much in medicines.
> *Judge:* Do you wish to contest?
> *Plaintiff:* Sir, do what you think is fair.

(1.5) *De Bonis Non*

The Indian Supreme Court argued that the "situational particularities" of the Bhopal case legitimated both the way the case was structured and the final judgment. Structuring the case in terms of parens patriae was said to be necessary because victims lacked the resources to represent themselves. The final judgment—a hasty settlement—was also justified through reference to the details of gas victims' lives: gas victims were poor, so the court was obligated to settle the case so that resources could be distributed sooner, rather than later. Ironically, however, the medical data on which the settlement was based limited the welfare responsibilities of the Indian government and rationalized victims into categories amenable to bureaucratized rehabilitation schemes, but with little diagnostic or therapeutic relevance. The "situational particularities" of the Bhopal case were left off the record, for Bhopalis to deal with the best way they could.

A pamphlet produced by the BGIA for the sixth anniversary of the disaster dealt with the "situational particularities" of the Bhopal case differently. Rather than reducing disaster to a single legal judgment, *Voices of*

Bhopal highlights the multiplicity of disaster on the ground in daily encounters with hospitals, bank clerks, and myriad categorization schemes. The attempt was to rewrite the record, detailing what the law ignored.

VOICES OF BHOPAL, 1990
BHOPAL GROUP FOR INFORMATION AND ACTION

Interview with Natthibai, age 55, resident of Rajendra Nagar
My husband's name was Dukhishyam. He got a lot of gas in him. On the night the gas leaked, both of us ran towards the forest. He remained sick afterwards. He used to get breathless, cough and his eyes would get very big. He could not see properly after the gas. Twice he was admitted to the hospital. The second time he was admitted, he never came back. He died in the MIC ward. I gave an application for Rs 10,000 [$400] in interim relief, but they haven't done anything about it yet. Last year, he died in Kunwar [autumn]. They haven't yet told me whether I will get Rs 10,000 or not. I gave them all the medical prescriptions of my husband with my application.

I stay sick. I have come back from the hospital on 13th of this month [November 1990]. I was there for one and a half months. I never got breathless before the gas; I used to work as a laborer. Now I get badly breathless and my chest pains. I was in the hospital during the Festival of Lights. This gas has destroyed us completely.

Interview with Sabra Bi, age 40, resident of Kazi Camp
I have been in and out of several hospitals since the gas disaster. In 1986, I was told that they were registering the claims of the gas victims. So I took my children and stood in the queue to get my claim form filled. It was a long queue and there were at least 250 people before me. When my turn came up, the fellow who was filling the claim forms said that he would not fill claims for children. He said only people over eighteen years could file claims. When I insisted, he asked me to put the medical prescriptions of the children along with their father's claim form. But that did not work. Later, when people were receiving notices to get themselves medically examined, there were no notices for my children. So their medical examinations were not done.

I was in the hospital when the people who were carrying on the [Tata Institute] survey came to my place. They took down the names that were listed in the family ration card. But all the names were not shown there. The ration card was issued fifteen years back and only three of my six children were listed on it. When the claims forms were filed, my daughter Afroz was twelve years old. Gulnaaz was ten years old. Mehenaaz was nine years. Neelofar was seven years and the youngest, a son, Firodous, was two years. The government fellows did not put any of these children's claims in their register.

Interview with Badruddin, age 50, resident of Pulbogda
I was asked to report to the Identification Center on 23rd of last month [October 1990]. There they told me that the names of my children that were on the notices did not match with the names they had on their records. The names of my two sons and one daughter were wrongly recorded. I went to

the Collector's office to get the names corrected. I was made to go from one office to another. It took three days to get the names corrected. And my daughter's name is yet to be corrected. I showed them all her medical papers, even her affidavit, but they are yet to correct the name.

Interview with Shaheem Bano, age 30, resident of Budhwara
Three of my children have yet to file their claims. All three were born before the gas disaster. The oldest, Samad, is 12 years old; then Malka is 8 years old and the third son, Amjad, is 7 years old. When they were filling the claims forms, I told them to file the children's claims. But they said such young children couldn't file claims. "We will put these children's claims along with their father's papers," they said. But they did not even write down the names of these children. Earlier, when the survey people had come, I told them the names of all my children. But they took down names from the ration card. Our ration card is twenty years old. It does not have the names of all the children. I told those government fellows that my children have been left out. All they say is "We will see, we will see."

Interview with Ramkishan, age 40, resident of Chhola Road
I used to work at the Formulation plant in Union Carbide's factory. I had been working there ever since the sixth month of the year 1973. When I joined, I used to work as a casual worker. For six years, I worked as a casual worker. They made me a permanent worker in the third month of 1980. I was working in the Formulation plant on the night the gas leaked. The tank of MIC which leaked was only 400 feet from the Formulation plant. When the gas started leaking, some people cried out "Run, Run" and we left our work and ran towards the west.

Later, the factory was closed down. There was nothing for me to do. The government offered me jobs but they were all away from Bhopal. I was given jobs in Mhow, Rajgarh and Indore, none in Bhopal. My wife had taken a lot of gas and she was pregnant at the time. So I could not stay away from Bhopal. Now I work as a daily wage laborer. I get jobs 15 to 20 days in a month and make about 20 to 25 rupees in a day [$1]. There are quite a few Carbide workers who could not find employment after the factory was closed down. I personally know about one hundred of such workers. After the factory was closed, I was given six months salary as compensation, nothing else.

I was just a worker there, how could I know what poisons were stored in there? I was never told that there were such dangerous chemicals inside the factory. If I knew, I would not have worked in that factory. The plant used to smell awfully at times, but we were just workers; how could we know? When we worked there, our eyes used to hurt and our skin itched but whoever knew that such a disaster could happen?

Interview with Jaya Mane, age 28, resident of Rajendra Nagar
I get breathless when I walk and now my head aches so badly that I cannot do any sewing or reading. The doctors had done my medical examination. But later they sent a notice that said they found me to be only temporarily injured. I do not know whether we can ask them to do another medical examination. I have started to get interim relief but the bank is quite far. I have to spend Rs 20 [$1] to go to the bank.

Interview with Sher Khan, age 45, resident of Chhola Road
I work at the railway coach factory. I have been living in this house for the
last 50 years. This year, they announced from a jeep that they would demol-
ish the houses on both sides of the road. People in my community and the
tenants who had rented shops were very troubled. Quite a few cried their
hearts out. Those who protested got arrested. Some people who tried to argue with the government officials were beaten up by the cops. And the bulldozers went on demolishing house after house as we watched in silence and
sorrow. After my house was demolished, I was so sad I could not eat food
for four days. It was raining then and my children were crying. I cried too.

I have made a tiny shelter out of whatever was left after the demolition. I
still have the registration papers for my house. This house had been there for
the last sixty years and they never told us that we had encroached upon government land. This anti-encroachment drive is a lot of bunkum. It was as if
the government had declared war on the people. All day they would carry
out the demolitions and at night they rested till they blew the bugle the next
morning. They cut off the water connections, the electricity connections and
turned us homeless. They have not given us any compensation or any land.
First, we were attacked by Union Carbide gas and then by the government's
bulldozers. Where are we to go?

Interview with Sahodra Bai, age 55, resident of Lakherapura
My husband Shantilal died on 12th May 1990. After the gas, he had difficulty breathing. He never went to work after the gas. He couldn't earn any
money. The children earned something by doing odd jobs. I cannot see properly and I get breathless. I cannot do any work. Union Carbide is responsible
for my husband's death. I should be given relief money of Rs 10,000 [$400]
and should be given enough compensation so that we have enough to eat and
get ourselves treated.

Interview with Puniya Bai, age 65, resident of Chhola Road
This was an old house. My husband's parents built it a long time back. And
now the government has demolished it. They did not give us any notice before breaking it down. It has been six months since the bulldozer demolished
my house and they have not talked about giving any compensation. I do not
want money; I should be given a house somewhere. They broke down the
house in the middle of the rainy season. The children had no shelter and
there was no place to light a fire or to cook food. The house had an identification number, which was a proof that the gas had affected us. Once we are
shifted away from this place, we will lose that proof. The government fellows will not believe us when we tell them we have been exposed to the
gasses from Carbide.

Interview with Chhotelal, age 50, resident of Barkhedi
I used to work as a porter for transport companies. Since the gas, I have not
been able to work for a single day. The gas killed my daughter; she died in
the morning after the gas leak. I am breathless all the time and I cough
badly. My eyes have become weak, too. I have been admitted to the MIC
ward more than 5 times since 1987. Last year, I was there for 9 months at
a stretch. This year, I have come home after eight months.

Interview with Ganga Bai, age 63, resident of Rajendra Nagar
I have started getting interim relief of Rs 200 [$8] per month. But there are
lots of problems. To get my photograph taken, I would have to go to Indra-
puri, which is twelve kilometers away. The bank is fifteen kilometers away,
in Abiragarh. My son took me on his bicycle to the bus stand and I had to
change two buses to reach the bank. And I had to make three trips to the
bank before I got the money. Now they have brought the bank a little closer.
Even then I had to spend Rs 24 [$1] on an auto rickshaw. The banks are very
crowded and the queues are long. Once I fell down on the ground while I
was waiting in the queue.

Interview with Suleman, age 45, resident of Shanajanabad
I am in Bhopal for the last 10 years. Before the gas disaster, I used to sell
vegetables on a pushcart. I have become too weak to work now. After I was
exposed to the gas, I tried pushing the cart but I became so breathless it was
impossible to move. Some time back, I went to sell vegetables but was so ill
afterwards that I had to be admitted to the hospital. Five bottles of intrave-
nous medicine were given to me because I was so sick. Since then, I have
not tried to take the vegetable cart around. For a few days, I worked as a
watchman but that, too, was difficult. Now I have given up. I stay in my sister's place; her family arranges for my food. All the treatment that I have
taken until now has done no good. Factories like Union Carbide's should not
be allowed anywhere in the world. Not even if they build it far from human
settlements.

Interview with Mohammad Nafees, age 25, resident of Budwara
My children are named Assu and Sharik. One is about 7 years old and the
other is 6 years old. When the gas leaked, my elder son was eight months
old; my younger son was born four months after the disaster. Both of them
have difficulty in breathing and they cough a lot. The doctor says that they
have tuberculosis. They have been taking treatment for tuberculosis for the
last four years. Even now they are under treatment. I get tablets and capsules
for them from the hospital. Sometimes I have to buy capsules from the store;
one capsule costs one and a half rupee. You tell me that one should not take
drugs for tuberculosis for so long. But what do we know? We do what the
doctor tells us. No one in my family ever had tuberculosis. Both my wife
and me also have breathing trouble. Before the gas I used to spend a lot of
time working at the bakery. Now I can't sit close to the oven.

Interview with Narayani Bai, age 35, resident of Mahamayee Ka Baug
This is the sixth time I have been admitted to the MIC ward. I have been
here since the last month of 1985. When I feel a little better, the doctors send
me home but I can't stay there for long. My breathlessness becomes acute
and my husband has to bring me back to the hospital. The doctors say that
the gases have damaged my lungs badly. They say nothing can be done
about my disease. Before the gas, I had never seen the insides of a hospital.
And now, I have spent most of the last five years on this hospital bed. I used
to work as an assistant at a day care centre and now I cannot do any work.
My husband Kaluram also cannot go to his job. He used to carry loads. My
son works as a tailor; he is the only one earning in the family.

Interview with Mohammad Ajez, age 22, resident of Vidhan Sabha
I am 22 years old and the notice for interim relief that came in my name said
I was 46 years old. When I went to get my passbook made, they told me to
get my age corrected on the notice. I went back to them two days later and
they said if you pay us Rs 50 [$2], we will get the age corrected. Then they
asked me to pay them this bribe after two days. My school examinations
mark-sheet says I am 23. I have got certificates that have my date of birth
and yet I am being harassed in this manner.

Interview with Ajeeza Bi, age 30, resident of Kazi Camp
Ever since the gas, my head aches 24 hours a day. I have pain in my stomach
and sometimes feel giddy. My daughter, Nasreen, cannot see properly, cannot
thread a needle and she is only eleven. My other daughter, Sofia, also stays
sick and she is eight. I have three children from before the gas disaster and
after the gas I have aborted thrice. All three times it happened in the hospital.
Once I was six months pregnant; the second time I was seven months pregnant and the third time I carried the baby for eight months. They were all born
dead. All with black skin like the color of coal and all shrunken in size. The
doctors never told me why such things are happening to me.

Interview with Shahida Bi, age 25, resident of Bharat Talkies
A few months after the gas disaster, I had a son. He was alright. After that I
had another child in the hospital. But it was not fully formed. It had no limbs
and no eyes and was born dead. Then another child was born but it died
soon after. I had another child just one and a half months back. Its skin
looked scalded and only half its head was formed. The other half was like a
membrane filled with water. It was born dead and was white all over. I had
a lot of pain two months before I delivered. My legs hurt so much that I
couldn't sit or walk around. I got rashes all over my body. The doctors said
that I would be okay after the childbirth but I still have these problems.

The pamphlet *Voices of Bhopal* can be characterized in legal terms as an
attempt to produce "the duty of care" that must be established to fulfill the
doctrinal requirements of the law of negligence. Establishment of the "duty
of care" requires proof that the responsible party ought reasonably to have
certain persons "in contemplation" when "directing the mind to the acts or
omissions which are called into question" (Cassels 1993, 62–63). In effect,
the pamphlet sought to establish a "public sentiment of moral wrong doing
for which the offender must pay." In this instance, the primary offender
was not Union Carbide, but the government of India. The disaster was not
confined to the mismanagement of a storage tank of toxic chemicals. The
disaster was extended to encompass both the form and the content of official response—as embodied in the settlement of the Bhopal case and the
medical categorization scheme that justified the settlement.

Circulation of *Voices of Bhopal* among journalists challenged how the
Bhopal disaster had been constituted as an object of official concern. It

demonstrated that the settlement could be considered "just" only if the situational particularities of life in Bhopal were ignored. The government of India was the accused—for negligence, for failing to contemplate the acts or omissions in question. The details of the Bhopal disaster are *de bonis non,* "of the goods or property not dispersed or taken care of," as when the executor of an estate dies before an estate is settled and an administrator must be appointed to complete the task.

(1.6) Idealic Images

Responding through law to the Bhopal disaster was complicated. Basic questions haunted every move, challenging all definition of what constitutes justice, representation, and advocacy itself. Fluency within the contradictions was undercut by lack of expertise and other resources—including time. But fluency was also undercut by ideal images of what advocacy should be.

Advocacy, conventionally conceived, embodies modernist ideals. The advocate is imagined to be sure and consistent in his beliefs.[7] He leads by providing a unifying language for movement beyond the ills of the past. Virtue consists in the performance of a focused, unified self, undistracted by what doesn't count—whether it be personal desires, secondary issues, or simple doubt. The good advocate can be defined in essentialist terms; we judge him by the match between standard and practice.[8]

Belief that the universal and the particular should confirm and harmonize with each other operates on many registers in received conceptions of advocacy. Behavior should match ideals. World should match theory. The micro and the macro are not considered at odds. Organic metaphors are pervasive.[9] The interruption of the ideal whole marks a problem, not potential. Movement must be forward or it does not count.

The goal is to "take the Capital," in all senses of the term. A new center must be installed to draw everyone and everything together.[10] Victory goes to those best able to assimilate others into their scheme. It is truth that gets you there and provides the currency for the new order of things. Reason continues to operate as a stable, shared source of authority.

Imagine, for example, filmic representations of Gandhi.[11] He possesses the truth, through which he draws people of every caste and creed into a shared program. His own behavior is a model, based on unquestionable values. While he did not seek to "take Delhi," he did seek to reinstate the village as the center of the new order.

Gandhi is, of course, renowned for emphasizing the means of politics, rather than merely the ends. But teleology remains intact. Gandhi knows where he is going and how to get there. The Idea is bigger than the particularities. The challenge is to make the world match theory; the world's capacity to interrupt all theorization is cut. We see Gandhi saddened, but never riddled with doubt. He is driven by belief. Nothing, and no one, is allowed to interrupt its purity.

Gandhi's virtue is derived from separation from what he critiques. He doesn't wear the dress of the British or eat their food. He doesn't sully himself with wagers for state power. He fasts—refusing to participate—until everyone promises to get along.

Desire does not contaminate Gandhi. The resources at his disposal are limited, but more than adequate. He is stubborn, but never impatient. He stands firm, focused, and sure of his next move. He is cast as the ideal advocate because he harmonizes the world. Different peoples and cultures are blended together. Conflict, ultimately, is supposed to be overcome—in an independent and unified India.

Idealized portraits of advocacy represent a certainty that is resolutely at odds with how environmental problems materialize on the ground, in continuing negotiations over what is real, what is past, and what is to come. Described in ideal terms, the advocate is never seen enmeshed in discrepancies, ambiguities, and paradox. Nor is he seen trying to force fit the world into available political ideologies. Discourse is able to digest everything the ideal advocate encounters. Nothing escapes conceptualization. Idioms for ethics without full knowledge remain undeveloped. Disaster is not confronted as disaster.

A drive for total and "well-justified" explanation drives idealized conceptions of advocacy, operating in environmentalism with ironic force. Pollution, after all, is about dirt—matter out of place—things we think of as waste. To oppose pollution is to oppose what has been defined as waste and what has been discounted as noise.[12] Purity is the goal of environmental advocacy, to be achieved through systems that contain or exclude what is noxious—well-integrated and balanced systems that harmonize the diversity within them. Harmonization links everyone and everything together through a shared logic. Diversity is subject to a unifying vision. Aspiration for purity encourages exclusion; aspiration for harmony encourages effacement.[13]

(1.7) Aesthetics and Expertise

When I came to Bhopal, I wrote slowly, valorizing patience, carefully crafted nuance, and repeated revision—which often took my words to higher and higher (most often ridiculous) levels of abstraction. Advocacy at the site of disaster demanded a different approach. Densely empirical accounts were essential to support clear statements of problems and alternative solutions. And they had to be produced in an afternoon, or by the following morning. The farthest horizon always seemed to be the next anniversary of the disaster, when journalists could be expected in Bhopal, or a promised, but repeatedly delayed decision from the Indian Supreme Court. Rushing and waiting became ways of being; writing on cue structured my days, and my understanding of the many registers on which disaster could operate.

Double binds proliferated. I learned languages of law and bureaucracy, while learning how badly these languages represent everyday life. I learned to speak in terms of environmentalism, while learning how badly environmentalism represents the Third World poor. I learned the many truths of theoretical critiques of representation, on the ground—while producing one representation of Bhopal after another.

With time, the double binds produced by the Bhopal disaster would become an important research focus, as a way to track new subject positions emergent from the fields of force and contradiction we call globalization. In 1990, I knew only that my methods of studying Bhopal could not be abstracted from the continuing disaster at the local level. Advocacy would have to become a way of fieldwork; I would have to work within what disaster exposed.

Ready, or not, I would have to see what it meant to play a double game, strategizing ways to represent the Bhopal disaster, aware of the inadequacy of all available idioms.[14] Competing demands would structure the work: Demands to acknowledge the unfigurability of disaster alongside demands for categorization—thrown into high relief in all attempts to render the disaster into the chronologies and comparisons needed to demonstrate responsibility. Demands to acknowledge both the contingent particularity of example and the universally valid—visible whenever stories of the disaster were turned into lessons for other places and times, rendering the experience of victims significant by emptying it of specificity. Demands for words that upheld entrenched regimes of power, alongside demands for words that disassemble—highlighted in every move to secure justice within the law, which works by discounting what cannot be translated into constitutional equivalencies.[15] Demands to respect both past and future, embodied in the need to remember Bhopal so that we may forget, staging a future less determined by the force of repetition.

The problem was always a matter of choice, between things of equal importance.[16] I could not insist that the preservation of particularity was more virtuous than legal argument. I could not forget how legal argument works to settle disaster, both literally and figuratively. I could not resist calls to recognize the Bhopal case as an important precedent for determining how the risks and rewards of globalization would be distributed. I could not forget the danger of idealizing disaster, troping gas victims as signs of something beyond themselves, accomplishing allegory but losing disaster's excessive specificity.

These contradictions could not be resolved. I had to work within them, oscillating between their demands, temporalizing what counted as proper response.[17] In some instances, the law was damned for being insufficiently lawful—because it failed to operate as a proper system, driven by an internal logic, immune to external influence. In other instances, the law was damned *because* of its systemic exclusions, which obviated any match between law and justice. Both these critiques were valid. But they called for very different responses. Following one meant discounting the other.

Oscillation between incongruent obligations was continual and exhausting. Established ethical paradigms provided little guide. Knowledge could not dictate action. Inaction could not be justified. Consistency and universality were ideals removed from practice. One always responded, on time—in ways tuned to a specific situation, at odds with unified theory, aware that interpretation could not stop. The desire for ethics was inconsolable—but was always deferred.

Learning to live within the double binds of the Bhopal disaster has not been easy. Patience for paradox is a scarce resource in cultures shaped by modernist images of ethical agents sure of themselves and at home in the world. But it was through such patience that advocacy within disaster acquired a tensile strength and a creative, even if hard, edge. Listening for and living within contradictions did not inspire confidence. To the contrary. But listening for contradictions did put advocacy *into* the world, even if into a world too complex to fully understand.

The most important challenge was aesthetic. Advocacy within disaster—particularly against the law—called for a change in what I valued and experienced as good. Virtue had to be gauged in terms of resilience, rather than in terms of principled uprightness.[18] Being well versed *in* the world became much more important than having an intellectual hold *on* the world.[19]

Two

Happening Here

(2.1) Cynicism and Contradiction

[handwritten: hmm, like so many others]

1984 was a complex year. Ronald Reagan swept into his second term, already credited with reinventing environmentalism by his contempt for it. Apple introduced the Macintosh personal computer, launching hopes for an information-rich, democratically driven future. Bruce Springsteen sang "Born in the USA" to fans around the world, cutting apart the American dream, while news of escalating national debt drummed in the background. America the Beautiful was in hock, and every citizen owed his or her fair share. Meanwhile, inflation was high, as was unemployment. The cost of medical care had climbed to the highest level since records were kept in 1935 (Hoffman 1997). Signs of increasing social stratification were visible, as were signs that the U.S. government could not be trusted—to look out for the common good or to provide reliable information through which citizens could pursue the common good themselves.

Reagan's solution had been simple from the outset: defederalization—remaking government by unmaking it. The Environmental Protection Agency (EPA) was a prime target for budget cutbacks and regulatory reform. The effects were contradictory.

By the end of 1983, the "gang of three" installed by Reagan to defederalize environmentalism had been ousted, but they left their mark. James Watt, head of the Department of the Interior, was forced to resign in October 1983 because of his use of racial and ethnic slurs while discussing the composition of a commission appointed to investigate his coal leasing policies (Vig 1997, 100). Watt had already alienated Congress by his efforts to open virtually all public lands and offshore coastal areas to mining and oil and gas development. Before leaving office, he either transferred or sold over 20 million acres of federal land to states that promised to turn responsibility for stewardship over to ranchers. The "wise use movement" began, offering conservatives a strategy for opposing environmentalism that could carry them through the next decade and beyond. But Watt's approach had side effects. He had hoped to unite the "hook and bullet boys" against the "daisy sniffers" (Dowd 1994, 68). Entrenched commitments to free enterprise and the right to bear arms were to provide the scaffolding. Watt's approach backfired, propelling extraordinary growth in mainstream environmental organizations. Staff of these organizations would come to play major roles in the interpretation of new flows of data on environmental disaster, shifting expertise out of established centers of power.

Rita Lavelle, head of initiatives to slow Superfund activity, was fired in February 1983 and in December was sentenced to serve six months in jail and pay a $10,000 fine for perjury committed in hearings on Superfund enforcement (Hoffman 1997, 88). Lavelle perfected ways to undercut the force of law without actually removing legislation from the books. Retaining legislation, but without funds for enforcement, reemerged as an explicit strategy during the environmental backlash of the 104th Congress, under the direction of Newt Gingrich. But Lavelle's imprint remained most visible in the black humor of citizens who learned that the EPA was part of their problem, rather than an ally in efforts to respond to environmental risk. During the 1980s, citizens also learned that just because government experts say something is safe doesn't mean that it is. Continuing controversy surrounding Three Mile Island was not reassuring.

The "acid reign" of EPA Administrator Ann Burford Gorsuch ended after she, too, was accused of making "sweetheart deals" on Superfund cleanups. Gorsuch was found in contempt of Congress for refusing to disclose documents and was forced to resign in March of 1983, along with twenty other high EPA officials (Vig 1997, 100). Together they are credited with dismantling the institutional capacity of the EPA. Regulation of pesticides was particularly affected, despite disclosure of serious dangers over the previous few years.[1]

Before leaving office, President Carter issued an executive order controlling the export of hazardous pesticides. President Reagan rescinded the order in February 1981. Then, in August 1981, Vice-President Bush announced that the EPA's pesticide registration program was targeted for special review by his "Regulatory Relief Task Force." The promise was to balance the economic benefits of pesticide use against uncertain risks through industry-EPA negotiations. The time of adversarial relations was said to be over. Timeliness became a key goal, realizable by reducing data provision requirements. According to industry representatives, data on risk estimates could not keep current with advances in science, so they were not useful in evaluating "tolerance requests."

Between 1980 and 1983, the staff of the EPA's pesticide program was reduced from 760 to 540. Meanwhile the approval of emergency exemptions escalated, increasing from 180 in 1978 to 750 in 1982. Many chlorinated compounds—including DDT—secured exemptions (Wargo 1996, 98). The effects were both material and symbolic. Trust in expert opinion eroded still further. Cynicism about promises of progress intensified.

By 1984, however, Congress had rebounded, promising to protect the environment through law. The Resource Conservation and Recovery Act was amended, partly in response to Reagan's "Sewergate" scandal, propelled into visibility by journalists who had learned that environmental crisis is good copy. Angry citizens also played important roles. By 1984, the Citizens' Clearinghouse for Hazardous Wastes—organized by Love Canal's Lois Gibbs—had assisted over 600 grassroots groups (Szasz 1994). Love Canal, which hit the headlines in 1978, had changed the landscape. The story of a community of working-class people raising their children on top of a toxic waste dump—without knowledge of the risk and with no means to reduce it—galvanized the public imagination.

Then, in December 1984, the U.S. Congress learned of the Bhopal disaster. On the first day of the 99th Congress, Senator Frank Lautenberg (D–NJ) outlined some of the questions requiring response: What percentage of the U.S. public lives in close proximity to facilities that produce or use hazardous chemicals? Is it known what these materials are and what hazards they present to adjacent communities? How adequate are the emergency procedures established by the federal and state government to respond to environmental disasters? Does the national emergency response team set up under the Superfund law have the capacity to provide or coordinate essential services in the event of a disaster? How would victims of exposure be compensated? (Long 1985, 53).

The question that seemed to encapsulate most others was basic: Can "it" happen here? Union Carbide said that it could not. The rest of the chemical industry concurred. But it could no longer be argued that hazardous chemical production was a closed system. Chemicals clearly leaked out, both routinely and catastrophically.

Representative Henry Waxman (D–CA) emphasized this shift in perspective in Congressional hearings ten days after the gas leak in Bhopal:

> We're being told on the one hand that it's a sealed system. But on the other, all these chemicals are leaking into the air on a routine basis. I find that troubling. The federal government doesn't know anything about it and that's outrageous enough. The state government hasn't the ability to regulate. We rely

on you to regulate yourself. And if you are regulating yourself, it doesn't seem to me that your own people know why these chemicals are going into the air and what effect they're having on the public. (Quoted in Lepkowski 1984, 20)

The EPA reviewed Union Carbide's plant in Institute, West Virginia— considered the "sister plant" of the Bhopal plant—and found few breaches of environmental laws. But the EPA did discover twenty-eight accidental spills of methyl isocyanate over the past five years that had not been reported. According to the EPA, at least sixteen of the spills appeared to have involved quantities of less than one pound and thus did not require reporting. Union Carbide later admitted that there were 190 leaks over the past five years in the methyl isocyanate (MIC) unit, but insisted that none was large enough to require reporting. EPA officials claimed to be troubled by the unreported leaks and promised to continue their investigation.

Representative Waxman argued that it was a discredit to the EPA that it did not know what was going on, pointing out that "EPA didn't mention the fact that there are no standards because EPA hasn't set any. After 14 years it has regulated only eight toxic pollutants. Methyl isocyanate is not considered a hazardous pollutant. . . . Aren't 2,500 deaths enough to convince EPA that methyl isocyanate is hazardous?" Waxman also described the contradictory process by which hazards "count": "EPA doesn't call something a hazard until it's ready to regulate it and it doesn't regulate something until it calls it a hazard. EPA has been chasing its tail for far too long" (quoted in Long 1985, 56). The main policy point was that industry had to be required to provide more information. The logic was simple: EPA couldn't regulate an industry without knowing what it was regulating.

In January 1985, Waxman also released an internal Union Carbide memo that warned of a "runaway reaction that could cause a catastrophic failure of the storage tanks holding the poisonous (MIC) gas" at the Institute, West Virginia, plant (quoted in Draffan 1994, 14). The report on which the memo was based was completed in July 1984. When Jackson Browning, Union Carbide's vice-president of health, safety, and environmental affairs, responded to publication of the memo in January, he insisted that the report was not applicable to the Bhopal plant because the cooling systems in the Bhopal plant relied on a refrigerant rather than brine, a possible source of water contamination. He also insisted that while the internal safety report indicated a potential for a runaway reaction at the Institute plant, such a reaction was in no way imminent (Draffan 1994, 18).

But the cat was out of the bag. Clearly, the chemical industry was operating open systems and would face demands for open accounting for what leaked out, and what was being done about it.

Legislating citizens' right to know about environmental risk has been the way the Bhopal disaster has been responded to in the United States. Notification of risk potential was also central in proposals for responding to possible misconduct by U.S. multinational corporations in their foreign operations. Representative Stephen Solarz (D–NY) visited Bhopal in December 1984 to make it clear that "if one is from the U.S. one doesn't have to be an attorney to be concerned about the tragedy" (quoted in Long 1985, 59). Solarz said he was shocked to find out that the mayor of Bhopal had no idea of the potential dangers posed by the plant. In his view, Congress should require that U.S. firms notify foreign governments of the nature of substances produced and used at all facilities, and of all potential threats to health, safety, and the environment. They also should be told about U.S. regulatory standards for similar facilities. Solarz did not enjoy broad support. Just two weeks after the Bhopal disaster, the United States was the only country to vote against a United Nations resolution requesting the secretary general to prepare a list of products that are banned, severely restricted, or not approved for use in various countries. The State Department defended its three years of opposition to the list, arguing that it was duplicative and not amenable to use by serious scientific or responsible public officials. The State Department said that it would support turning the list into an index of 'potentially troublesome' substances and materials (Long 1985, 59).

Bills that would have required international notification were proposed. They could have stood alone or have been attached to either the Export Administration Act or the Overseas Private Investment Corporation's fiscal 1986 authorization bill. A group of about sixty national environmental and public welfare groups—including the Audubon Society, National Wildlife Federation, Environmental Policy Institute, and Natural Resources Defense Council—went further. Operating as the Global Tomorrow Coalition, they sought to respond to both Bhopal and famine in Africa with a package of legislative proposals to address worldwide environmental problems. Their proposals were solid, but beyond what Congress would support. One proposal was for technical assistance programs to help developing countries train workers and managers in the use of hazardous chemicals. Another proposal was for assistance to improve the capabilities of environmental ministries in developing countries. Another proposal called for independent audits conducted by a team of government and environmental experts, available on request by any country in the world. The audits would have been available to the public to deter governments from just sitting on them (Long 1985, 59).

Calls for organizational reform that would redistribute technical expertise were not built into law. Legislative response to the Bhopal disaster focused almost exclusively on notification—on the right to know—which would provoke organizational change, but not change of the sort that built third-party technical expertise. Corporations would face new reporting requirements, but the accounting would be done internally. The public would be told more about what came out of "the end of the pipe," but would not have authority over measurement techniques. The effect was twofold: an unprecedented amount of data on pollution became available; environmentalism became a negotiation over how things would be counted and categorized.

In the mid-1980s, citizens around the United States had a desperate need and desire for environmental risk information. They also had solid grounds for distrusting the information made available to them—and for distrusting the implicit promise that information, somehow, is power. Citizens therefore became environmental advocates in large numbers, in contexts rife with cynicism and contradiction. Environmental risk information had to be demanded, alongside arguments that the right to know is far removed from the capacity to change.

(2.2) Safer Systems

On August 11, 1985, a cloud of chemicals used in the manufacture of aldicarb (Temik) spread over four communities in Institute, West Virginia. Residents were concerned, but confused. After the leak in Bhopal, they had been assured that any hazardous emission from the plant would be detected by a new, $5 million, computerized leak-detection system, allowing plant operators to warn exposed communities immediately following any release. The system, known as "Safer," was supposed to be able to identify instantly the speed and direction of a toxic leak. Carbide, however, had failed to program the system to detect the specific chemical combination spilled. As a result, plant workers thought that the gas cloud was hovering above the plant itself and not migrating over surrounding communities. The release included aldicarb oxime, which contains trace amounts of MIC, and methylene chloride, a nervous system toxin and a suspected carcinogen under special EPA review.

For twenty minutes, no siren sounded. The eyes of plant workers were glued to a computer screen while the world outside their window was being covered with a cream-colored cloud of chemicals. Meanwhile, thousands of nearby residents were exposed, and 135 were headed for hospitals. Officials later warned that food grown near the Institute plant might be contaminated and should not be eaten. The Occupational Health and Safety Administration (OSHA) cited Union Carbide for "willful neglect" and fined Carbide $1.4 million for 221 violations of 55 federal health and safety laws. Labor Secretary William Brock stated, "We were just surprised to find constant, willful, overt violations on such a widespread basis." In March 1986, Carbide negotiated a settlement with OSHA, agreeing to pay $408,000 for "five serious violations" in return for an agreement to correct the violations immediately (Dembo, Morehouse, and Wykle 1990, 84).[2]

In interviews, residents and workers of the Institute area referred to the month following the incident on August 11 as "toxic hell month." On August 18, a tank truck accident leaked sulfur trioxide, requiring the evacuation of thirty homes. On August 26, a cloud of water and hydrochloric acid leaked and drifted over South Charleston, in the vicinity of a Chubby Checker concert with 35,000 people in attendance, whom officials decided "it was better not to warn than panic." On September 7, an undisclosed amount of dimethyl disulfide and Larvin leaked from the Institute plant, also without any alarms sounding. On September 8, a leak of methyl mercaptain in nearby Nitro hospitalized three workers. On September 11, the South Charleston plant leaked liquid monomethylamine, but did not report the release until three days later (Draffan 1994, 18).

Community response to these incidents was vehement, and multidirectional. The most immediate response was a parade by about 400 Carbide employees to show support for Union Carbide. Their chant was clear: "Hey! Hey! We believe in Carbide all the way!" The wives of plant workers were particularly vocal. One woman spoke to camera crews as she marched alongside her husband: "He worked at Carbide 31 years and look at him. Ain't nothin' wrong with him." Another woman took the microphone and explained that "I wasn't worried about him then, and I'm not worried now. His father worked at Carbide, too. And, by golly, there's not a better employer in the valley and never will be." Someone nearby is overheard saying that "almost heaven would be almost hell without Carbide." All the demonstrators caught on camera were white. A meeting held the next day on the campus of West Virginia State College had more diverse attendance. West Virginia State College shares a fence line with Union Carbide, as does an African-American neighborhood.

The meeting at West Virginia State College was convened by a newly formed group called "People Concerned About MIC." The meeting was attended by Union Carbide's new CEO, Robert Kennedy.

The first speaker was an African-American woman named Donna Willis. She started with an even tone, but quickly crescendoed: "You fumble and stumble, and cause our lives to be turned upside down over things you misplace, over 500 gallons of this mixture. Now I can see misplacing one or two gallons of gasoline around your house. . . . Who owes us for our gardens that we couldn't eat? And to tell you the truth, I wouldn't eat it now.

Whose garden got checked? Ya'll from Institute, raise your hands. Ya'll going to eat that mess? It wasn't the emergency whistle. It was the fire whistle that woke me up. Firemen were running for a fire, while I was running out of Institute with my two kids. Trying to save the lives that I put on this earth, not for Carbide to take away!" The next speaker was a white woman named Savannah Evans. She said that the birds stopped singing in Institute after the leak, and they didn't come back for a week. And she was mad about it, and about having to live where she had to worry all the time. The next speaker was a young woman named Cheryl Whitened. She explained, "Last Sunday I got up on the interstate and they detoured me, so I got off right at the gates of Union Carbide. There was a blue light flashing; I realized there was a gas. I'm pregnant. I'm two months pregnant. I called everywhere to find out what I had been exposed to. And I still don't have any answers." Next was Claire Smith: "The great big corporation. Yet I'm across the street and don't even hear a damn signal. By the time my family got up the gas was all in our house. I'm not fortunate enough to have central air conditioning. So the fans just brought it right in."

Robert Kennedy, CEO of Carbide, took the next turn. First, he acknowledged that some people in the audience seemed to want Union Carbide to shut down the plant. He said that he didn't think they could shut down the plant because "we aren't an organization of quitters. We don't want to admit that we aren't capable of managing our affairs. I had a dog once that was overly aggressive. He bit a mailman once. . . ." Then Kennedy was interrupted. Someone from the audience shouts, "Why don't you listen? I don't want to hear your dog story. We're talking about people. . . . And my payroll number is *****. Fire me if you want. But until you do I'm going to be one of those inside making *sure* you're running a safe plant."

Payroll ***** exhibited one of many identities that have emerged in Institute since the Bhopal disaster. Many are angry. Many are defensive. Many are confused. Some can't separate what's good for the chemical industry from what's good for West Virginia. Charleston City Councilman Ernest Layne was neither atypical nor representative, explaining that MIC-based Sevin was "the best pesticide on the market. It saved the cotton crop in Egypt two or three different times. It was used in India to mass-produce their wheat crop, and saved thousands of people from starving to death. And I know it's not harmful. I participated in a test. A crop was sprayed with the pesticide, then I ate it." [3]

People Concerned About MIC has helped configure alternative identities. But their efforts have been constrained by the extraordinary authority Union Carbide has in West Virginia. The sale of Union Carbide's agricultural products division (including the MIC unit) to Rhone-Poulenc in 1986 made things even more complicated. Many West Virginians interpreted the sale as a sign of disloyalty, which deserved no further loyalty in return.

Many people in the Kanawha Valley feel duped, both by Union Carbide and by the broad promise of democratic participation in risk decisions. One Institute resident told me that "while it looks more democratic, it really is a child's game in which we all pretend that everyone has a grown-up role." Many say that they live in even more fear of disaster than before and that the only function of the new "right-to-know" laws has been to inform them of danger, without allowing them to do anything about it. Some residents, however, have turned right-to-know laws into a powerful tool for acquiring new kinds of risk information—information that confirms that a Bhopal-like disaster can indeed "happen here."

(2.3) Rights to Know

A family of environmental regulations sometimes referred to as "Bhopal's Babies" has transformed the chemical industry by significantly extending required disclosure of risk. "Disaster preparedness" plans have become part of the right to know, as has information about routine emissions. The effect has been widespread recognition that hazardous chemical production is not a closed system. Even industry seemed surprised by the data released. When the first round of data for the Toxics Release Inventory was submitted in 1988, the president of Monsanto was so taken aback by the figures disclosed that he pledged to reduce emissions by 90 percent over the next five years. The next year the Chemical Manufacturers Association initiated its Responsible Care program.

The right to know was attached as Title III to the Superfund Amendment and Reauthorization Act (SARA), passed in 1986 to expedite the cleanup of hazardous waste sites, to augment the "Superfund" (through taxes on the petrochemical industry), and to strengthen mechanisms for recovering cleanup costs from responsible parties.[4] By most accounts, the Superfund portion of the bill was so contentious that the sections including provisions for emergency response and public access to information were not considered at length (Hadden 1994).

Title III initiated four different reporting requirements intended to provide public access to information about chemical risk. The first requirement was for provision of an emergency evacuation plan to local rescue personnel— beginning community awareness of "worst-case scenarios." The second requirement involved emergency notification whenever a "reportable quantity" of a broad set of chemicals was released—beginning community awareness of the frequency of small releases and "near mishaps." Third,

unity right to know provisions" required companies to provide in-
on on all chemicals stored or used in excess of certain threshold
quantities—beginning the circulation of Material Safety Data Sheets,
which describe the physical characteristics and health effects of individual
chemicals. Fourth, a "Toxics Release Inventory" required companies to
provide the EPA with extensive information on releases and disposal of
about 350 chemicals—allowing citizens as well as the EPA to track and
evaluate emissions standards.

Implementation of Title III had two primary effects. Most obvious was
the increased circulation of information regarding existing and possible
hazards. Less visible, but equally important was the creation of institutional
structures for citizen participation in environmental risk assessment and
mitigation—through Local Emergency Planning Committees (LEPCs).
LEPCs were modeled on the Community Awareness and Emergency Response program established by the Chemical Manufacturers Association
(CMA). Each committee had designated seats for company representatives,
for emergency response personnel, for elected officials, and for ordinary
citizens. The statutory duty of an LEPC was to write an emergency response plan that coordinated all involved parties based on a risk-benefit
analysis of each site where hazardous chemicals were stored or used.

The effectiveness of LEPCs has been uneven. It has not helped that Congress has not appropriated funds for implementation of Title III, leaving
localities, states, and even the EPA to pull money from other programs.[5]
Nor has it helped that emergency response personnel often seem to understand the right to know as important for their own planning, but of little
relevance to citizens. And who counts as a citizen is always at issue. Members of the local chamber of commerce often play this role. Labor unions
do not have a seat at all.

Similar problems have beset the CMA's Responsible Care program. "Responsible Care" is a "public commitment" to run clean and safe plants—
voluntarily and "beyond compliance" with the law (Union Carbide Corp.
1989). One mechanism for fulfilling this commitment was the creation of
forums for citizen participation in discussions about plant risks, which duplicated similar forums created by Title III.

Responsible Care was initiated in 1989 under the CMA presidency of
Carbide CEO Robert Kennedy. It was a response to a study conducted by
CMA's Public Perception Committee to gauge public distaste for the industry following the Bhopal disaster. The results were devastating, and
CMA decided that Responsible Care was the answer. CMA provided the
framework, with each member company responsible for setting its own
specific goals in each of the Management Practice Codes.

The initiation of Responsible Care demonstrated how Bhopal would be responded to on two registers. First, it spoke to the fact that in the wake of Bhopal the chemical industry would present a unified front. Carbide would not be scapegoated as the "bad company," unlike Monsanto, Dow, or DuPont. Instead, Carbide would be salvaged as a sign of survival. Union Carbide CEO Robert Kennedy would lead the charge, not be subject to it.

Second, Responsible Care established the institutional structures through which public concern about chemicals would be articulated. In his introductory speech, Kennedy described Responsible Care as a "major performance-based initiative" to "work together in ways that the public sees as responsive to their concerns." In addition to pledging to support the industry's responsible management of chemicals, the program instituted "Public Advisory Panels composed of environmental, health and safety thought leaders" who would be asked to comment on all aspects of management, including the ways products are developed, made, transported, used, and disposed of. Carbide's brochure describing the program states that "beyond specifics, Responsible Care signals the Chemical Industry's commitment to keep faith with America and the world" (Union Carbide Corp. 1989).

The Public Advisory Panels Kennedy promised have become Citizen Advisory Panels (CAPs). They have been constituted in many communities where chemical companies have major facilities, often existing alongside LEPCs, also promising inclusive risk decision-making. Both CAPs and LEPCs have enrolled diverse parties in discussion, if not decision-making. Industry, however, has remained in control of how the discussion proceeds. The focus has been on managing information—itself considered a high risk. The risk of chemicals has been largely deferred.

Right-to-know legislation has nonetheless been immensely productive, as has been the right to comment, as provided for by LEPCs and CAPs. Citizens around the United States have participated, and their activism has acquired focus. The right to know has also given citizens data for comparing and consolidating the concerns of different communities. A national network of grassroots environmental groups is one result. Most important, however, has been the simple acceleration of the desire to know. The Bhopal disaster has become a prompt to know about what was not known before.

(2.4) Sheltering in Place

Right-to-know laws are an imperfect response to the Bhopal disaster. The effects of right-to-know laws have nonetheless been enormous—sparking environmental initiatives within corporations, in the communities where hazardous production facilities are located, and by national and international environmental groups. The potential—and contradictions—of the right to know is particularly clear in the Kanawha Valley.

In 1992, Pam Nixon, a founding member of People Concerned About MIC, requested the LEPC to provide "worst-case scenarios." Her request would turn into a major initiative that would finally bring national attention to West Virginia. The Kanawha Valley would be cast as a model for national implementation of new provisions of the Clean Air Act of 1990, which significantly extended the right to know by mandating the disclosure not only of what is produced and stored, but also of risk management plans and "worst-case scenarios." [6]

Nixon's request for worst-case scenarios turned into the Kanawha Valley Hazard Assessment Project, organized around the theme of "Safety Street." A DuPont process safety expert explained its significance: "What's exciting about participating in this whole activity in the Kanawha Valley, [is that] it clearly is at the leading edge of pending regulations at the federal level that are three or four years away. . . . [We] are going to have a positive impact on how federal regulations will be written, developed, and implemented" (Jackson 1994, 69). EPA's Carl Matthiessen concurred, encouraging participants in Safety Street not to wait for the EPA, because "we don't have all the answers. You have all the answers. And I think most of what you're doing is already state-of-the-art" (Chemical Manufacturers Association 1994).

Staged jointly by the fourteen chemical plants in the valley, Safety Street was organized to share data on the potential results of a plant mishap. The event's banner read: "Safety Street: Managing Our Risks Together." The first part of the event, staged at the Charleston Civic Center, included nine presentations explaining different roles played in risk management. Presenters included industry representatives, emergency planning officials, regulators, and "ordinary citizens." After lunch, formality ended, and the audience could approach presenters at convention-style booths. The following day, the relay of information was moved to a local shopping mall, where passersby could scan graphs that carefully discriminated between "worst-case" and "more likely" scenarios.

The pamphlet distributed at this event that described risk management for MIC is instructive. A "worst-case scenario" is identified as "sudden loss of contents of the underground storage tank when completely full (253,600 lbs.)." The accompanying graph draws two concentric circles around the plant. The inner circle, identified as ERPG3, has a radius of nine miles, within which concentration levels would be 5 ppm. The outer circle,

Worst Case Scenario

Scenario Description:
Sudden loss of contents of the underground storage tank when completely full (253,600 lbs.).

ERPG* 2 distance = 28 miles   ERPG 3 distance = 9 miles
ERPG 2 concentration = 0.5 ppm.   ERPG 3 concentration = 5 ppm.

Methyl Isocyanate (MIC)

RHÔNE-POULENC'S PREVENTION

Process Design

- Emergency dump tank available for safe transfer of product from leaking vessel
- Scrubber will destroy MIC from any storage tank
- Flare tower will destroy MIC vapor from process vents
- Storage tank heat exchanger fluid will not react with MIC
- Backup control room instruments

- Automatic MIC isolation valves stop leaks
- Diking and spill collection sump
- Fire deluge system
- MIC leak detection alarms
- Safety relief valves protect vessels from over pressure
- Diesel generator for backup power
- Sealless pumps for pure liquid MIC

- Fire protection for pipe rack transfer lines
- Independent nitrogen supply to prevent cross contamination

Equipment Design

- Double-walled underground storage tank
- Pressure vessels coded by American Society of Mechanical Engineers
- Blast mat protection on above ground MIC storage facilities
- Stainless steel construction
- Pipelines over roads protected by barriers
- Double-walled pipelines with leak detection analyzers on critical transfer lines

Safety Reviews

- Process Hazard Analysis completed every 5 years
- Ongoing safety reviews for design changes
- Operational reviews completed for all process changes
- Safety review team includes safety experts, engineers, union operators and union maintenance personnel

Procedures

- Strictly enforced inventory limits
- Annual review of operating procedures

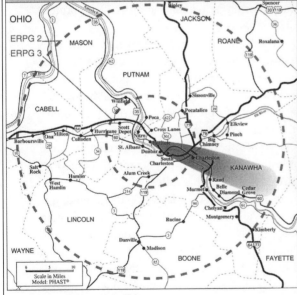

* ERPG = Emergency Response Planning Guideline

FIGURE 2.1 WORST-CASE SCENARIO   A Rhone-Poulenc pamphlet describing a worst-case scenario for methyl isocyanate, the bulk chemical released in Bhopal. The Rhone-Poulenc plant in Institute, West Virginia, that this pamphlet describes stores more than four times as much methyl isocyanate as was released in Bhopal. Like the Bhopal plant, the Institute plant relies on a large-tank storage system.

identified as ERPG2, has a radius of twenty-eight miles, within which concentration levels would be 0.5 ppm. The graph is surrounded by detail on process and equipment design, safety review procedures, and emergency response plans. The next page describes a "more likely scenario," during which 118 pounds of MIC would escape through a 0.8 mm–diameter hole. The graph here doesn't show the circles, but the numbers indicate that the risk is substantially less. The area designated as ERPG3 has a radius of only half a mile; ERPG2 has a radius of only three miles (Rhone-Poulenc 1994).

Nowhere within this pamphlet are readers told what it would mean to be within an ERPG3 area at the time of a catastrophic release of toxic gas. An article in the journal *Corporate Environmental Strategy* provides the explanation. An ERPG is an "Emergency Response Planning Guide," administered by the American Industrial Hygiene Association. ERPG3 designates an area within which "it is believed that nearly all individuals could be exposed for up to one hour without experiencing or developing life-threatening health effects." ERPG2 designates an area within which "it is believed that nearly all individuals could be exposed for up to one hour without experiencing or developing irreversible or other serious health effects or symptoms that could impair an individual's ability to take protective action" (Jackson 1994, 73).

In other words, if you don't get out of an area designated as ERPG3 within an hour after a worst-case release, you're dead. If you don't get out of an area designated as ERPG2 within an hour, you aren't going anywhere, and, at best, you'll develop "irreversible or other serious health effects." Such, at least, might be the scenario if you are the generic "individual" designated by the ERPG. It is even less clear what happens to a child, an elderly person, an ill person, or some other differentiated category. Institute is twelve miles east of Charleston, which has a population of 60,000. A quarter of a million people live in the Kanawha Valley.

This overview is bad enough at the theoretical level. On the ground, it is truly catastrophic. The Kanawha Valley is hemmed in by mountains, with one interstate leading out of town. There are a few back roads through the hills, but they are often difficult to access. One such road, adjacent to the Institute plant, is gated and locked. Industry representatives say they will be happy to open it upon necessity, though it is unclear whether emergency planning procedures assign anyone to this task. The road empties out near an African-American community that shares a fence line with the plant, as does the campus of West Virginia State College. Nearby there is a rehabilitation facility that houses severely disabled patients.

Industry recognizes the problems associated with any attempt to evacuate the Kanawha Valley. Its solution is to encourage residents to "shelter in place." Instructions are distributed on small, house-shaped magnets that can be attached to the family refrigerator:

1. Go inside.
2. Turn off air moving equipment.
3. Turn on radio or television.
4. Tune in emergency broadcast system.
5. Stand by for telephone ring-down.
6. Stay calm.

The friendly, house-shaped reminder neglects to explain how residents need to cover doors and windows with plastic, taped to walls with duct tape. It also forgets about the lime green sign residents should post in a window, visible from the street, once they are locked inside. The sign reads: WE HAVE BEEN NOTIFIED.

Residents of the Kanawha Valley are divided in their evaluation of "sheltering in place" and of the Safety Street Initiative that helped bring it to public attention. Some are harshly fatalistic, insisting that in the Kanawha Valley the chemicals will eventually "getcha." One resident insisted that this is almost natural: "First they killed the Indians; now they're killing the hillbillies." Others are more strategic, pointing out that it is not insignificant that local chemical companies have to replace their own stock of gas masks so frequently: workers steal them and take them home to their families. Still others cooperate with initiatives such as Safety Street, playing roles in instructional videos, serving as company diplomats at school and civic functions, attending meetings that "bring everyone to the table" and demonstrate "Responsible Care."[7]

In 1995, when I was in the Kanawha Valley, industry representatives were confident that their safety plans were sufficient. Management experts who wrote about Safety Street confirmed their sense of security, reminding their readers that "worst case scenarios don't have to be ugly, don't have to be scary. Like planning for your own funeral, it is a necessary activity that can be looked upon with fear and loathing, or it can be an eye-opening activity" (Jackson 1994, 69). Nowhere in the article where this encouragement is deployed are residents quoted or otherwise given space to air their own views.

Worst-case scenarios provide important information. They do not provide tools for interpretation. Comparison is a device of understanding that dilutes the import of its examples. Individuals are homogenized and wrested out of their geographic specificity. There is only one way in, and

FIGURE 2.2 WE HAVE BEEN NOTIFIED COLLAGE   The "We Have Been Notified" sign residents are to post in a window visible from the street while sheltering-in-place during a hazardous chemical emergency, overlaid with the house-shaped magnet to be posted on a refrigerator with instructions on how to shelter-in-place; with the logo for "Safety Street," the event at which worst-case scenarios were released to the public in Charleston, West Virginia; and with the cupped-hands emblem of Responsible Care. The "We Have Been Notified" sign and the instructions on the back were distributed to residents by the Kanawha County Emergency Services. Collage by Kim Fortun.

out, of the situation; alternate routes are gated and locked; no one identifiable has been assigned the task of securing access. The final message: shelter in place, stay calm, prominently display the sign: WE HAVE BEEN NOTIFIED.

Worst-case scenarios imply that the problem is fear of the unknown—an information deficit. The problem is corrected through notification. No further movement is necessary, or possible. The answer—practically as well as conceptually—is to shelter in place. Trade unions in the Kanawha Valley don't like this answer, posing an important question: Safety Street, or Death Row?

Safety Street
or
Death Row?

That depends on the efforts made by all to prevent accidents from happening.

The ACT Foundation commends local chemical industry leaders for finally releasing some of their accident scenarios, but they shouldn't stop there.

The next step must be the release of corporate risk reduction plans and safety audits. *All* of these are key to accident prevention.

Some questions you should ask:

☐ **What plans are there for reducing the risks identified by these scenarios?**

☐ **Are the workers who build, repair and maintain these plants skilled?**

☐ **Are emergency plans based on these scenarios?**

☐ **What commitments are there for new investment in each plant?**

☐ **What plans are there for increased or decreased employment at each plant?**

☐ **What effort is being made to replace the most hazardous chemicals with less dangerous ones?**

If we are to manage our risk together, we need to start right now as partners.

Safety Street cannot be a One Way street.

Companies must commit to an open planning process to move forward. We need a plan that includes decreasing risk, reducing pollution to our air, water and land, new investment and expansion, and a commitment to hiring local qualified people for the jobs that exist.

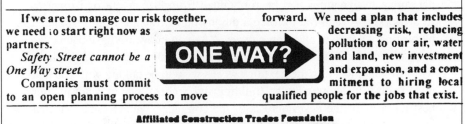

Affiliated Construction Trades Foundation
523 Central Avenue, Charleston, WV 25302

FIGURE 2.3 SAFETY STREET OR DEATH ROW? Affiliated Construction Trades Council poster, 1994.

(2.5) New Vehicles of Meaning

Steve Midkiff refuses to say the name "Union Carbide." "It's Carbide; there ain't nothin' union about it," he insists. Midkiff was a pipe fitter in Carbide's construction unit for fifteen years. In 1992, Carbide eliminated its in-house construction crew—and then hired Brown & Root, a Houston-based construction firm, to build a new unit. Rhone-Poulenc and DuPont already had contracts with Brown & Root for projects in West Virginia. Brown & Root "is known as a fighter when faced with a union challenge. It hires the best lawyers—and it has a history of winning" (Sixel 1993, 10C). Brown & Root has never lost a union election in the United States.

One response in West Virginia was the formation of ACT—the Affiliated Construction Trades Council—to consolidate the efforts of smaller unions and to develop an organizing strategy specifically tuned to the timing of construction work. By 1995, twenty-five different union locals were members of ACT, which got 25 cents for each hour a union member worked—creating a yearly budget of almost $1 million (Woodrum 1995). Midkiff, who went to work for ACT after Carbide laid him off, explained that union organizing on construction sites can be particularly difficult because jobs are often completed before organizers really have a chance to "set." Cycles of working and being laid off are common, as is migration between jobs in different places. So Brown & Root had a significant advantage, claiming to be going through normal hiring and layoff cycles. ACT claimed Brown & Root was bringing in workers from out of state to sabotage a union vote.[8] And workers that signed a union card were laid off.

Brown & Root employees asked the National Labor Relations Board to hold an election on whether they and their coworkers wanted to be represented by ACT in early April 1993. The next step was to be a hearing to determine if signatures had been obtained from 30 percent of Brown & Root workers—required to force a union vote. Brown & Root responded by posting a notice to all its employees: "As you know, we have been talking with you the last few weeks about the many good reasons for working with Brown & Root. In the next few weeks we will be talking with you and giving you facts about why the Building and Trades Council is not in your best interests or the best interests of the company. We are confident that once you have the truth, you will agree with us and you will vote no in the election." The next Sunday's papers in Charleston included new Brown & Root ads asserting its commitment to "improving the environment and enhancing the quality of life for people throughout the world." The ad told how Brown & Root had helped clean up San Francisco after the earthquake, had donated its service to repair a county pool, and had adopted a

FIGURE 2.4 ACT: WEST VIRGINIANS WANT TO WORK   Affiliated Construction Trades Council poster, 1994.

local highway. These activities were said to "represent our employees' involvement in environmental matters worldwide" (Ward 1993, 1A).

By 1994, the controversy had escalated into "an all out advertising war." Brown and Root ran full-page ads extolling its safety record and the quality of its construction.[9] ACT responded with ads of its own. One set of television ads used the slogan "It's cheaper for them, dangerous for you." Plants explode in the background while a roster of Brown & Root workers killed on the job scrolls across the screen. The Bhopal disaster was also a recurrent reference. At one rally, in May 1994, protesters outside Carbide's South Charleston offices carried signs that said, "Don't Repeat Bhopal. Hire from the Union Hall."

Union protesters also said that they were upset that Carbide was building new plants in Canada and Kuwait and paid CEO Robert Kennedy $3.6 million at a time when local Carbide employment was dropping. ACT Director Steve White told reporters that "[p]eople here not only made the money [for Carbide], they went over and fought to keep Kuwait free and then our jobs go over there." Thad Epps, longtime regional spokesman for Carbide, said that "this kind of demonstration is going to discourage construction and economic development in the state." He also said that Carbide had no authority over the hiring practices of Brown & Root or any other contract company. Ken Ward, a reporter who covered the rally, ended his article with a question for Epps posed by Bubby Casto, ACT's business manager: "You're telling me that if you hire somebody to do a job you can't tell them how to do it?" (Ward 1994a).

ACT has relied on many strategies for building union strength, legislative support, and public education. Its advertising has been creative, and very funny. A poster examining the nature of expertise is illustrative. It insists that expertise of many different kinds is important. And it asks us to think comparatively, drawing out the connections between seemingly disparate social practices.

In 1995, race cars became another part of the ACT package, built by pipe fitters who once demonstrated their skill on chemical plant construction crews. Steve Midkiff, his brother Randy, and Tim Jones, who also used to work at Union Carbide, started the venture—called M&J Racing. ACT was their biggest sponsor. Two important possibilities were harbored within their new welds: In an era of declining support for unions, a new way was found to "rev up" support. ACT literally created new vehicles for disseminating labor advocacy. ACT's racers, one for a dirt track, one for pavement, also harbored, or garaged, new social possibilities. The garage in which the racers were built and maintained became a gathering spot,

FIGURE 2.5 ACT: GYM TEACHERS MUST   Affiliated Construction Trades Council poster, 1994.

FIGURE 2.6 STEVE MIDKIFF AND TIM JONES   Steve Midkiff and Tim Jones with their
union racer. Photograph by Mike Fortun, 1995.

where union members strategized their public performances, in the legislature, at new construction sites, and at tracks around the state. The goal
was to win new audiences by opening up pathways along which people
could move without giving up their positions. A crash course remained
possible, but was guarded against by purposeful disclosure of the techniques of construction. Steve Midkiff, who both built and drove the cars,
pointed this out by having me inspect all the weld points, all painted white
so that their strengths, and possible cracks, were thrown into the highest
possible visibility. Midkiff didn't start from scratch. He worked with existing forms, retrofitting them so that they responded in the best possible way
to prevailing road conditions. One of the cars appeared to have started as a
Camaro.

(2.6) Covering Events

In November 1984, just two weeks before the gas leak in Bhopal, a "raw
material" used to produce the pesticide Furadan spilled at an FMC plant in
Middleport, New York. "Vapors from the spill entered the ventilation system of a nearby grammar school, requiring the evacuation of 500 children
and the hospitalization of 9 of them, along with 2 teachers, due to eye
irritation and respiratory difficulty. Later the city fire chief complained

about the lapse of time between the actual spill and the notification of out-side support agencies, school authorities, ambulance, fire and evacuation personnel" (Weir 1988, 32). Later still Middleport residents found out that the chemical was MIC.

It was only in the context of the news coverage of Bhopal that Middle-port residents became aware that a lethal chemical was being shipped and handled in their community. Following the spill, OSHA found FMC guilty of four "serious violations" and labeled it the most seriously deficient of all five U.S. plants using MIC. OSHA also confirmed that the plant did not have the necessary safety systems to prevent an uncontrolled chemical re-action like the one that led to the leak in Bhopal.

At the time of the spill, FMC provided 170 jobs and 20 percent of Middleport's tax base. Thus, even after the almost simultaneous leaks in Middleport and Bhopal, residents continued to offer widespread support. When the plant was temporarily shut down after the Bhopal leak, the *New York Times* quoted a shop steward as saying that "[i]t'll be like Christmas if we ever see a train full of MIC pulling in here" (quoted in Weir 1988, 34).

Some residents remained concerned about the health of their children. Diane Hemingway, the mother of two children who attended the exposed school, polled parents and found that many local children complained of many lingering health effects, including headaches, skin rashes, and cold-like symptoms. Parents then demanded medical exams to assess long-term effects on their children. Finally carried out in June 1985, seven months after the spill, results proved "inconclusive." Diane Hemingway moved to another town.

Incidents like the one in Middleport happen, but don't count as "events." And it takes "events" for things to change—even though "events" don't often happen in western New York, or in Charleston, West Virginia—un-less it's Love Canal. Hugh Kaufman, the EPA official who helped expose the Love Canal story, provides a blunt explanation: "Love Canal was a catastrophe, but it wasn't the worst catastrophe in the country, or in New York state, or even in Niagara County." But "it was Love Canal, Hooker Chemical and Niagara Falls, honeymoon capital of the world. Editors all over the country had orgasms. So Love Canal became catastrophic, not any of the other 10,000 hazardous dump sites across America" (Hoffman 1997, 163, citing Keating and Russel 1992, 33).

Business historian Andrew Hoffman refers to Kaufman's interpretation of Love Canal to explain how events operate as catalysts for institutional change. Bhopal is on his list of significant events, as are *Silent Spring,* the Santa Barbara oil spill, fire on Ohio's Cuyahoga River, the first Earth Day, the formation of EPA, Sevesco, termination of Ann Burford Gorsuch, the ozone hole, Chernobyl, the Mobro 4000 garbage barge, medical waste on East Coast beaches, and Exxon Valdez. These events, in Hoffman's analysis, catalyzed public concern about environmental risk, and programmatic corporate response.

Events don't operate in isolation. They require what Hoffman calls "social surrogates," within which events operate as catalysts. A catalyst "is a compound that, through its mere presence, facilities a more rapid chemical reaction. It does not force a reaction that would not occur otherwise. In the same way, events can serve to hasten the development of institutional issues, without causing them to form" (Hoffman 1997, 165). Love Canal thus became "a landmark opportunity, empowering environmental activists to wage institutional 'war,' given the conveniently sensational participants, the emerging awareness of something called hazardous waste sites, and the event's coincidental timing with Jimmy Carter's presidential campaign" (Hoffman 1997, 163).

Analyses that examine events as part of complex systems are useful. What is missing in Hoffman's account, however, is analysis of how many incidents fail to ever achieve the status of "events"—"incidents" like the one in Middleport in December 1984. The problem is exacerbated by standard modes of media coverage. According to Lee Wilkins, "event analysis" by the mainstream media circumscribes problems to the local level, leaving out significant background information that would indicate translocal relevance (Wilkins 1987). Places like Middleport are left out of the loop. So are many other places where tension between jobs and safety is high. The contradiction is harsh: events matter, but events are events because they are encapsulated in space and time.

In 1984, MIC was used at five sites in the United States: Middleport; New Iberia, Louisiana; Woodbine, Georgia; La Porte, Texas; and Institute, West Virginia. Institute was the only place MIC was produced. Despite this fact, Institute remains "offset" from the Bhopal story. Union Carbide has insisted that "it can't happen here." Event analysis by the media implicitly concurs, even if only by failing to mention structural connections. So production and storage of MIC in Institute have never gained the status of an event. Nor has Institute enjoyed the supposed effects of Bhopal becoming an "event." In 1984, 52,000 pounds of MIC were released in Bhopal. In 1984, 400,000 pounds of MIC were stored at Union Carbide's plant in Institute. By 1994, storage of MIC at the plant (bought by Rhone-Poulenc in 1986) was down to 240,000 pounds. And Bill Frampton, head of the MIC unit, could say that the Bhopal tragedy had been overplayed. According to the *Charleston Gazette,* Frampton insists that Bhopal "has nothing to do with the way we operate. It has no relevance anymore to me" (Charleston Gazette 1994, 11A).

(2.7) Rewriting Company Towns

Muriel Rukeyser traveled to West Virginia in 1936 at the age of twenty-two. The poem she published in 1938, "The Book of the Dead," is often compared to the work of James Agee and to Walker Evan's *Let Us Now Praise Famous Men*. "The Book of the Dead" is multivocal, integrating excerpts from congressional hearings, from interviews, from letters, from printouts of financial information obtained from the Stock Exchange— with Rukeyser's own struggle for words to represent disaster. "The Book of the Dead" helps explains the imprint Union Carbide has made on West Virginia.

THE BOOK OF THE DEAD
Statement: Philla Allen

—You like the State of West Virginia very much, do you not?
—I do very much, in the summertime.
—How much time have you spent in West Virginia?
—During the summer of 1934, when I was doing social work
    down there, I first heard of what we were pleased to call
    the Gauley tunnel tragedy, which involved about 2,000 men.
—What was their salary?
—It started at 40¢ and dropped to 25¢ an hour.
—You have met these people personally?
—I have talked to people, yes.
  According to estimates of contractors
  2,000 men were
    employed there
    period, about 2 years
    drilling, 3.75 miles of tunnel.
    To divert water (from New River)
    to a hydroelectric plant (at Gauley Junction)
  The rock through which they were boring was of a high silica
    content.
  In tunnel No. 1 it ran 97–99% pure silica.
  The contractors
    knowing pure silica
    30 years' experience
    must have known danger for every man
  neglected to provide the workmen with any safety device . . .
—As a matter of fact, they originally intended to dig that
    tunnel a certain size?

—Yes.
—And then enlarged the size of the tunnel, due to the fact
        that they discovered silica and wanted to get it out?
—This is true for tunnel No. 1.
    The tunnel is part of a huge water-power project
        begun, latter part of 1929
        direction: New Kanawha Power Co.
            subsidiary of Union Carbide & Carbon Co.
        The company—licensed:
        to develop power for public sale.
        Ostensibly it was to do that; but
        (in reality) it was formed to sell all the power to
            the Electro-Metallurgical Co.
            subsidiary of Union Carbide & Carbon Co.
                which by an act of the State legislature
                was allowed to buy up
                New Kanawha Power Co. in 1933. . . .

Absalom

I first discovered what was killing these men.
I had three sons who worked with their father in the tunnel:
Cecil, aged 23, Owen, aged 21, Shirley, aged 17.
They used to work in a coal mine, not steady work
for the mines were not going much of the time.
A power Co. foreman learned that we made home brew,
he formed a habit of dropping in evenings to drink,
persuading the boys and my husband—
give up their jobs and take this other work.
It would pay them better.
Shirley was my youngest son; the boy.
He went into the tunnel
    My heart   my mother my heart   my mother
    My heart   my coming into being
My husband is not able to work
He has it, according to the doctor.
We have been having a very hard time making a living since
        this trouble came to us.
I saw the dust in the bottom of the tub.
The boy worked there about eighteen months,
came home one evening with a shortness of breath.
He said, "Mother, I cannot get my breath."
Shirley was sick about three months.
I would carry him from his bed to the table,
from his bed to the porch, in my arms.
    My heart is mine in the place of hearts,
    They gave me back my heart, it lies in me.
When they took sick, right at the start, I saw a doctor.
I tried to get Dr. Harness to X-ray the boys.

He was the only man I had any confidence in,
the company doctor in the Kopper's mine,
but he would not see Shirley.
He did not know where his money was coming from.
I promised him half if he'd word to get compensation,
but even then he would not do anything.

I went on the road and begged the X-ray money,
the Charleston hospital made the lung pictures,
he took the case after the pictures were made.
And two or three doctors said the same thing.
The youngest boy did not get to go down there with me,
he lay and said, "Mother, when I die,
"I want you to have them open me up and
"see if that dust killed me.
"Try to get compensation,
"you will not have any way of making a living.
"when we are gone,
"and the rest are going, too."
   I have gained mastery over my heart
   I have gained mastery over my two hands
   I have gained mastery over the waters
   I have gained mastery over the river
The case of my son was the first of the line of lawsuits.
They sent the lawyers down and the doctors down;
they closed the electric sockets in the camps.
There was Shirley, and Cecil, Jeffrey, and Oren,
Raymond Johnson, Clev and Oscar Anders,
Frank Lynch, Henry Palf, Mr. Pitch, a foreman;
a slim fellow who carried steel with my boys,
his name was Darnell, I believe. There were many others,
the towns of Glen Ferris, Alloy, where the white rock lies,
six miles away; Vanetta, Gauley Bridge
Gamoca, Lockwood, the gullies
the whole valley is witness.
I hitchhike eighteen miles, they make checks out.
They ask me how I keep the cow on $2.
I said one week, feed for the cow, one week, the children's flour.
The oldest son was twenty-three.
The next son was twenty-one.
The youngest son was eighteen.
They called it pneumonia at first.
They would pronounce it fever.
Shirley asked that we try to find out.
That's how they learned what the trouble was.
   I open a way, they have covered my sky with crystal
   I come forth by day; I am born a second time,
   I force a way through, and I know the gate
   I shall journey over the earth among the living.

He shall not be diminished, never;
I shall give a mouth to my son. . . .

The Disease

This is a lung disease. Silicate dust makes it.
The dust causing the growth of

This is the X-ray picture taken last April.
I would point out to you: these are the ribs;
this is the region of the breastbone;
this is the heart (a wide white shadow filled with blood).
The windpipe. Spaces between the lungs.
     Between the ribs?
Between the ribs. These are the collar bones.
Now, this lung's mottled, beginning, in these areas.
You'd say a snowstorm had struck the fellow's lungs.
About alike, that side and this side, top and bottom.
The first stage in this period in this case.
     Let us have the second.
Come to the window again. Here is the heart.
More numerous nodules, thicker, see, in the upper lobes.
You will notice the increase: here, streaked fibrous tissue—
     Indicating?
That indicates the progress in ten months' time.
And now, this year—short breathing, solid scars
even over the ribs, thick on both sides.
Blood vessels shut. Model conglomeration.
     What stage?
Third stage. Each time I place my pencil point:
There and there and there, there, there.
     "It is growing worse every day. At night
     "I get up to catch my breath. If I remained
     "flat on my back I believe I would die."

     It gradually chokes off the air cells in the lungs?
     I am trying to say it the best I can.
     That is what happens, isn't it?
     A choking-off in the air cells?
Yes.
     There is difficulty in breathing.
Yes.
     And a painful cough?
Yes.
     Does silicosis cause death?
Yes, sir.

The Disease: After-effects

This is the life of a Congressman.
Now he is standing on the floor of the House,

the galleries full; raises his voice; presents the bill.
Legislative, the fanfare, greeting its heroes with
ringing of telephone bells preceding entrances,
snapshots (Grenz rays, recording structure) newsreels.
This is silent, and he proposes:
    embargo on munitions
to Germany and Italy
as states at war with Spain.
He proposes
  Congress memorialize
the governor of California: free Tom Mooney.
A bill for TVA at Fort Peck Dam.
A bill to prevent silicosis.

This is the gentleman from Montana.
—I'm a child, I'm leaning from a bedroom window,
clipping the rose that climbs upon the wall,
the tea roses, and the red roses,
one for a wound, another for disease,
remembrance for strikers. I was five, going on six,
my father on strike at the Anaconda mine;
they broke the Socialist mayor we had in Butte,
the sheriff (friendly), found their judge. Strike-broke.
Shot father. He died: wounds and his disease.
My father had silicosis.

Copper contains it, we find it in limestone,
sand quarries, sandstone, potteries, foundries,
granite, abrasives, blasting; many kinds of grinding,
plate, mining, and glass.

Widespread in trade, widespread in space!
Butte, Montana; Joplin, Missouri; the New York tunnels,
the Catskill Aqueduct. In over thirty States.
A disease worse than consumption.

Only eleven States have laws.
There are today one million potential victims.
500,000 Americans have silicosis now.
These are the proportions of war.
    Pictures rise, foreign parades, the living faces,
    Asturian miners with my father's face,
    wounded and fighting, the men at Gauley Bridge,
    my father's face enlarged; since now our house

    and all our meaning lies in this
    signature: power on a hill
    centered on its committee and its armies
    sources of anger, the mine of emphasis.
      No plane can ever lift us high enough
    to see forgetful countries underneath,
    but always now the map and X-ray seem

one country marked by error
and one air.

It sets up a gradual scar formation;
this increases, blocking all drainage from the lung,
eventually scars, blocking the blood supply,
and then they block the air passageways.
Shortness of breath,
pains around the chest,
he notices lack of vigor.

Bill blocked; investigation blocked.

These galleries produce their generations.
The Congressmen are restless, stare at the triple tier,
the flags, the ranks, the walnut foliage wall;
a row of empty seats, mask over a dead voice.
But over the country, a million look from work,
five hundred thousand stand.

"The Book of the Dead"
Muriel Rukeyser 1994 [1938], 32–33

Rukeyser wrote about what is now referred to as the Hawk's Nest disaster, the most deadly industrial disaster in the United States. At Hawk's Nest in the 1930s, somewhere between 700 and 2,000 mostly African-American workers died while building a hydroelectric tunnel for the New Kanawha Power Company, a subsidiary of Union Carbide.[10] The tunnel provided more than a water route. It also contained very high purity silica that could be used in ferro-alloy production. By the 1930s, silicosis was well documented. Fibrosis of the lungs caused by inhalation of silica dust was established by British physicians in 1860. Routine precautions included wet drilling and particular attention to ventilation. According to workers at Hawk's Nest, Union Carbide procedures only occasionally included these precautions. Black workers were put at particular risk, working inside the tunnel in far greater numbers than white workers (Dembo, Morehouse, and Wykle 1990, 25).

The total workforce on the tunnel was about 5,000. Less than 20 percent of the workers were local. Many were brought in from out of state, particularly from the South. Carbide's own records indicate that 65 percent of the workers were black, while the population of Fayette County, West Virginia, was 80 percent white. Of those employed to work on the tunnel, 1,115 black and 379 white workers were assigned duties only inside the tunnel; 1,129 black and 359 white workers had assignments both inside and outside the tunnel; and 952 black and 952 white workers had duties only outside the tunnel (Dembo, Morehouse, and Wykle 1990, 23–24; Cherniack 1986). Those who worked longest inside the tunnel died the fastest.

According to Prince Ahmed Williams, there was a time when things were different. Williams explains that "in the old days, West Virginia was an area that was rather free, beyond the hills. Mr. Cato was white and he owned slaves but he married one of them. They really were the very early settlers in this valley. One of the grandchildren sold the first three acres to the State of West Virginia for a school, for negro children, and children of mixed parentage, in this area. . . . It was very beautiful. It was challenging. There was a time when you could find, from any corner of the state, black representation" (Pickering and Johnson 1991).

Things changed when the chemical industry moved in, during World War I. The hills around the capital city of Charleston filled with boom-towns. The war effort drew Union Carbide into the manufacture of helium for dirigibles, ferrozirconium for armorplating, and activated carbon for gas masks.[11] In 1920, Union Carbide built a plant to manufacture synthetic organic chemicals—the first petrochemical plant in the United States, near Institute (Dembo, Morehouse, and Wykle 1990, 13). African Americans were not given work in the plants—even though many lived in nearby communities. Carbide's plant in Institute shares a fence line with West Virginia State College—until recently a predominantly African-American school.

Mildred Holt came to Institute at age seventeen, as a freshman at West Virginia State College. She thought Institute was a smelly, dirty place where it was hard to breathe. Holt says Carbide didn't even try to be an attractive neighbor, installing tanks and "other plant stuff" right next to houses. Holt has wondered if Carbide would have done the same thing had it been a white neighborhood. "Maybe a low income white neighborhood," she says, "but not to people of affluence." But, until recently, no one from the college "would have dared speak to the corporation or even contact them. I dare say, even the President of the college wouldn't have done it. It wasn't the way things were done. So, we just endured" (Pickering and Johnson 1991).

At one time, Union Carbide was the largest single employer in West Virginia. Carbide's headquarters have remained out of state, like most of the other chemical companies and the coal companies that supply them. Unemployment in West Virginia always ranks among the highest in the country. Carbide jobs were always "the good jobs." So Carbide paid little in taxes and was not pressured to clean up its act. If questioned, Carbide had a habit of suggesting that it could move on to more hospitable environments (Dembo, Morehouse, and Wykle 1990).

The 1970s were a particularly bad period. Eight out of ten South Charleston City Council members were working for Union Carbide, or had previously (Dembo, Morehouse, and Wykle 1990, 58; Gerrard 1871, 130). In 1976, six workers from Union Carbide's South Charleston plant died of angiosarcoma of the liver, a form of cancer. Together they constituted 10 percent of the total number of known cases of angiosarcoma worldwide. Vinyl chloride workers at Carbide's South Charleston plant also had four times the expected rate of leukemia and twice the expected rate of brain cancer. Union Carbide continued to fight against stricter regulations on permissible levels of vinyl chloride (Dembo, Morehouse, and Wykle 1990, 63).

According to a study reported by the *New York Times* in 1979, cancer rates for the Kanawha Valley as a whole were also high. Out of a sample of 819 death certificates for the period between 1965 and 1978, there were 9 cases of multiple myeloma rather than the 2 that were statistically expected, 12 cases of kidney cancer instead of the 3 to 4 expected, and 9 cases of central nervous system cancer rather than the 3 to 4 expected (Dembo, Morehouse, and Wykle 1990, 63). Studies such as this one did provoke change. West Virginia is visibly cleaner, and it's even possible to swim in the Kanawha Valley. But the hold the chemical industry has over the state remains visceral.

In 1985, there were twenty chemical plants in the Kanawha Valley, employing 10,000 people earning an average weekly salary of $601 and an average annual salary of $31,000 (Draffan 1994, 16). What prosperity there was came from chemicals. Union Carbide in particular enjoyed extraordinary loyalty. By the 1990s, the loyalty was breaking apart. Because the Bhopal disaster provoked distrust. Because the chemical industry provides far fewer jobs than it once did. Because the history of West Virginia is being rewritten, by a new generation of local reporters in particular.

The fact that African Americans never enjoyed the prosperity promised by chemicals has risen to the surface, as has the fact that company unions have not protected worker health. The impact on West Virginians' sense of the everyday has been profound. Most remain staunchly patriotic, but a little unsure of what patriotism means. Their attitude about the chemical industry is even more ambivalent.

The cultural transformation in West Virginia has been profound. Environmental advocates—in government, in journalism, in labor unions, and in residential communities—have become prominent figures—sophisticated, articulate figures who can draw out the contradictions in which they live. The future of West Virginia will depend on them.

THREE

Union Carbide, Having a Hand in Things

(3.1) Things Fall Apart

The press meeting scheduled for December 3, 1984, at Union Carbide's corporate headquarters in New York had been planned well in advance. Corporate executives had promised to provide a detailed overview of their initiatives to restructure the company for entry into markets shaped by new information technologies, new financial instruments, and the opening of trade regimes. Instead, Union Carbide had to relay news of the Bhopal disaster (Lepkowski 1994).

Times were indeed out of joint. 1984 was also the Golden Jubilee of Union Carbide India Limited (UCIL). The commemorative brochure presented UCIL as a competent and loyal participant in the national development of India. It integrated the company into national history, which was folded into world history. UCIL showed itself to be a good citizen of India, and of the world.

Hexagon, The Commemorative Issue: Golden Jubilee
Union Carbide India Limited, June 1984

A TIME FOR NOSTALGIA

This is the story of a Company that has a unique personality and a special work ethos.

The Company, beginning its operations in India at the dawn of this century, has ever since shared the country's dreams and aspirations and contributed to the nation's growth.

From initial sales of mere Rs. 500 to Rs. 210 crores; from a small one-plant operation to a corporation with five operating divisions, fourteen manufac-turing units and a subsidiary company in Nepal; from a handful of employ-ees to over nine thousand, Union Carbide has grown indeed.

Today, the Company has 24000 shareholders, 26000 debenture and fixed de-posit investors and a network of dealers and retailers who reach UCIL prod-ucts to consumers all over India.

Union Carbide has much to be grateful for to the people of India, and has much to be proud of.

UCIL believes that a company ensures its own growth only when it serves the needs of the people; and that some things are worth doing in themselves, without thought for profit or gain.

1905–1915

The first Eveready battery came to the country almost at the beginning of this century, in 1905. Total sales amounted to less than Rs. 500. It took eight years to double that figure.

These were times of tensions, turmoil and fierce nationalism. The Indian National Congress launched its struggle for self-government; the Provisional Government of India was established in exile, at Kabul; Bengal was parti-tioned and the Morley-Minto reforms instituted.

The First World War broke out in 1914. The following year, Mahatma Gandhi returned to India, to join the struggle for Independence.

In these years also, the seeds of Indian industry were sown: Jamshedji Tata established a steel plant, and thus launched a new era in India's indus-trial history.

1916–1925

During this period, the Montague-Chelmsford reforms were instituted, the Rowlatt Act passed, the Treaty of Versailles signed.

World War I ended, the massacre at Jallianwala Bagh took place and Ta-gore returned his knighthood. His Royal Highness the Prince of Wales ar-rived in India on a state visit, but his tour was boycotted.

The Non-Cooperation Movement spread; C. R. Das and Motilal Nehru joined hands to form the Swaraj Party.

And amidst the strife, tension and political uncertainty, one company—Union Carbide—saw enough promise to look upon India as a manufactur-ing centre, and not just a market for imported products.

For, in these years, the demand for Eveready batteries had been growing steadily.

1926–1935

As the Independence movement gathered strength, the Simon Commission was met with hostile demonstrations. At the Lahore session of the Congress, for the first time, the demand for complete independence was raised.

Carbide owned Eveready

The Chittagong armoury riots took place; the first round table conference was held; Mahatma Gandhi led the march to Dandi, where the Salt Satyagraha was launched; the civil disobedience movement started; J. R. D. Tata flew from Karachi to Bombay, and started India's first domestic air service; the talkie era began in the world of cinema.

The Eveready (India) Company started manufacturing activities at the Canal Road Works, Calcutta, with an installed capacity of six million cells a year.

Within nine years, the Eveready Company (India) Limited was incorporated, and the installed capacity at the Canal Road Works was increased to fifteen million cells a year.

1936–1945

United Battery Distributors Limited was formed, and appointed sole selling agent for The Eveready Company (India) Limited; India's first modern battery plant was established at Camperdown, Calcutta, with an installed annual capacity of 40 million round cells.

World War II broke out, and the greater portion of Camperdown's output was directed to meet defense needs.

The Congress passed a resolution calling upon the British to leave, and the "Quit India" movement began; Subhash Chandra Bose formed the Indian National Army, to fight for the Country's liberation.

World War II ended; United Battery Distributors Limited merged with The Eveready Company (India) Limited; the new company was named National Carbon Company (India) Limited, and sales depots were established at Cochin, Dacca, Vijaynagar, Chheharta, Allahabad and Nagpur.

UBDL + TECL = NCCL

1946–1950

Perhaps the most momentous period in the nation's history.

India gained Independence from the British Empire; the Country was partitioned into India and Pakistan; the Reserve Bank was nationalized; C. Rajagopalachari became India's first Governor-General, Lord Mountbatten the last British Viceroy; Mahatma Gandhi assassinated.

India became a democratic republic, adopted a new constitution, and set up a Planning Commission; production commences at the Chittaranjan Locomotive Works.

Installed capacity at National Carbon Company's Camperdown Works was expanded to 100 million round cells; a manganese crushing and grinding mill was installed; the Company commenced distribution of a wide range of chemicals, plastics, and allied products, and became the distributors for Union Carbide Corporation.

1951–1955

The Government took steps to consolidate India and chart a path for progress: the railways were nationalized; the first five-year plan was launched; the first general elections were held, and the socialistic pattern of society was adopted by the ruling Congress party as the country's goal.

Computer technology came to India; Mount Everest was scaled by sherpa Tenzing Norgay and Edmund Hilary; National Carbon Company established a second battery plant at Madras, introduced the first India-made mini-max radio batteries and set up an Industrial Products Division.

1956–1960

The first man-made satellite was launched into space by the USSR. India's second five-year plan was launched, and the second general elections were held.

Oil was struck near Baroda; the first TV station in the Country commenced telecasts; India's first atomic reactor was commissioned; life insurance was nationalized and the Life Insurance Corporation of India was formed; linguistic pressures led to the split of Bombay state into Maharashtra and Gujarat.

National Carbon Company created history when it offered 8 lakh [100,000] shares to the public: registers were kept open for exactly one minute. In that one minute, the issue was oversubscribed 17 times.

The Eveready Flashlight Case plant was inaugurated at Lucknow, and National Carbon changed its name to Union Carbide India Limited.

1961–1965

The Country's peace was disturbed by two wars, one with China and the other with Pakistan. Both times, Union Carbide India worked round the clock to produce special batteries required for the nation's defense; India acquired her first aircraft carrier.

Three special landmarks during this period were the liberation of Goa, the adoption of Hindi as the national language and the passing away of Pandit Jawaharlal Nehru. Lal Bahadur Shastri succeeded Pandit Nehru, but his term was destined to be a short one.

The Carbide Chemicals Company, a division of Union Carbide India, began operations at Trombay, and India's first film and pipe unit went into production. The Carbon Products Company, another division, commences production of cinema carbons and midget electrodes for the first time in India; and the Company's R&D efforts resulted in a photo-engraving plate that was to become the industry standard.

1966–1970

The Paradeep Port was inaugurated; fourteen commercial banks were nationalized; and construction work began on the giant steel plant at Bokaro.

India made huge strides in technology, and prepared to enter the space age, even as Neil Armstrong became the first man to walk on the moon.

A time of major expansion and growth for Union Carbide; the Company's third—and most modern—battery plant was set up at Hyderabad; the Pesticides and Formulation plant went on stream at Bhopal; at Chemco, India's first naptha cracker was installed, ushering in the development of India's petro-chemicals industry; UCIL commissioned the first sewage water reclamation plant; and the EMD plant at Bombay made India the second country in the world to produce Electrolytic Manganese Dioxide.

1971–1975

A quarter century after Independence, the country helped another new nation, Bangladesh, gain independence.

Privy purses were abolished; India's first nuclear device was imploded at Pokhran, thus ushering in a new era; India's first rocket was launched from Thumba.

Mazagon Dock built the first Indian-made trawlers; oil was struck at Bombay High; India's first satellite, the Aryabhatta went into orbit, and SITE programmes commence.

During this period, UCIL began its Marine Products operations, with the arrival of its first two trawlers; the foundation stone was laid for a massive Research and Development centre near the Company's Bhopal plant.

1976–1980

Bombay High went commercial, and a major expansion programme began for the peaceful uses of atomic energy; the first steps were taken to create a national power grid; subscriber trunk dialing linked far flung cities, and a pin code system was adopted to speed up postal services.

Political turmoil rocked the country; the first non-Congress government was installed as the Janata Party came into power; it was replaced a few years later by a new Congress government.

UCIL developed indigenous process technology for the manufacture of Carbaryl pesticides and the Sevin Carbomylation unit went on stream, followed by the battery plant at Taratola, Calcutta. The Rs. 2 crore R&D facility for agricultural chemicals was inaugurated at Bhopal, and the Rs. 2 crore electro-chemical R&D centre was inaugurated at Calcutta.

1981–1984

With the arrival of low-cost microcomputers, the computer era began in a big way in India; planning began for a national computer network.

India got recognized as the eleventh largest industrial nation in the world; Insat 1B was launched into space by the US space shuttle, and successfully began operations; India's first astronaut conducted scientific experiments in space.

The Country's importance as an exporter of finished products, services, and know how grew; an export-import business was set up to facilitate even faster growth.

UCIL completed fifty years since its incorporation as an Indian company, the Company's fifth battery plant went into production in Srinagar, a subsidiary company, the Nepal Battery Company was formed.

The brochure begins with personal memories, which seem addressed to others in the UCIL "family." Photographs of former directors are encased in cameo-like frames, followed by warm greetings and walks down "memory lane." The historical overview reads like an exam, demonstrating comprehensive knowledge of the history of India. Events and place names function metonymically, connoting knowledge of a whole much larger than what is actually stated. No mention is made of Indira Gandhi or the Emergency she had so recently imposed. Little mention is made of the Green Revolution. Two particular contributions are highlighted: to war and to the development of Indian media.

The story assumes a cooperative posture, highlighting India's national accomplishments, to which the company had contributed. But the tone is somewhat forced, hinting at contradictions and split allegiances. In his opening statement, Chairman Keshub Mahindra says that "at the present moment some of you may have concern and doubts on the future of our Company, especially in view of the news of our recent decision to rationalize some product segments. I would like to stress that such decisions, though extremely difficult, are essential steps if we are to maintain the viability of our products, and meet the challenges of a changing economic and business environment." R. Natarajan, director of UCIL and vice-president of Union Carbide Corporation (UCC), follows with the insistence that "as a Carbide man, I can state quite categorically that the bond between Union Carbide Corporation and India is very strong indeed. All of us would very much like to see Carbide India grow and will extend all the necessary assistance to ensure such growth."

The last page of the brochure reiterates the company's loyalty to the nation of India and hints at another dependency: "Our destiny: tied in with our country's future. For fifty years, one goal has guided the business policies, plans and actions of Union Carbide India Limited: to serve and to live up to the trust of the people of this country. Identifying itself with the mainstream, UCIL has been contributing in its own way to the socio-economic development of the country." UCIL appears caught within multiple, potentially conflicting dependencies. The relationship between UCIL and the nation of India seems forced, as does the relationship between UCIL and UCC. The complexity of these relationships was, of course, exacerbated by the complexity of the times. The 1980s was a far from straightforward decade.

In the early 1980s, Union Carbide was the seventh largest chemical company in the United States, with almost 100,000 employees working in forty countries. Both assets and annual sales approached $10 billion. But the company was not making money. Paul Shrivastava explains:

The metals and mining division, which supplied products to the declining American steel industry, saw its pretax profits fall from $291 million in 1981 to $31 million in 1984, with little chance predicted for recovery. In 1981, the company introduced a cheaper process for the manufacture of ethylene derivatives that were used as raw materials for polyester, antifreeze and polyethylene plastics. Instead of driving the competition out of the market, this strategy backfired by creating a worldwide oversupply of ethylene derivatives and driving prices down. The division manufacturing the product dropped from a $131 million profit in 1981 to a $39 million loss in 1982. Even in the normally lucrative area of real estate, Union Carbide had bad luck. In 1977 it lost money by selling its corporate headquarters at one of the few times in history when the Manhattan real estate market was depressed. (1987, 41–42)

The 1980s would spiral out of control for Union Carbide. The confidence expressed in advertising during the 1960s and 1970s began to erode, with a bitterly ironic edge. Carbide's advertising during the 1960s was built around a memorable theme: "A Hand in Things to Come." One hand sends Carbide ambassadors around the world. Another snaps of power, advertising the portable batteries that made UCIL profitable. Another reminds us that without air, life stops. Carbide takes care of the children as they sleep. Yet another ad shows how Union Carbide brings science to India. Luminescent red liquid pours from a test tube. In the foreground, an Indian man plows. In the background, there is the Gateway to India, the Ganges River, and a chemical plant.

By the 1980s, Union Carbide itself was a complex tangle of remotely related parts. The company was both tightly coupled and dispersed. It had grown into a global giant with far-flung subsidiaries and inadequate feedback mechanisms. The problem of relating parts to the whole was literally as well as figuratively "out of hand."[1] Bhopal continues to be a site where this problem works out.

The relationship between UCC and UCIL has never been settled. Nor has the relationship between UCIL and India, much less between UCC and India. The "function" of corporate subsidiaries, particularly in the Third World, remains undefined. New relationships between corporations and their various "stakeholders" have begun to be articulated. But effective modes for connecting stakeholders (parts) to the corporation (whole) remain undetermined. And then there is the problem of the significant event (part) in history (whole). Union Carbide has had to figure out what to do with Bhopal in relation to itself as a historically legitimated totality. Corporations are legitimated by "their record." Establishing what counts as part of the corporate record is the work of corporate advocacy.

TOP ROW (left to right): Australia, Switzerland, Great Britain, India, Mexico, New Caledonia, Venezuela, Panama, Italy, Japan, Puerto Rico, British Guiana, Canada, France, Ghana. MIDDLE ROW: Thailand, Malaya, Philippines, South Africa, Brazil, Pakistan, Hong Kong. BOTTOM ROW: Argentina, Norway, Indonesia, Greece, Sweden, New Zealand, Colombia, Nigeria.

Meet the ambassadors

Around the world, Union Carbide is making friends for America. Its 50 affiliated companies abroad serve growing markets in some 135 countries, and employ about 30,000 local people. ▸ Many expressions of friendship have come from the countries in which Union Carbide is active. One of the most appealing is this collection of dolls. They were sent here by Union Carbide employees for a Christmas display, and show some of the folklore, customs, and crafts of the lands they represent. "We hope you like our contingent," said a letter with one group, "for they come as ambassadors from our country." ▸ To Union Carbide, they also signify a thriving partnership based on science and technology, an exchange of knowledge and skills, and the vital raw materials that are turned into things that the whole world needs.

A HAND IN THINGS TO COME

FIGURE 3.1 CARBIDE'S AMBASSADORS   In the 1950s and early 1960s, Union Carbide ran a series of ads with the slogan "Having a Hand in Things to Come" in magazines such as *Fortune, Scientific American, National Geographic,* and *U.S. News and World Report.*

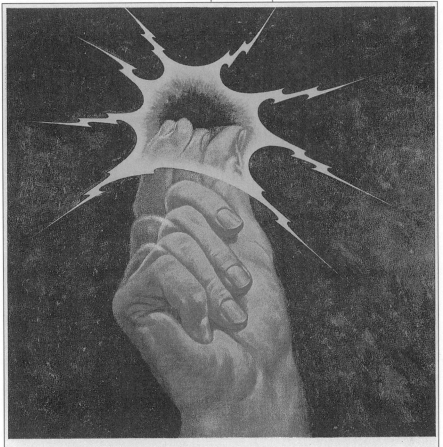

Instant portable power... any time, any place

In this battery-sparked new world of portable convenience, hand tools are driven by their own re-chargeable batteries . . . toys perform their tricks by remote control . . . a hearing aid with its button-size power cell can be slipped into the ear . . . cordless radios and television sets are lively companions in the home or outdoors . . . missiles and satellites are guided through the vastness of space. ▶ Developments like these have brought more than 350 types of EVEREADY batteries into use today, 73 years after Union Carbide produced the first commercial dry cell. Ever-longer service life and smaller size with power to spare are opening the way for batteries, such as the new alkaline cells, to serve hundreds of new uses. ▶ For the future, along with their research in batteries, the people of Union Carbide are working on new and unusual power systems, including fuel cells. And this is only one of the many fields in which they are active in meeting the growing needs of tomorrow's world.

A HAND IN THINGS TO COME

UNION CARBIDE

LOOK for these other famous Union Carbide consumer products—
LINDE Stars, PRESTONE anti-freeze and car care products, "6-12" Insect Repellent, DYNEL textile fibers.
Union Carbide Corporation, 270 Park Avenue, New York 17, N. Y. In Canada: Union Carbide Canada Limited, Toronto.

FIGURE 3.2 CARBIDE'S PORTABLE POWER   A Union Carbide advertisement for Eveready batteries. In 1984, over 50 percent of Union Carbide India Limited revenues came from this product.

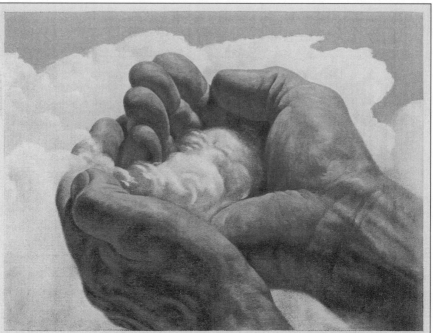

Without air, life stops

... for you and for industry

SEEING A CLOUD is probably the nearest we come to 'seeing' air, because air is a mixture of invisible gases.

Life-giving oxygen comprises about 21 per cent of the air. We all know how it helps sick people get well, but few of us realize that steel and other major industries could not operate without the same oxygen in tremendous quantities. About 78 per cent of the air is nitrogen. Food processors use it as an atmosphere to protect freshness and flavor of food.

The remaining one per cent of the air is composed of the little-known yet vital "rare" gases — argon, helium, krypton, neon, and xenon. These gases are essential in making incandescent light bulbs, in electric welding processes, and in refining new metals such as titanium.

For fifty years, the people of Union Carbide have been separating the gases of the air and finding new ways in which they can help make a better life for all of us.

FREE: *Learn how many of the products you use every day are improved by research in alloys, carbons, chemicals, gases, plastics, and nuclear energy. Write for "Products and Processes" booklet M.*

Union Carbide Corporation, 30 East 42nd Street, New York 17, N.Y. In Canada, Union Carbide Canada Limited, Toronto.

UNION CARBIDE

UCC's Trade-marked Products include

LINDE Oxygen	CRAG Agricultural Chemicals	EVEREADY Flashlights and Batteries	ELECTROMET Alloys and Metals
SYNTHETIC ORGANIC CHEMICALS	PREST-O-LITE Acetylene	PRESTONE Anti-Freeze	HAYNES STELLITE Alloys	Dynel Textile Fibers
BAKELITE, VINYLITE, and KRENE Plastics	PYROFAX Gas	NATIONAL Carbons	UNION Calcium Carbide	UNION CARBIDE Silicones

FIGURE 3.3 CARBIDE'S WITHOUT AIR, LIFE STOPS   Bhopal gas victims could not but agree: without air, life indeed does stop. The primary cause of death immediately following the gas leak in Bhopal was pulmonary edema.

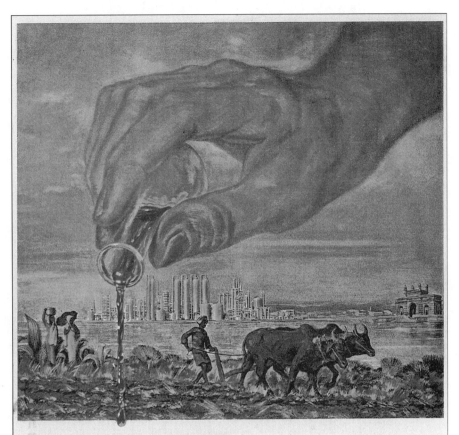

Science helps build a new India

Oxen working the fields . . . the eternal river Ganges . . . jeweled elephants on parade. Today these symbols of ancient India exist side by side with a new sight—modern industry. India has developed bold new plans to build its economy and bring the promise of a bright future to its more than 400,000,000 people. ▶ But India needs the technical knowledge of the western world. For example, working with Indian engineers and technicians, Union Carbide recently made available its vast scientific resources to help build a major chemicals and plastics plant near Bombay. ▶ Throughout the free world, Union Carbide has been actively engaged in building plants for the manufacture of chemicals, plastics, carbons, gases, and metals. The people of Union Carbide welcome the opportunity to use their knowledge and skills in partnership with the citizens of so many great countries.

A HAND IN THINGS TO COME

UNION CARBIDE

WRITE *for booklet B-3 "The Exciting Universe of Union Carbide", which tells how research in the fields of carbons, chemicals, gases, metals, plastics and nuclear energy keeps bringing new wonders into your life.*
Union Carbide Corporation, 270 Park Avenue, New York 17, N.Y.

FIGURE 3.4 CARBIDE BRINGS SCIENCE TO INDIA   A Union Carbide advertisement run in *Fortune* magazine in April 1962. Understandably, this particular advertisement has been reprinted often since December 1984.

(3.2) Assuming Moral Responsibility

In August 1986, Jackson Browning, vice-president of Union Carbide, wrote a letter to the *New York Times* explaining that an article published the previous month about the Bhopal disaster "points to one of the continuing tragedies in that city: the apparent decision by the Indian Government to make health concerns subservient to litigation concerns." The letter goes on to bemoan the lack of "third-party evaluation of health effects of the gas release," supposedly stymied by the Indian government and creating a situation in which "enlightened decisions on medical treatment—now and in future years—cannot be made." Nonetheless, Carbide has continued to offer "major humanitarian assistance—with no strings attached and no ties to an ultimate litigation award. . . . India has refused most of these offers apparently in the mistaken belief that acceptance would reduce ultimate compensation." The letter closes with the insistence that "immediately following the gas release, Union Carbide said that it would take moral responsibility and let legal responsibility get sorted out later so that the immediate needs of the victims could be met. We believe that the Government of India should make the same kind of commitment to the people of Bhopal" (Browning 1986, 11A).

Browning's letter reiterates Union Carbide's standard line on the Bhopal disaster. The insistence that the company accepts "moral responsibility, but no liability" was put into circulation immediately following the gas leak and has been the most repeated response ever since. Warren Anderson, CEO of Union Carbide in 1984, revisited the argument in an anniversary statement in 1990, following the out-of-court settlement of the case. Anderson's emphasis is on humanitarianism, which is opposed to legalism; on local responsibility, which is opposed to corporate liability; and on victims, who are set in opposition to their own government, rather than in opposition to Union Carbide.

BHOPAL: WHAT WE LEARNED
Warren M. Anderson
Former Chairman, Union Carbide Corporation

The main lesson from the Bhopal tragedy? When you have a disaster in which your company is involved, make everyone understand that the needs of the victims come first. Don't get lost in the morass of lawyers, politicians and public relations managers. . . .

On this anniversary of the settlement, I'd like to review the Bhopal disaster, how it unfolded, and some of the lessons we believe we learned from it.

The tragedy occurred on the night of December 2, 1984, at a plant in Bhopal, in central India. The plant was owned by Union Carbide India Limited, or UCIL. Union Carbide Corporation owned 50.9% of the stock of UCIL.

Another 22% was owned by the Indian government and the balance was widely held by Indian citizens. At the time, UCIL had been in existence for almost 50 years and had 14 plants throughout India. For many years, it had been a wholly Indian operation; the last American employee left the Bhopal plant two years before the release.

For decades, at Carbide, safety has been a paramount concern and even a single critical injury or death has always been considered a major calamity which immediately receives urgent attention at the highest levels of the company. You can imagine our horror, then, when we began to receive initial reports from India that hundreds of people had died in a gas release from UCIL's Bhopal plant. When the full magnitude of what had happened became known that day, the entire company was shaken to its core.

Despite our shock, I believe our instincts that day were correct, and if anything, the history of the last five years has repeatedly confirmed that view.

We saw Bhopal for what it was—a terrible tragedy involving real people who had either lost family members or had suffered injuries, in some cases serious injuries. They needed medical relief, prompt aid in any form possible, and an early settlement which would help restore their lives and bring long-term relief. They didn't need what they ultimately got—armies of lawyers and politicians who spent years claiming to represent them and deciding what was in their best interests. We saw Bhopal in moral—not in legal—terms. Although we had good legal defenses—the plant wasn't ours and it later was established that the tragedy had been caused by employee sabotage—we didn't want to spend years arguing those issues in court while the victims waited. We therefore said immediately that Union Carbide Corporation would take moral responsibility for the disaster.

I decided that day to go to Bhopal myself because I believed that if I was there in person, I could work with the Indian authorities to bring relief to the victims and get them to see that it was in the best interests of the people of Bhopal that the matter be resolved quickly. I also believed that being there myself would help to focus everyone's attention on the fact that what we were dealing with was a human tragedy which required a human response, not a media event or a protracted legal battle.

Politics took over, as it would so many times afterward in the history of the tragedy. State elections were imminent in Madhya Pradesh, the state where Bhopal was located. In an effort to help win those elections, the state's chief minister had me arrested and sent back to the United States.

The arrest meant the loss of a major opportunity to resolve the matter quickly and it set the stage for a long legal and political battle, which did not benefit the victims at all. . . . However, I believe that my trip did an enormous amount of good. By my going, we were saying to the people of Bhopal, and to the world, that we were willing to take moral responsibility for the tragedy and that we would do everything in our power to help. . . . We then set about trying to do two things, bring immediate relief to the victims, and effect a prompt and fair settlement of the matter. . . . All in all, we gave or offered on the order of $20 million.

Unfortunately, virtually all of these offers were rejected by the Indian government. By then we were enmeshed in litigation with the government, and it apparently felt that allowing Carbide to offer aid was somehow inconsistent with its litigation position. The aid we offered was always on a "no-strings attached" basis and we couldn't understand why it wasn't accepted. What was clear was that it wasn't going to be accepted.

Frustrated, we then provided funds to independent organizations who could then provide aid on their own. In one we provided over $2 million to Arizona State University to build and operate a rehabilitation center in Bhopal. The center was built, and operating well, but when the state government learned that Carbide money had funded it, bulldozers were sent in to knock the building down.

Our efforts to settle the matter, if anything, proved even more frustrating. In April 1995, the Indian government passed a law that gave it the right to represent all of the victims in court and to settle the case on their behalf. For the next four years, we made constant efforts to effect a settlement through every conceivable channel we could think of.

Oddly enough, the problem was not money. We believe that the government negotiators realized early on that the amount on the table was very generous and more than adequate to fully compensate all of the victims. Notwithstanding, they still wouldn't settle.

The problem was political, both at the government level and on an individual level. Multinationals such as Carbide are widely disliked in India, as is the case in much of the Third World. For obvious reasons, after the tragedy, Carbide was even more disliked than most. There was therefore a large body of opinion in India that was opposed to any settlement with us, irrespective of the amount. In addition, it was also clear to the government that—politics being politics—the opposition would inevitably claim that the amount of any settlement was far too low. During the last few years, the Gandhi government was in serious political trouble for other reasons, and it was afraid to settle because of the potential political repercussions.

On an individual level, this fear had similar consequences. The Indian government is one of the largest bureaucracies in the world and it became impossible to find an individual official or group of officials willing to accept the responsibility for the settlement. . . .

What ultimately settled the matter was the courage of the Indian Supreme Court. Every judge who had overseen the litigation—either in the U.S. or India—had recognized immediately that it cried out for settlement, and they all tried to effectuate a resolution. . . . The reason the case settled was because Chief Justice Pathak and the rest of the panel were willing to take full responsibility for the settlement, which eliminated the major obstacle to a resolution from the Indian government's point of view. The judges knew that they would face tremendous criticism from activist groups—and they did—but they were willing to face it because they also knew that in the end the case was about poor people who needed relief, and not about politics, lawyers, grand legal doctrines or multinationals in the Third World. The Indian Supreme Court understood what we had been saying immediately and they deserve most of the credit for the settlement.

What have we learned from Bhopal?

Most importantly, the tragedy caused all of us in the industry to take a hard look at ourselves. We believed we'd been making good progress and that a tragedy on the order of Bhopal's was impossible.

Now we know differently, and that knowledge has changed the way we operate forever. . . . We've also learned something a little less obvious—that when you have a tragedy or a disaster with which your company is involved, you have to keep focusing everyone's attention on what's most important. Certainly, your public relations and your legal defense are extremely important. Clearly, there's a vital need to satisfy your various constituencies: employees, shareholders, regulators and the public. And it's also critical that management not be diverted from running the company.

But it's easy to get lost in the morass of lawyers, politicians and media managers and to forget the fact that a tragedy like Bhopal is primarily about victims and their needs. If you remember that, and keep telling people that, you'll do the right thing and in the long run you'll be much more respected by everyone involved.

Circulated in Bhopal, April 1990

The argument that Union Carbide assumes moral responsibility but no liability for the Bhopal disaster has had legal implications. In 1985, it was the implicit basis of Union Carbide's plea for dismissal of the Bhopal case from U.S. courts, on grounds that the parent company, headquartered in Danbury, Connecticut, had no jurisdiction over the plant operations in Bhopal. This claim was substantiated with a highly publicized sabotage theory, wherein blame for the disaster was located within a five-minute interval of purposeful action by a "disgruntled worker." In sum, Carbide argued that the preemptive knowledge that could have prevented the disaster was beyond the organizational capacity of the corporation. There was intentional misconduct, but it was not their own.

(3.3) Bhopal Is History Now

Carbide's sabotage theory was first fully presented at an independent conference of chemical engineers in London during the spring of 1988. The paper "Investigation of Large-Magnitude Incidents: Bhopal as a Case Study" was presented by Ashok Kalelkar, a representative of the engineering firm of Arthur D. Little. Kalelkar's primary goal was to challenge the water-washing theory put forth by journalists and by India's Central Bureau of Investigation, a theory constructed through the testimony of plant workers.

The water-washing theory states that around 8:30 on the evening of December 2 a routine maintenance procedure was carried out to clear blocked lines in the relief valve vent header piping configuration, downstream from the methyl isocyanate (MIC) storage tank. Despite Carbide's own maintenance regulations, the procedure was carried out without the insertion of slip binds to keep water from backing up into the storage area. MIC storage tank 610 was not holding pressure, so water was able to pass the pressure control valve into the tank.

Carried with the water that flowed into tank 610 were iron rust filings from corroding pipe walls, residue of the salt compounds that had blocked the lines being washed, and other contaminants. The entry of water plus contaminants into the MIC storage tank set off an exothermic reaction that caused catalytic trimerization, a runaway reaction causing massive rise in pressure and temperature that resulted in the release of the tank's contents to the atmosphere.

The water-washing theory implicates Carbide management both for decisions immediately prior to the disaster and for long-term processes of plant design, maintenance, and personnel training. In his refutation, Kalelkar insisted that "salient, non-technical features" be investigated, requiring "an understanding of human nature in addition to the necessary technical and engineering skills." Kalelkar then refuted the water-washing theory with a technical analysis grounded on the claim that "there is a reflexive tendency among plant workers everywhere to attempt to divorce themselves from the events surrounding any incident and to distort or omit facts to serve their own purposes" (Kalelkar 1988, 5).

Because of this reflexive tendency among workers, Kalelkar based his investigation on the accounts of peripheral witnesses. A primary witness was Mr. Rajan, an instruments engineer, who claimed that on the morning following the leak he noticed that the pressure gauge on tank 610 had been removed, leaving an opening through which water could have been introduced from a nearby hose. It took Rajan over one year to remember this detail, after which he was relocated to Bombay. The other primary witness was a twelve-year-old tea boy, retrieved after much effort from his native Nepal. The tea boy is said to have been on duty the night of the disaster and describes a tense atmosphere just preceding the leak, thus "verifying" that all workers on-site were involved in a conspiratorial cover-up.

Carbide's sabotage theory pins the cause of the disaster on a five-minute interval when a "disgruntled" worker attached a water hose directly to a storage tank filled with MIC. No mention is made of the reason there were no safety mechanisms to prevent unauthorized inputs. No mention is made that, according to Carbide's own regulations, the tank should only have been three-quarters full and an adjacent tank empty to allow for transfers in case of emergency. No mention is made that there was no early warning system that would have tracked subsequent rises in temperature, indicating that disaster was imminent. No mention is made that four of five major safety systems failed to work due to indifferent maintenance. No mention is made that these systems were all underdesigned to accommodate mass escape of gas at rates up to 720 pounds per minute.

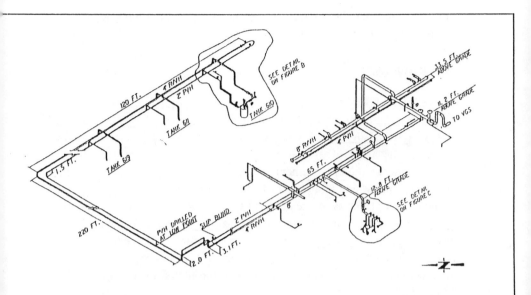

Source: A. S. Kalelkar, Investigation of Large-Magnitude Incidents: Bhopal as a case study (I. C. E. S. Series No. 110, 1987)

FIGURE 3.5 KALELKAR'S PIPING CONFIGURATION   A drawing of the piping configuration of the Bhopal plant included in A. S. Kalelkar's report titled "Investigation of Large-Magnitude Incidents: Bhopal as a Case Study." The report "refutes" the water-washing theory of the cause of the disaster and argues instead that the disaster was caused by worker sabotage. The circle on the right is drawn around the lines that were water-washed the night of December 2, 1984, circumscribing where the disaster originated in the water-washing theory. The circle on the left, drawn around the MIC storage unit, delimits the disaster even further, showing where a "disgruntled worker" would have attached a water hose directly to the MIC storage tank, setting off the exothermic reaction that led to release of the contents of the tank into the atmosphere.

The drawings Kalelkar provided to support his argument clearly delineate his circumscription of the disaster. Even if the disaster could be linked to the water-washing theory, a circle could be drawn around the specific place on the piping configurations where slip binds should have been inserted. If people had been drawn in, a handful of workers and a supervisor would have been included—all of whom were Indian, as Carbide so often points out. The sabotage theory is even more reductive, allowing a circle to be drawn around the MIC storage tank itself. If people were drawn in here, only one worker would be included. He would appear "disgruntled." The reasons for his grievances would not be relevant.

Robert Kennedy, CEO of Union Carbide following Warren Anderson, offers the final interpretation: "The Bhopal incident has not affected our ability to do business. It was a terrible human tragedy along with the sabotage that led to it. The tragedy had an indirect consequence in the form of GAF's takeover attempt a year after Bhopal. But Bhopal is history now"

wow—
Not a silver lining.

(quoted in Reisch 1990, 9). Kennedy's history lesson is straightforward: the Bhopal disaster can be circumscribed in time and space. It is finished.

(3.4) Psychosis?

In 1989, Union Carbide's public affairs strategy group circulated a memo concerning the environmental group convened by Lois Gibbs, famous as a key voice from Love Canal. The memo warned that the Citizens' Clearing-house for Hazardous Waste (CCHW) is "one of the most radical coalitions operating under the environmentalist banner . . . with ties into labor, the communist party, and all manner of folk with private/single agenda." The agenda of CCHW was to "restructure U.S. society into something unrecognizable and probably unworkable. It's a tour de force of public issues to be affecting business in years to come." Amidst the media coverage of Earth Day 1990, the author of the memo sent an apology to Gibbs after the *Washington Post* threatened to publish the memo in full (Lepkowski 1994, 38).

Just eight days prior to circulation of the memo defaming Lois Gibbs, Robert Kennedy gave his opening speech as elected chairman of the Chemical Manufacturers Association. This speech, taking up the challenge of regaining the "trust and confidence of the public," insisted that "there is a growing need for predictable, consistent environmental stewardship from nation to nation and region to region around the world. The chemical industry has a great opportunity in the years ahead to make new history." Kennedy followed his speech with the initiation of the CMA's Responsible Care program (Lepkowski 1994, 38).

How can this concurrence be drawn out? How—in the space of eight days—could Carbide manage this shift? How was old-fashioned paranoia about environmentalism entangled with proactive embrace of environmental stewardship? Carbide—like many other chemical corporations in the 1980s—was caught up in a time warp, exhibiting postures of times past concurrently with postures demanded by the future. The "corporation" was undergoing an identity crisis that brought divergent personas to the surface.

By law, corporations are persons enjoying the rights promised by the 14th Amendment to the U.S. Constitution. Corporate identity, legally defined, has many of the attributes of the Enlightenment subject: while constituted through multiple bilateral contracts with employees, vendors, customers, and others, the overarching responsibility of a corporation is singular—maximization of shareholder wealth. In other words, the multiplicity of social relations constitutive of the corporation is reduced to a single objective—profit maximization.

The legitimacy of the "shareholder primacy principle" began to be challenged in the 1980s by legal and management theorists, as well as by lawyers and citizens active in social movements. A fundamental redefinition of the corporation was under way. Corporations were still trying to figure it all out.

The stakeholder concept of management debuted in 1984 with the publication of Freeman's "landmark" book titled *Strategic Management: A Stakeholder Approach* (1984). According to management theorists Jerry Calton and Nancy Kurland, "[T]he problematic nature of this concept is reflected in a history of efforts by Freeman and others, to reconcile the conventional assumption of managerial agents' fiduciary responsibility to serve a unitary organizational purpose (i.e. profit maximization) with the expectation of agent responsiveness to the counterclaims of non-owner stakeholder groups" (Calton and Kurland 1995, 155). Nonetheless, by the late 1980s, the stakeholder concept of management was in wide circulation—prompting new demands by workers and communities inhabited by corporations, as well as by shareholders themselves.

In the late 1980s, socially conscious investors moved into visibility by filing large numbers of environmental proxy resolutions at shareholder meetings. Only three such resolutions were filed in 1989; forty-three were filed in 1990; an average of almost sixty resolutions was filed per year between 1990 and 1993 (Hoffman 1997, 110). Some of these resolutions were filed at Union Carbide meetings, demanding changes in the company's approach to the rehabilitation of gas victims in Bhopal.

PROCEEDINGS OF THE ANNUAL MEETING OF SHAREHOLDERS
UNION CARBIDE CORPORATION
APRIL 1989
Hyatt Regency West Houston
Texas Ballroom

. . . While I'm on the subject of rules and procedure, I should inform you that we expected a group of three Bhopal victims or survivors here with us in the meeting this morning. Their patrons are either shareholders or had proxies. There were other groups with them who had a press conference in front of the hotel approximately at 9:00 o'clock this morning. When they sought to enter the hotel, they were asked to leave outside placards and handouts and bills that they apparently were prepared to hand out to people inside the hotel.

The hotel manager met them at the door, talked to them for about 10 minutes, repeatedly advised them that it was a rule of the hotel—not of Union Carbide, but a rule of the hotel—that demonstrators and people carrying placards were not to be allowed in with placards or handouts. They were welcome to come in, and they were told they were welcome to come in if they would set aside their demonstration materials at the door.

After about ten minutes, there was some pushing and shoving, and the hotel manager asked the police if they would restrain those people who were the most aggressive. I believe they were detained and have been escorted away. I also believe that plans are underway immediately for their release.

And I would repeat that they are welcome. Anyone having a proxy or a guest of a proxy or, of course, a shareholder is welcome at this meeting and is welcome to make statements as long as the rules of procedure are followed, in fairness to all present. . . .

Our third item of business today is the stockholder proposal regarding South Africa. The proposal is sponsored by the New York State Common Retirement Fund, the American Baptist Home Society, the New York State Teachers' Retirement System, the General Board of Pensions of the United Methodist Church, and the Society of Jesus, Detroit Province. . . .

MR. ROBERT D. KENNEDY: Back in the center aisle. Andy Smith.

REVEREND ANDY SMITH: Yes, good morning Mr. Kennedy. I am Reverend Andy Smith from the American Baptist Home Mission Society. We are a co-sponsor of this proposal.

I second the proposal and would like to comment on it.

I returned on April 6th from a six—from a three-week trip to South Africa. I went there in order to find out what the situation was today, having been there already once before in 1985.

I found out that by talking with many people in the townships—religious leaders, local people living there and as many people as I could—that in fact many cosmetic changes have occurred. If you go there and go into the big cities and don't go into the townships, you would think that everything is okay. But once you probe under the surface, you find that the situation is literally unchanged from before.

I also recognize that there is a real moral dilemma for South Africans in supporting sanctions because the economy is in terrible shape and it is hurting all, including the blacks. We often hear that. And yet at the same time many, many people told me that they had this moral dilemma because they knew it would hurt, but they saw no other option.

I spoke with a member of parliament who said the only way that the nationalist members of parliament are going to do something is under pressure, under force. Therefore, I came away feeling that sanctions are indeed the only alternative.

We approve of the fact that Union Carbide has in fact left all its profit or dividend from the operations there in the Hexagon Trust. The Hexagon Trust is beginning to aid some development of blacks. However, all of the development of blacks that has occurred until this point has not ended apartheid. So, no black can vote, no black can live where he or she wants; and the basic thing is that the blacks are not accepted as equal human beings. I heard in a number of instances from people: All we want is to be accepted as people, human beings.

Therefore, I would ask: What is indeed this company doing to bring the kind of pressure on the South African Government that is necessary in order to force it to change, because it will not change through rational dialogue?

I believe that the company is doing little and in fact is supporting the government more through its taxes than what it is doing through the Hexagon Trust. I would, therefore, call on shareholders to support this proposal.

MR. ROBERT D. KENNEDY: Thank you, Reverend Smith.

We now have a resolution and it has been seconded. Is there further comment or discussion?

(Pause)

There are, of course, many thoughtful opinions on the subject of investment or disinvestment in South Africa. Our position which we took in 1986 and had held prior to 1986, but it was in 1986 that we moved to take and separate our earning from the issue and stand aside and say: This is not an economic issue for Union Carbide. The issue is whether we stay and have some voice in what happens or whether we walk away and just really wash our hands of it.

We have elected to stay and in every way that we can, as a participant in the community, by creating minority businesses, by giving scholarships which we now have given to a 104 students who are in bursaries, in mixed race universities, by conducting our business in a way that meets the highest standards of what were formerly called the Sullivan Principles. We believe you have a voice. We believe that staying is better than running. . . .

We believe that some day the situation will change and that South Africa will continue to be a very profitable and very important, strategic place for us to be. So, it's really kind of a long-term view. . . .

. . . Our fourth item of business is a stockholder proposal regarding Bhopal. The proposal is sponsored by the American Baptist Home Mission Society and the Sisters of Charity of Saint Elizabeth.

I would like to introduce the Reverend Andy Smith of the American Baptist Churches, USA, who will present the resolution.

REVEREND ANDY SMITH: Thank you, Mr. Chairman.

We feel that the company's intention in this resolution is to make the proxy statement appear moot and unnecessary. Such a contention is misleading in the extreme. Our corporation wants its shareholders to believe that the case is settled. It is not. There is no final settlement in Bhopal. . . .

The company has misled the shareholders in terms of what its moral responsibility is. It has said all along it is taking moral responsibility. We feel that its actions have indicated otherwise at this point in time. . . .

DR. CLARENCE DIAS: I hold a proxy for the Sisters of Charity. . . . The victims are not bargaining for higher compensation alone. They are seeking guarantees that their health care needs must be met over the coming decades as their health continues to deteriorate drastically.

You, the stockholders of Union Carbide, are you willing to deny these victims such health care and thereby condemn them to a lingering and painful death?

One slight amplification in conclusion about the factual direction that your chief executive officer, Mr. Kennedy, made. The Bhopal victims are not in this room. They chose not to be here. They chose not to be here for several reasons. First, because they attempted to meet with some of the Carbide board of directors and explain their concerns, their problems, their priorities with them before this meeting. They were denied access to such directors.

They are not in this room because they refused to submit to proceedings dictated by Carbide, which will not allow them to raise the questions that they consider to be important to raise.

They are not in this room because they were arrested this morning outside this hotel for trying to share with you a document that contained information crucial to this proposal. Two of the victims who have not been arrested are in the lobby of the hotel outside and are willing to meet with any of you. Unfortunately, in more senses than one, this meeting has been an arresting one.

I urge you to live up to your moral responsibility and vote "yes" to the proposal. Only then can we, the stockholders of Carbide, live down the shame epitomized in the slogan that environmentalists are raising repeatedly these days, the slogan that "Exxon spills, Carbide kills."

Thank you.

MR. ROBERT D. KENNEDY: Dr. Dias.

We now have a proposal that has been put forward and seconded. I think I will ask for other comments at this time on the proposal before I respond on the part of management.

MR. GARY COHEN: (Indicating)

MR. ROBERT D. KENNEDY: Over here in Section D.

MR. GARY COHEN: Hello. My name is Gary Cohen. I'm with the National Toxics Campaign, which is a shareholder of Union Carbide.

I believe it's critical for the company to support this resolution providing immediate relief to the victims of the Bhopal accident.

I work with many environmental groups, local civic groups, and public health groups around this country and in other countries concerned about environmental problems. And the feeling among all these groups is that our company, Union Carbide, is an outlaw company, is a criminal company, is a company that poisons people for profit and does not pay the full consequences. Despite all the glossy public relations efforts of our company, this attitude damages us. . . .

So, I would—I would ask you right now, Mr. Kennedy, in front of everyone whether the board of directors and especially the outside board of directors—Mr. Jordan, who is a civil rights leader; Mr. Train, who is an environmental leader; Ms. Rivlin, who is an environmental leader—to meet with these victims in Washington a couple of weeks from now and to discuss this, and also to distribute this report to your shareholders who were not given the benefit of this information because the victims were arrested. . . .

MR. ROBERT D. KENNEDY: . . . We share the sense of frustration and anger that has been expressed here that in four and a half years the tragedy of Bhopal has been politicized and so little has been done for the victims, and that in fact is the second tragedy of Bhopal.

Most of the things proposed or suggested that Union Carbide do to fulfill its moral responsibility for the victims were in fact proposed by Union Carbide when Mr. Anderson went to Bhopal in the days immediately following the tragedy and was detained and a week later expelled from the country. . . .

I think it is well-known to most, if not all, of the people here that all of the proposals put forward by Union Carbide, indeed many of which we tried to act on unilaterally or through third parties—money through the International Red Cross, funds for a rehabilitation center through Arizona State University, other efforts were turned down, spurned. In the case of the rehabilitation center, actually bulldozed.

It became clear that since the government by act of parliament, also in March of 1985, had made itself the sole source of remedy and representation for its people, that it was up to us to deal with the government. It will be recalled that the government aggressively pursued a litigation strategy, came to this country and for a year and a half pursued that route in this country, in the courts of the United States. . . .

No hearings have been held on the substance of what happened in Bhopal. Bhopal, we must remember, was the act of a single, misguided, unhappy employee, who we believe had no idea what kind of turmoil he would turn loose by introducing water into that tank.

Today it's not a question of who is to be punished or what the law should do. The failings of the law in this are serious and evident. The real question is, as it has been from the beginning, is to get compensation to the victims without further discussion and debate. . . .

The proper place, we think, to address these comments is to the representative of the people of India; and that is the government. For Union Carbide to become further involved at this time would constitute something we have very seriously avoided from the very beginning, would constitute some kind of political involvement on the part of the Union Carbide in the internal affairs of India.

Beyond that, in real terms, it is quite clear that the emotions and sentiment surrounding the issues of Bhopal that an effort on our part in India to do any of these things today unilaterally would simply serve as a lightening rod for those who want to bring attention onto the multinational corporations. . . .

Is there any further discussion or comment on this resolution?

MR. WARD MOREHOUSE: (Indicating)

MR. ROBERT D. KENNEDY: Yes. In the first row. Section G.

MR. WARD MOREHOUSE: Thank you, Mr. Kennedy.

My name is Ward Morehouse from the Bhopal Action Resource Center in New York. I'm here as a proxy for the Sisters of Charity.

There are many misstatements of fact, I regret to say, in the remarks you have just made. I shall not here indulge your patience by pointing them out, but I do invite any shareholders who are interested to meet with the victims outside the meeting after it is over.

I know how anxious Mr. Kennedy and all the rest of you are to get on with the remaining agenda of this meeting, so I shall be brief.

I'm acutely conscious of my responsibilities in representing at this meeting a religious order committed to the sanctity of life. I, therefore, ask all of you to join me in one minute of silence in memory of the 10 victims who will die this week in Bhopal as you contemplate how you will vote on this resolution.

MR. ROBERT D. KENNEDY: The chair will accept that very well intentioned motion.

Thank you, Mr. Morehouse.

(One minute of silence)

MR. ROBERT D. KENNEDY: Are there any further comments at this time on this resolution?

MR. JAMES BARNES: (Indicating)

MR. ROBERT D. KENNEDY: In Section E.

MR. JAMES BARNES: I'm Jim Barnes. I'm a retired Carbider.

I'm standing to request that you defeat this resolution. I think Mr. Kennedy has explained what Carbide has tried to do and has done. Every one is sorry that it happened. It was not the company's fault. It was an employee's fault.

I feel like these groups that each year try to condemn our country—our company are not helping this country or the people in this country. I just made a quick figure of the stock that they held, which comes to—today's price of our stock—about $400,000. I believe that those people could use that stock money to help our poor and needy better than they could do to come here and disrupt a stockholder's meeting.

I thank you. . . .

MR. O. J. ROMARY: I have the preliminary results of the voting. . . .

On the stockholder proposal regarding South Africa, the vote was 12,960,000 shares for withdrawal, 76,380,000 shares against withdrawal, and 29,600,000 abstained. The vote against this resolution represents approximately 85 percent of the total shares voted.

On the stockholder proposal regarding Bhopal, the vote was 5,220,000 shares for, 81,270,000 shares against, 32,460,000 shares abstained. The vote against this resolution represents approximately 93 percent of the total shares voted.

On management's proposal for the formation of a holding company, the vote was 95,350,000 shares for, 1,108,000 against, 22,490,000 shares abstained. The vote for this resolution represents approximately 68 percent of the total shares outstanding. . . .

Prepared by Richer, Bernhart & Probst/Houston, Texas.

The scene at Union Carbide's 1989 shareholder meeting illustrates what a pluralization of corporate identity means. Many at the meeting insisted that the Bhopal case was settled. Others *within* Carbide said it was not. The definition of "moral responsibility" was questioned and diversified. The possibility that corporate environmentalism could be turned toward the future—infected by the past—was opened up.

(3.5) Responsible Care?

In March 1990, the Texas Industry Chemical Council elected Union Carbide's Seadrift plant the safest plant in Texas. On March 5, 1991, Assistant Secretary of Labor Gerard Scannell announced that Union Carbide's Seadrift plant had been approved for participation in Occupational Safety and Health Administration (OSHA) STAR program, one of the agency's Voluntary Protection Programs for companies with exemplary safety and health programs. On March 12, the plant's No. 1 ethylene oxide production unit exploded. The fireball could be seen from ten miles away. John Resendez, a contract worker, was killed. Twenty-six others were injured.[2]

Environment and Development in the USA: A Grassroots Report for UNCED

OUR NIGHTMARE
Melonie Masih, Goliad, Texas

Our nightmare began early on March 12, 1991. We were awoken from a very sound sleep by a tremendous explosion. The roof of our home appeared to lift several inches from the wall. Later, we would discover cracks in walls, broken windows, and pictures on the floor. The very earth underneath us shook with terrible vibrations. We thought we would be thrown from our very bed. It was shortly after 1:00 A.M. but it appeared to be daylight due to the enormous fire resulting from the explosion. In our minds we felt certain an earthquake or the end of the world approached.

We immediately ran to our children ages 10 years and 12 years. Their safety being paramount in our concern. Our 12 year old suffers from a handicapping condition that results in grand mal seizures. Under stress her seizure activity increases.

My husband soon discovered Union Carbide once again was the culprit for disturbing our rest. We had ceased to count the episodes of lost sleep resulting from penetrating odors coming from Union Carbide.

We telephoned the emergency number Union Carbide had provided for us at an earlier "near neighbor" meeting. This number had remained posted on our refrigerator for easy access. Following previous instructions given by Union Carbide we asked to speak to the emergency director on call. We were told, "he is busy." Next, we asked what had happened. We were informed, "lady, there has been an explosion." No joking, an idiot could have ascertained that bit of information! Before we could inquire as to what exploded the Union Carbide spokesperson hung up on us.

Then, we dialed the sheriff's department—no answer—dialed again—no answer. During this time we were trying our best to reassure and calm two very hysterical children. The children were literally shaking with fright and crying. It was months after the explosion before they would sleep alone or be free of nightmares. Even today loud noises upset them.

We could see this horrendous, murky cloud approaching our home. Our home was located half a mile from Union Carbide–Seadrift Plant, on the northeast corner of their property. The wind that early morning was from the south-southwest. We smelled a very suffocating, nauseous odor. It seemed to take our very breath away.

During the next few minutes a second and third explosion occurred. We had visions of being completely annihilated. Our decision was made we must evacuate. We telephoned our 80-year-old grandmother, who lived alone and closer to the plant, to inform her we were on our way to get her. Then, we telephoned our parents to let them know we were evacuating. Our daughter began to seizure. It took several minutes to place her in the car. All we could think of was that we would be found dead when this nightmare finally ends.

The odor was horrible, terrible, suffocating and terrifying. Our children were begging and crying to put our two dogs in the car. There was no room. The dogs sensed danger so they tried repeatedly to get in the car. Finally, we pulled away from our home and pets not knowing if we would be alive at daybreak.

Our grandmother was waiting at her back door. We hurriedly placed her in the car and drove away quickly. She stated she had been up some time vomiting.

Once reaching the main highway there was tremendous traffic for such an early hour. All types of emergency vehicles were heading toward the Union Carbide Plant. There also were numerous vehicles heading away from the plant. We later learned these were employees and neighbors seeking escape.

On the way to our small town of Port Lavaca located 11 miles east of the plant we stopped twice to question sheriff and police personnel. The stops proved futile. They had no information. In fact, they knew less than we did. One policeman even told us he would probably be dead by morning. They could not even tell us how far to go to be considered safe or what we had been exposed to.

During this time our daughter began to seizure again. We had to administer her emergency medication. We decided to seek medical attention and drove to the hospital in Victoria some 30 miles away. . . .

Upon arriving we learned that the hospital had received no information. In fact, they thought Saddam Hussein had launched a scud missile. Soon after our arrival word came over the radio that ethylene oxide was the chemical involved and to expect numerous casualties.

Our children became weak, nauseous and had diarrhea. We gave them Coke to try and settle them. The physician who saw us said all he could do was treat symptoms and advised us to see our family physician during office hours that day.

Just as we were beginning to recover somewhat from the nightmare earlier, statements and fear again began to enter our minds. A previous statement made at a "near neighbor" meeting by the county judge, emergency management director and Union Carbide Plant Manager that in case of an emergency we would be notified within 90 seconds. NO ONE EVER CAME OR CALLED!! A statement by the emergency management director that he could be reached at anytime and anywhere. Yet, he was the last official to report to the courthouse due to the fact he was unable to locate his pants! Statements regarding safety, cancer concerns, fear of chemical plants, and their emission. Why had Union Carbide lied to us?!?

Today, as we sit here recounting this nightmare of events we know the answers. Answers we have had to learn the hard way. As we recall this information our heart rate increases, tension develops, fear begins to consume and yes, anger. Righteous anger that a company such as Union Carbide can be left to prey upon innocent people without fear of censure. The answers remain simple and point to one all consuming fact: Union Carbide decided a long time ago that a dollar was more important than human life or planet earth.

> In conclusion, doctors tell us it may be years before we learn the extent of damage to our bodies resulting from the nightmare of the explosion and our close proximity to the plant. We have been forced to sell our home, a five-generation farm, and relocate in order to protect our families from further harm. Our resolve will continue to be to do everything in our power to protect this God given home we call EARTH!

In July 1991, *Chemical Week* reported that there was a 92 percent satisfaction rate with emergency response by those living within two miles of the plant. Melonie Masih has a different story. So does Diane Wilson:

> People were listening to scanners and all you heard was pure chaos. They didn't know how to stop the fires. There were these big oxide tanks sitting close by. They were sitting there watching them expand; they didn't know what was in them. It was just, you know. Half the people supposed to be in the control room were down in Seadrift, having taken off on a tear . . . but according to the local media the explosion went so well they oughta have another one next year—just to show how great this county is at handling explosions. Yet, the workers who went in to handle it didn't even have protective gear on. (Bedford 1988)

Wilson later obtained internal OSHA documents stating that Carbide management had prevented government investigators from questioning workers without company lawyers present. Only one worker was willing to sign a statement.

In September 1991, OSHA proposed a fine of $2,803,500 for 112 willful violations of health and safety regulations.[3] Among the willful violations cited were 106 instances of fire and explosion hazards, 3 instances of inadequate fire water supply, and 3 instances of locked gates and blocked emergency exits. OSHA also revoked its approval of the Seadrift plant for participation in the STAR program and said that Union Carbide has a "significant" history of workers' safety violations.

Robert Kennedy, CEO of Carbide, recalled Scannell saying on March 5 that "the quality of your [Seadrift plant] safety and health program is impressive. . . . Your commitment to the safety and health of your workers is commendable." Kennedy insisted that Carbide agreed with Scannell's March conclusion and that "the evidence will confirm that Seadrift is a fine plant, operated safely by individuals committed to safety." Kennedy also said OSHA should not have "abandoned" Carbide as a STAR member just because of the accident (LaBar 1991, 37).

Union Carbide eventually paid $1.5 million to OSHA in fines. In March 1992, Union Carbide agreed to pay $3.2 million to the widow and two children of Resendez (Draffan 1994, 254). Diane Wilson still had questions. So she made a citizens' request to meet with Carbide's Seadrift plant manager, as provided for under the Responsible Care program.

On March 16, 1992, a group of eleven local and national environmental activists met at the Seadrift plant to attend the scheduled meeting. The UCC spokesman said he would not meet with them, saying that he would be prepared later in the week—which would have been after outside experts had left the area. The discussion held just outside the plant gates proceeded as follows:

> *Fred Millar, Friends of the Earth:* Is there a policy about not talking to people from outside the community?
> *UCC:* As I told Diane, we prefer to talk to local people, to keep it local.
> *Ramona Stevens, Louisiana Action Network:* The main thing is that Diane has the ability to bring in technical people to go through the documents so that they can discuss them with your technical people.
> *UCC:* I'm local; Diane's local. We'll talk.
> *Millar:* When you have the meeting with Diane will you be bringing engineers and people like that?
> *UCC:* We'll bring the right people.
> *Millar:* Experts?
> *Stevens:* What about Diane's experts? It's not fair for ya'll to gang up on her. Ultimately, that's what you're doing.
> *Reverend Andy Smith, Director of Social and Ethical Responsibilities, American Baptist Churches, USA:* This message is going to shareholders, that Carbide is not willing to allow people to come in that might know what the data you are giving them is, and be able to interpret it. This message is going to go out loud and clear to all the shareholders—that Carbide is not doing what it says it will do under Responsible Care. (Bedford 1988)

In late 1992, Union Carbide formed a Citizen Advisory Panel for its Seadrift facility. Diane Wilson was not asked to join. Since then, the company insists she has been asked to join twice. Wilson insists she has only been asked once and forfeited the invitation when she was unable to attend the first meeting. Wilson also insists that Carbide has refused to provide her with data on groundwater contamination and hazardous waste disposal practices. The plant's community affairs manager, Kathy Hunt, says Carbide has given Wilson the data and repeatedly asked to meet with her to discuss it. Wilson says, "That's a bunch of bull." In reporting this story, *Chemical and Engineering News* (C&EN) correspondent Lois Ember noted that it is "difficult for an outsider to make heads or tails of divergent stories. But Hunt [Carbide's plant community affairs manager] promised to send C&EN a list of names and phone numbers of CAP members. C&EN never received the list, even after repeated requests" (Ember 1995, 10).[4]

(3.6) Designing Disaster

Union Carbide Corporation began distancing itself from the operations of its Bhopal plant immediately following the 1984 gas leak. Arun Subramaniam, a journalist for *Business India,* understood the connections differently, explaining that

> how close the relationship between the two companies actually was can be seen from the fact that four senior executives of UCC's regional division, Union Carbide Eastern (UCE), including its chairman, were members of UCIL's board of directors. UCIL's budgets, major capital expenditures, policy decisions and company reports had to be approved by UCC corporate headquarters. The Bhopal plant formed an integral part of UCC's agricultural products division (APD), and was directly under the control of the director, APD, at the UCE headquarters in Hong Kong. The director, APD, in turn occupied the position of executive vice president at UCC. Thus the chain of command stretched all the way from Bhopal to corporate headquarters in Danbury, Connecticut. (1985, 44)

Subramaniam's picture is confirmed and elaborated on from within the ranks of Carbide executives. Edward Munoz, once managing director of UCIL, provided details in an affidavit and later interviews (Munoz 1985).[5] Munoz, a chemical engineer, began working at Carbide in 1958. First, he was based at corporate headquarters in New York City, and then he became a regional manager and spent some time in Mexico. Eventually, Munoz became a project development manager in charge of studying the viability of new projects. He argued—to a skeptical management—that India was a viable place for developing the production and distribution of agricultural chemicals. So he was sent to India. Munoz intended to stay in India one or two years, but ended up staying nine; if he had stayed longer, he would have been subject to India's tax laws.

Munoz insists that we must admit that the government of India isn't the easiest government to work with. But, in the 1970s, agricultural chemicals were considered an important part of the effort to increase agricultural production. Thus, the government of India itself approached Carbide to ask that the company develop the market. Carbide decided that in order to promote chemicals it would have to formulate them itself, but it was not a sector open to foreign companies. UCC's first permit, for an experimental plant in Bhopal, was to formulate only twenty tons of pesticides. Then a man approached them from Bhopal who had a license from the Madhya Pradesh government for a formulation plant, but no money to build it. Collaboration developed, and within six months, Carbide was building a large plant in Bhopal.

Munoz says that managers at UCC's New York office were aloof about the plant. In part, this was due to the shifting of expertise around the corporation structure. In earlier years, UCC's headquarters was filled with engineers, many of whom had moved up through the ranks. Munoz says that by the 1970s "they were all lawyers and MBAs, who didn't know anything about building plants or engineering." UCC headquarters did, however, approve plans for building the plant, as was required for capital expenditures in excess of $500,000.

In his affidavit, Munoz says that "in 1973, the Management Committee approved $20,000,000 for the erection of manufacturing facilities for MIC and MIC-based pesticides" (1985, 2). After the initial approval, the Engineering Division in South Charleston, West Virginia, was assigned primary responsibility for engineering and construction of the plant. Its duties included the conceptual design work, the preparation of the design report (including operating manuals), the appointment of a project manager, and the training of UCIL employees who were to constitute the start-up team. UCIL, with permission from the government of India, reimbursed UCC for engineering expenses and paid a "technical service fee" for "use of UCC technology, patents, trademarks as well as continuous know how and safety audits" (Munoz 1985, 3).

Changing configurations of expertise in New York gave Carbide's Engineering Division in West Virginia significant authority. Munoz describes the Engineering Division as a "sort of Mafia, a very inbred group of buddies, who are very jealous of their prerogatives and do things the way they want." And, in the case of the Bhopal plant, they wanted to build big tanks. Partly because they were paid according to the size of the projects they designed and oversaw. Partly because a large bulk storage design could be patterned on a Carbide plant at Institute, West Virginia.

Munoz says that there was controversy over the design of the plant from the outset. First, Carbide began believing "its own propaganda about market needs"—propaganda created both to preempt competition and to legitimate its licensing requests to the Government of India. Munoz explains that

> plans to begin with should have been for two million pounds, not ten million pounds. We believed our own propaganda. We were cornering the market by asking for ten thousand tons. We knew that ten thousand tons was something far into the future, and was unrealistic for the next five years or so. But we knew that if we asked for ten million pounds, Bayer would ask for another two million pounds, and then we would have competition. So we ask for ten million pounds. The engineering department said "well, if we're supposed to build a ten million pound plant, we're going to build a ten million pound plant." I said, "Wait a minute, that's propaganda." They said that's what the

paper says, that's what the government has authorized, that's what the management committee has authorized. So we're going to build it like that. It was a childish comedy of errors. (Karliner 1994) *or intentioned?*

Munoz also describes controversy over the design specifications. Once it was decided to build a plant that would supply the MIC needs of ten thousand tons of carbonates, there was the choice *to store* MIC and then choices about *how* MIC would be stored. Munoz says that he did not think that MIC should be stored and that "South Charleston knew my position well." Bayer Chemical, for example, used a closed loop process. Munoz explains that "MIC is a very unstable chemical. It trimerizes without notice, and with the evolution of a lot of heat, the trimerization can be of a very explosive nature. Which cannot be controlled beyond a certain point. It is a very dangerous product to store" (Karliner 1994).

Munoz finally took the position that the Bhopal plant should have only token, small-drum storage, based solely on downstream process requirements. Engineers in South Charleston overrode him, despite his provision of detail on the location of the plant, adjacent to residential colonies and barely two kilometers from Bhopal's main railway station. Munoz says the decision was analogous to planting a bomb near where people live and children play.

Munoz's description of the design decisions leading to the Bhopal disaster indicates a highly integrated corporate structure within which all major decisions were made or approved by the parent company. The government of India's reputation as "license raj" hovers in the background. But disastrous design decisions made by Union Carbide are more than evident—as are problems with the organizational structure within which Carbide engineers and managers made those decisions. Carbide's efforts to circumscribe responsibility for the disaster to a few isolated actions in the Bhopal plant are dismantled. Munoz—a voice from within Carbide—corrodes Carbide's dominant line. The corporation is caught by its own complexity.

(3.7) Old Dogs, New Tricks

The Bhopal litigation was settled out of court in February 1989. That same year, Carbide doubled its profits from its chemicals and plastics business. According to business analysts, these improvements were not due to improvements in management. When shares plummeted immediately following the disaster, Moody's lowered Carbide's rating to the lowest investment grade, quoting "fundamental weaknesses" in the firm's day-to-day business operations. Carbide's rise after the settlement was seen not as a reversal of this weakness, but merely as a market correction—an effect, rather than a cause. One cause of Carbide's rise was a massive restructuring in response to a takeover bid by GAF Corporation. Resisting the takeover justified huge debt to finance recapitalization. Financing the debt legitimated the sale of entire divisions: Carbide's Battery Products division, the world's leading flashlight and small appliance battery business, was sold to Ralston-Purina. The Home and Automotive Products division, which carried consumer brands such as Prestone Antifreeze and Glad Bags, was sold to employees. Critics claimed these were moves to immunize the company from consumer boycott.

Another, less direct cause of the rise in Carbide's investment ratings was a general market cycle that promoted the commodity chemical business. One analyst, quoted in the *New York Times,* insisted that "the so-called turnaround at Carbide is a result of the improvements in petrochemical markets and not much more." The same analyst nonetheless described the settlement of the Bhopal case in euphoric terms, saying that "psychologically, it's terrific. Financially, it's reasonable. This relieves the pressure on Carbide and the stigma" (Hicks 1989, 7A).

By 1994, the tenth anniversary of the disaster, the prognosis was even better. A March 1994 profile article in *Forbes* began with the claim that "like Mark Twain, Union Carbide thwarts the obituary writers" (Moukheiber 1994, 41). Despite a multimillion-dollar lawsuit resulting from the Bhopal disaster and a takeover attempt by GAF Corporation that required the sale of key assets and encumbrance of enormous debt, "Carbide refused to die." An insert quote argued that "people who say you can't teach an old dog new tricks haven't examined the tremendous progress made in the past few years by that old mutt Union Carbide." The article goes on to describe the method of renewal. In short, Union Carbide downsized, shrinking from a conglomerate with $9 billion in sales in 1985 to $5 billion in sales in 1990.

In 1994, industry analysts pointed out that the price peaks of 1989 were again evident and were expected to continue to rise, promising growth in operating earnings from 15 to 30 percent through 1995. On the tenth anniversary of the Bhopal disaster, the chemical industry was also benefiting from lower fixed costs. According to an economist from the Chemical Manufacturers Association, "[T]he industry has been through a ten year makeover, shedding old capacity and repositioning businesses to emphasize more profitable specialty chemicals over commodity brews. . . . The results are beginning to pay off" (Moukheiber 1994, 41).

The repositioning of Union Carbide was complex and radical. Two shifts of orientation seem particularly significant. First, Union Carbide overturned its diversification strategy of earlier years, focusing on becoming a low-cost producer of one product, ethylene glycol, used to make polyester fibers and antifreeze. Second, Union Carbide undertook major initiatives to utilize new information technologies to cut costs and raise efficiency.

The chemical industry used to be thought of as one of the most internally focused U.S. industries. In the 1980s, the focus began to shift. From L. L. Bean, the mail-order catalog company, Carbide learned how to centralize global customer orders into one computer center in Houston. Insight on computer tracking of inventory to adjust production to demand was borrowed from retailers such as Wal-Mart. Global distribution processes were revamped according to the model proven by Federal Express, to streamline the routing of products, rather than privileging lowest-cost forms of transport, whatever the circuitous route they took.

Carbide's initiatives to systemize production and distribution allowed both order accuracy and "just-in-time" production scheduling—which meant that Carbide reduced its finished goods inventory by 20 percent. The work processes to carry out these correspondences were redesigned by employees and resulted in a 30 percent reduction in required personnel for carrying it out. Overall, selling and administrative costs dropped 27 percent between 1989 and 1994; fixed costs at Carbide plants dropped 18 percent, and sales per employee rose 77 percent. The result: operating earnings jumped 27.5 percent during 1993, to $227 million, despite a weak market for polyethylene and ethylene glycol, which made up nearly 30 percent of Carbide's revenues. These earnings were a 45 percent gain over 1992, when sales were 5 percent higher (Moukheiber 1994, 41–43).

The business press offers an incomplete story of Union Carbide. But it does reveal a contradiction that has catalyzed the company's legitimacy in the years since 1984. When Union Carbide is "covered," Bhopal is not a pivot point for descriptions of a corporate identity crisis provoked by environmental disaster. Instead, "Bhopal" is a moment in the restructuring of Union Carbide that began long before December 1984, driven by the demands of a globalizing market. The Bhopal disaster is an effect, not a cause. The contradiction is made clear in a statement by Arthur Sharplin, McNeese University professor of management: "Clearly, by any objective measure, UCC [Union Carbide Corporation] and its managers benefited from the Bhopal incident, as did UCIL [Union Carbide India, Ltd.]. They were politically able to close a burdensome plant, take aggressive actions to restructure both companies, and enhance management benefits. . . . It is ironic that a disaster such as Bhopal would leave its victims devastated and other corporate stakeholders better off" (quoted in Lepkowski 1994, 30).

Four

Working Perspectives

(4.1) Knowing that You Don't Know

On the night of December 2, 1984, T. R. Chouhan was not working. His official role in the methyl isocyanate (MIC) unit of the Carbide plant was being played by someone else. Chouhan was at home, sleeping with his wife and two small children. When the family woke up coughing, the house was filled with a white mist. Soon Chouhan decided they must leave. He didn't own a vehicle so the family left on foot. His wife carried his son; he carried his daughter, who was too young to have ever left the house previously. Outside there was total hysteria, and the fear of being trampled matched that of being unable to breathe. Finally, they caught a ride on a truck headed for New Market, away from the poorer sections of Bhopal most proximate to the Carbide plant.

By morning, officials announced that the gas was released from a storage tank holding MIC. Shortly thereafter MIC became the only reference of possible medical response, despite knowledge that the ruptured storage tank still contained over 26,000 pounds of by-products, possibly identifiable through Union Carbide's own assertion that the thermal decomposition of MIC "may produce hydrogen cyanide, nitrogen oxides, carbon monoxide and/or carbon dioxide" (Union Carbide Corp. 1976, 27). Further, it was well established that MIC, hydrolyzed by water or water vapor, forms mono-methylamine, a derivative of ammonia with a stronger base strength. Many of the possible reaction products of MIC trimerization, which led to the explosion of the storage tank, had been researched in depth to explore toxicological effects and remedial therapies. Meanwhile, little was known about the acute and chronic effects of exposure to high levels of MIC, since most available data were related to low-dose occupational exposure.

The mist that filled Chouhan's house was not MIC. He knew the smell all too well from routine exposure, since small leaks in the plant were discovered through sense detection, despite an entry in Carbide's operating manual warning that MIC's "obnoxious odor and tearing effect cannot be used to warn of dangerous concentration, since this concentration is approximately two parts per million (ppm), one hundred times greater than its threshold limit value—the maximum permissible exposure limit—of 0.02 ppm" (Union Carbide Corp. 1976, 25). Moreover, Chouhan also knew that if MIC had been there in such intense concentration, his entire family would be dead. In his account, the reaction of MIC with water in the tank and its subsequent dispersal in the atmosphere led to a chemical breakdown, meaning that different sections of the city were exposed to different gases in different combinations. His own family and others living nearby did not suffer the immediate disabilities of those living in other areas. But he worries that they may be at high risk for long-term effects like cancer.

By mid-morning on December 3, 1984, Chouhan was confronted with a harsh reality: he knew that he did not know what chemicals were affecting the people of Bhopal with such deadly force—despite claims to the contrary by Union Carbide, the source of his own sense of authority. Thus began a process of discovery that transfigured Chouhan.

One material effect is Chouhan's book, *Bhopal, The Inside Story,* published in 1994. It is story about Union Carbide's negligence as far back as 1975, when Chouhan was first hired. It is also a story about Chouhan.

Bhopal, The Inside Story
T. R. Chouhan

INTRODUCTION

"Bhopal," the world's worst industrial disaster, can only be fully understood within the context of the managerial and technical situation at Union Carbide's Bhopal plant in the years preceding the disaster. Since the plant was established in 1969 the conditions leading to the disaster were building up. Unsafe design of the plant was made still more unsafe by such cost cutting measures as poor maintenance, use of poorly trained personnel and gross neglect of the most basic safety procedures. These factors turned the pesticide factory into a ticking bomb that could explode at any time. Workers individually and through their unions did not remain silent spectators as this bomb went on ticking. As a plant operator, I was one of the workers to articulate my concerns over the deteriorating condition of the factory. In the following pages, I will describe the processes that led to the tragedy and our attempts as workers to prevent the disaster from happening.

FIGURE 4.1 UNION CARBIDE'S BHOPAL PLANT   Union Carbide's Bhopal plant, seen from
an adjacent residential neighborhood. Photo by Mike Fortun.

> I must tell this story of Carbide's greed and negligence for two reasons:
> First, I must challenge Carbide's "sabotage theory," which slanders the good
> name of workers around the world. By placing blame on a "disgruntled
> worker," Carbide has tried to hide management culpability. This theory
> shifts the focus of attention from Carbide to a single person, who Carbide
> has identified as a "typical" worker—stupid, vindictive, prone to lying. The
> basis of Carbide's sabotage theory is not a reconstruction of facts gathered
> from those on the site, but a fiction created from stories pulled from periph-
> eral "witnesses," easily influenced by interviewers. Carbide's report justifies
> this unorthodox methodology with the claim that it was necessary due to a
> "reflexive tendency among plant workers everywhere to attempt to divorce
> themselves from the events surrounding any incident and to distort or omit
> facts to serve their purpose" (Kalelkar, 5).
>
>   My second reason for telling the Bhopal story lies in the belief that if
> evil is not exposed, it will continue to run rampant. Unless we publicize the
> gross negligence that made the Bhopal tragedy inevitable, similar tragedies
> are certain to happen in the future. Unless corporate negligence and greed
> are exposed, we cannot prevent future Bhopals. . . . [1]

(4.2) Culpable by Comparison

The expertise Chouhan has assumed responsibility for within the Bhopal
disaster has been constituted within competing messages. Union Carbide

taught him an authoritative mode of knowledge. Then Carbide itself failed to respect it. So Chouhan has turned Carbide's knowledge back on Carbide.

Chouhan has collected every Carbide document he could get his hands on and then cross-checked it with his own experience, and through interviews with other workers. Operating manuals, plant audits, design drawings—all have been pored over and restated. Chouhan keeps a collection of sketchbooks. He has drawn out what happened again and again—in Carbide's own terms.

Bhopal, The Inside Story
T. R. Chouhan

JOINING CARBIDE

In 1975 I was in the middle of my diploma in Pharmacy when there appeared in the newspapers an advertisement of job vacancy in the Union Carbide factory in Bhopal. Friends and relations said that I should try for a job in the Carbide factory since it was quite a good opportunity. My family and friends in the small town 150 km away from Bhopal which was my home thought that it was time I started earning. So I applied for a job as trainee operator. When I got it, I dropped out of my pharmacy studies. When I joined on 11th August 1975 at Rs200 per month, I knew little about the factory except that they produced insecticides. I didn't even know that those chemicals were dangerous. There were about 45 people in our batch of trainees and the going was good for the first months. We were told that as operators we would be placed in a rank above that of the general workers. So for the first six months when we took part in classroom training we had a notion of ourselves as pretty important people. The bubble burst when we were sent for training on the job.

We began to get treated as casual workers and were asked to carry loads, sweep the floor and do other such humble jobs. The factory canteen has separate sections for workers and supervisory staff, tables and chairs for the staff and hard benches for the workers. We were directed to the "workers side" of the canteen, and so we came to know our place. About twelve trainees left half way when they found that they were being treated as workers. I, too, was disappointed with the way we were being treated but could not quit as I came from a poor family. I couldn't afford to leave the job even if I wanted to.

During training we were told about the hazardous nature of the chemicals stored, used and produced in the factory. During on-the-job training in the Sirmet pilot plant, acid slurry fell on Achilles Gupta, a fellow trainee and he suffered intensive chemical burns. Another trainee, D. C. Mathur became allergic to chemicals and suffered from a painful skin disease. Later he was transferred to Utilities where he died of stomach ailments in 1982. After some months our training period was extended by six months and our stipend was raised to Rs300 per month. By this time we had realized that on completion of our training, we would be placed as hourly rated workers and most of us had resolved not to stay in Carbide for long. Of course, our first reason for outrage at being given hourly status was the prestige, the money, and lack of control of our workspace. Since, I've learned that there is a connection between these personal ends and our obligation to the community. Workers cannot fulfill their responsibilities beyond their own workplace if they are continually threatened with termination. Workers need the same tenure status as professors in universities. Professors were given secure job status so that they could be free to speak for the protection of society; workers have an even greater task, as theirs is both ideological and material; hence, it is essential that they be given security. This must become a goal of the unions, and it must be emphasized that it is not only to serve the interests of the individual workers but also to provide a context in which he or she can carry out their responsibilities to the community.

On completion of our training in February 1977, we were absorbed as hourly rated workers—confirming our apprehensions. I was placed as plant operator in Phase II Sevin plant that started working in July 1977. In this plant, Sevin insecticide was produced from methyl isocyanate (MIC) and x-naphthol. During our training we were told in detail about this plant with the help of a model. I was rather awed by the sophistication of technology and attention to safety matters, which this plant was supposed to have. Things turned out to be different when the plant started operating. In all systems of the plant—the MIC and x-naphthol charging station, reaction system, filtration system, drying system, solvent recovery system, packing system, vent gas scrubbing system, and refrigeration system—modifications were made in the design in response to operational problems. Soon after the plant was started, supply lines started getting choked, control instruments failed and equipment turned out to have inadequate capacity. All these required substantial changes in order to build up and maintain production volumes. Safety appeared to be the last consideration. I thus learned the urgency of looking beyond corporate appearances, especially regarding safety.

The modifications made in the x-naphthol charging system are illustrative of the management's approach. During our training by means of a model, we were shown the working of the x-naphthol charging system. It consisted of a jaw crusher that pulverized solid x-naphthol. The x-naphthol powder was then taken up by means of bucket elevators and dumped into a melting tank where it was melted through steam heating. Then the molten x-naphthol was pumped through the charging lines into the MIC reactor. On the model, the whole operation was pretty impressive and looked quite smooth and safe. In actual practice, the jaw crusher did not work properly so x-naphthol was crushed manually—by contract workers, who knew nothing about the damage x-naphthol dust could cause to their liver and kidneys. But the problems were not over. The molten x-naphthol started getting solidified in the line due to insufficient heating. That choked the valves, pumps and the lines, necessitating frequent steam lancing. This went on for two months after which the charging system was closed down.

Charging was made more "direct" after this. Contract workers were employed to pound solid x-naphthol with hammers in a shed specifically constructed for this purpose. Crushed x-naphthol was carried to the MIC reactor vessel and charged manually through the manhole. Opening the manhole cover would lead to escape of MIC, carbon tetrachloride, trimethylamine, x-naphthol and carbaryl dust. Thus, not only the casual workers but also operators like myself were routinely exposed to these toxic chemicals. The casual and contract workers who performed the most hazardous tasks were not considered to be members of the workers' union and could be laid off at any time.

Once a casual worker while charging x-naphthol accidentally dropped the empty sack containing pounded x-naphthol into the MIC reactor tank. Afraid that he would be punished for this error, the worker lowered himself into the tank and was trying to lift the sack off with his feet. The plant supervisor spotted him and this incident was used to terminate the service of Shyam Pancholi, the operator in charge who was also a militant union activist. The casual worker was also laid off but no changes were made in the obviously hazardous charging procedure.

Most of the modifications from the original design that were made in the Sevin plant consisted of changes from automatic and continuous to manual and batch processes. The solvent recovery system, for example, was designed for recovery of the solvent carbon tetrachloride from the filtration and drying unit. In actual practice, recovery was grossly affected by frequent choking problems. Due to impurities, carbon tetrachloride was found as semisolid tar that would cause choking in the rotary valve that was supposed to drain out the solvent (kettle product) at periodic intervals. To tackle the problem the rotary valve was eventually removed and solvent was periodically drained out by manual prodding—a procedure that would often lead to the solvent spreading all over the shop floor. This and other similar modifications in plant plans led to more hazardous work conditions, poor recovery of solvents and leakage of chemicals leading to atmospheric pollution and insufficient control over the process of production.

The operators of the plant brought forward the problems of plant pollution that were caused by the modifications through the workers' union. Also, letters were sent to the managers of the plant as well as to the Ministry of Labor, Government of Madhya Pradesh and the factory inspector. All our letters went unanswered and the management told us that we would have to work under "normal" pollution levels. However, in abnormal conditions (like when carbon tetrachloride leaks were heavy) we used to close down the plant and wait for conditions to become bearable. Around this time, I started suffering from problems of indigestion, acidity, and a constant choking sensation in the throat. My coworkers too had similar problems and quite a few of them started vomiting every time the pollution levels were high. Every six months the factory doctor examined us. Samples of our urine and blood were tested and chest x-rays were done on us but we were never given the reports of these examinations. Four of my coworkers who by then had developed pronounced health problems were shifted to the Utilities section and there was a rumor that this was because of the adverse test reports on their health condition.

In 1978, there was a big fire in the factory. I was working in the Sevin plant on the first shift. As I was coming out of the factory, I saw that all the four rear wheels of the fire truck in the factory had been taken off and the vehicle was resting on a jack. As we all were walking out, I commented to my coworkers about the dangerous possibility of a fire breaking out with the only fire truck being out of commission. My friends thought the management was much too stingy to invest in another fire truck to have at least one fire truck always in readiness. Our discussion ended with an ominous chant of "anything can happen in this factory."

When we reached the peer gate, about 3 km away from the factory, we saw fire trucks rushing toward the direction of the factory. Looking back, we saw smoke billowing out and we went back to the factory. There was a raging fire at the coke yard of the Carbon monoxide plant and fire fighters from the municipal corporation and the airport were battling to arrest the spread of fire.

There was a huge crowd around the factory watching the fire as a crowd watches any sensational event. The fire continued until midnight, when most of the material was burnt out. The newspapers next day carried a full account of the fire saying that the x-naphthol that was kept in the coke yard had caught fire. The next day when I went to work, I saw a number of police officials moving around and was told that they were investigating the fire. I was told that x-naphthol worth 6 crore had been burnt off but, oddly enough, the factory bosses did not show any sense of loss—they did not seem to care. Contrary to what had been reported, I gathered that it was naphthalene that was stored in that coke yard and not only x-naphthol. Many workers were also wondering why naphthalene (or even x-naphthol) should be stored in a place that was exclusively meant for storage of petroleum coke, particularly when there existed storage facilities for specific raw materials.

There was a strong suspicion among the workers that the bosses deliberately started the fire. It was rumored that the management was not being allowed by the government to import x-naphthol and given that the naphthol plant was not being run successfully, this was the only way the management could create a situation so as to make a case for the import of x-naphthol. Most workers held that the fire was set up to circumvent the import restrictions of x-naphthol laid down by the government. Whatever be the basis of these rumors, it is a fact that after the fire, the factory manager resumed the import of x-naphthol, which went on until 1984. The newspapers gave a lot of coverage to the fire and indicated the potential hazards of such a factory that was situated in the middle of human settlement. Neither the factory management nor the government officials came out with a report on the fire, no one was held responsible and most of the sensation died down in a week. By that time, Carbide bosses were throwing parties at the posh hotels of the city to thank government officials, city corporation officers and journalists for their cooperation in managing the fire.

The same year, the Sevin plant was closed down for six months and I was shifted to the x-naphthol plant that had started operation in 1978. Here again, a lot of modifications in the plant design were made in an effort to cope with frequent choking of the lines and improper control of the final product leading to increases in production costs. The overall impact of these modifications was once again a shift from continuous and automatic to manual and batch processes. The plant had a bad start and a worse run and finally was closed down in 1982.

WORKING IN THE MIC UNIT

In 1979, I came back to the Sevin plant when it was restarted after some modifications. I stayed there till 1982 when I got transferred to the MIC plant along with two other coworkers. We were rather surprised because we had known that to be an operator in the MIC plant one had to be either a graduate in science or hold a diploma in mechanical or chemical engineering. Among us workers, it was known that the MIC plant was the most dangerous and yet MIC plant operators' jobs were coveted because they meant higher grades and better salaries. And so it was with mixed feelings that the three of us accepted our jobs but we were assured by the managers that we would be given six months of training before we were put on the job. That helped matters.

Our training began on 12th March 1982. For one week, it was classroom training and then we had on the job training for one month. At the end of five weeks, the plant supervisor asked me to take charge as a full-fledged plant operator. I refused to do this and insisted on getting six months training, as I felt quite ill prepared to handle a plant known for its complexities and dangers. The plant supervisor was furious at my refusal to take orders and complained against me to the plant superintendent and manager and the manager of industrial relations. During my transfer to the MIC plant, I was given a letter mentioning a minimum training period of six months so when the manager of industrial relations D. S. Pandey called me in I showed him the letter. Two more officials, safety manager B. S. Pajpurohit and plant manager K. D. Ballal joined Pandey and the three of them tried to convince me that I was being unreasonable in insisting on further training. They said that I had been working for six years in a chemical factory and that was enough experience and that I did not need any further training. But I stood my ground and flashed the letter before them as often as I could. Finally, it was resolved that I would be given three weeks of training for handling the storage unit of the MIC plant and then I would take charge of the storage unit alone. So in June 1982, I took over as operator in charge of the storage unit in the MIC plant.

Work in the storage unit consisted of transfer of MIC from the storage tank to the Sevin plant and the unloading, transfer and storage of chlorine, monomethylamine and chloroform, all of which were brought from outside the factory. While working in the Sevin plant, I had on many occasions been told about the advanced technology and the efficient crew of operators in the MIC plant. But, after I joined, I saw the steady decline in the quality of personnel with more and more untrained staff coming in each category.

One of the reasons for such a state of affairs was the unfair manner in which some of the personnel were treated. In the matter of promotions, individuals with little experience but with an unquestioning loyalty to the bosses were invariably selected over others. For example, Mr. K. D. Ballal with 23 years experience of working in Carbide plants behind him was not selected for promotion while others with far less experience fared much better. S. P. Choudhury who joined as a plant superintendent in 1980 was promoted to production superintendent in 1982. This, many alleged, was due to his smart "handling" of the matter of death on the job of Ashraf Lala, a maintenance worker in December '81. Choudhury was the plant superintendent when Ashraf died of phosgene inhalation. Further, in 1983, he was made manager of the Sevin and MIC plant after Mr. K. D. Ballal resigned from the post and left Union Carbide.

Unlike the Sevin plant, most of the equipment and vital instruments of the MIC plant were imported from the United States. Senior plant personnel had been given training in the Institute plant in West Virginia. We were told that the MIC plant in Bhopal was based on 20 years of experience at the Institute plant and that nothing could go wrong. But early on I came to know of modifications made in the MIC plant. The modifications here were made essentially to cut down on costs. The pilot flame in the flare tower was lighted only when the carbon monoxide plant was running and alternate arrangements for using liquid petroleum gas were not made all the time. This, of course, was a serious breach of safety procedures.

In the same manner, the refrigeration plant was not kept running all the time as had been stipulated in the manual on safety procedures. And MIC was chilled only while it was filled into drums (meant for transport to the Temik plant so as to cut down on evaporation losses). As operators in the MIC plant, we also had to cope with corroded lines, malfunctioning valves, faulty indicators and absence of some key control instruments. Continuous monitoring of temperature of MIC in the tank was crucial for safety reasons, but the temperature indicator alarm never functioned properly and we did not record the temperature of MIC on the log sheets. As part of the economy drive that had been sweeping through the plant for some time, the caustic soda feed to the vent gas scrubber was modified into a batch process from the continuous process originally designed in order to cut down on caustic soda losses. The economy drive also meant that malfunctioning valves were not replaced, faulty gauges were not repaired and process monitoring was done inadequately. The economy drive had serious consequences on the lives and health of the workers.

On 26th December 1981, Ashraf Lala, a maintenance helper was fatally exposed to phosgene while working on a phosgene line. All the valves had been closed, the phosgene had been purged out and slip binds inserted near the valves. Ashraf and his coworkers were repairing the phosgene vaporizer. Work had been going on for two days and once it was over, Ashraf removed the slip binds, as was the procedure, to get the vaporizer back into operation. Suddenly, he was splashed over with liquid phosgene from the line. This could only happen if there was a malfunctioning valve along the line and possibly there were several. The shower of phosgene must have been chilling for Ashraf and in panic, he ripped off the mask he was wearing and thus inhaled a lot of phosgene gas that evaporated from his clothes. He was taken to the factory dispensary and from there to the Hamidia Hospital, where he died after 72 hours.

The managers blamed Ashraf for removing his mask and thus bringing about his death. The workers' union put the blame squarely on the company management. We pointed out that it was the malfunctioning valve that led to the accident, that Ashraf should have been informed about the possibility of such an occurrence and, further, that he should have been provided with PVC overalls.

Ashraf's death precipitated the feeling of fear and anxiety that had been growing among us workers. While the panic over Ashraf's death was still raging, another incident occurred on 9 January 1982. During the night shift, a valve in the phosgene line in the MIC plant broke off and thick clouds of deadly phosgene leaked out. Workers in the MIC and Sevin plant ran away in panic and soon the rest of the workers in the factory joined them. I was working in the Sevin plant at that time and I ran out to the fields behind the factory and went back to the factory in the morning. Twenty-four workers who were badly exposed to phosgene had to be admitted to the hospital that night.

In January 1982, the workers' union began an agitation over safer working conditions. We carried our campaign to the community around the factory and tried to make them aware of the hazards posed by the factory. We got fliers printed, organized rallies and held public meetings at the factory gate. After two weeks of much activities, we were told by government officials that a committee should be set up to investigate matters related to Ashraf's death. Meanwhile, our attempts to inform the community were met with such indifference that we gave up before we put more effort into it.

I wondered about the indifference of the community towards an issue that I thought was of immense relevance to their lives. It was quite plain that the people in the community were suspicious of the workers. They suspected that the workers were whipping up community support with an eye for getting a better deal for themselves.

There was another reason for the non-involvement of the community. Apart from the communities that have been there for hundreds of years, there were a few communities where a number of people have been living without paper land deeds or "pattas." Quite aware of their place in the social hierarchy of Union Carbide, these people were afraid that if they joined the workers in their demand for shifting of the factory, the government might instead drive them away.

The agitations of January 1982 that grew around the issue of safer working conditions were supported by almost every worker. Most had had nasty and frequent encounters with unsafe work conditions and Ashraf's death had shaken everyone up. And yet, we could not carry on with the agitation. The lack of support from the community added to the apprehensions of a section of the workers; they were afraid that the company might close down the plant in response to such agitations. Around the same time, the factory management terminated the service of two workers, Shard Shandilye and Bashaerullah, who were the most active union leaders and who had played key roles in all the agitations and negotiations during that time. This was the final blow and the agitation fizzled out before it could gain real momentum.

The workers' agitation of 19 January '82 and its disappointing consequences brought about a change in the union's response to the unsafe conditions in the factory. While earlier, the unions concerned themselves with
unsafe plant designs and work procedures, after January '82, the emphasis
shifted to personnel safety. Prior to and in the immediate aftermath of Ashraf's death, the unions asked for design modification or even shifting the
plant so as to ensure safety for the neighboring communities and workers.
The demands later focused on the supply of safety equipment for individual
workers. The potential danger to communities of workers and neighboring
residents was thus presented as hazards to individual workers.

Around the same time, the management set up a safety committee within
the factory in which workers' representatives were included. The safety
committee, however, deliberated on the supply and fixing of quotas of
gloves, mask and overalls and kept away from the more fundamental issues
of unsafe conditions of work. The removal of the two most vocal union
leaders by the management led to a comparative softening of the union's
stance on unsafe work conditions. The negotiations between the management and the union in 1983 were marked by the absence of any agitational
activity that was quite unlike earlier years. The chief features of the '83 negotiations were introduction of a five day week, rationalization of workers, and a raise in salary to make all this palatable. As a part of the rationalization plans, a voluntary retirement scheme was introduced and about
30 workers left the factory under the scheme.

The introduction of a five-day week in the factory brought about quite a
few changes. While earlier maintenance jobs were done in all three shifts
round the clock, the introduction of a five-day week coupled with job rationalization meant that major maintenance jobs could only be done during the
general shift during the five-day week. It also meant that for two and a half
days, work in the plants was comparatively less supervised. The rationalization of workers even extended to the safety crew within the factory. Prior to
'83, there were two squads of workers who were trained to respond to accidental situations—these were the fire squad and the rescue squad. After the
1983 negotiations, both the squads were merged with the emergency squad,
which had fewer numbers of workers. The quality of personnel was also
affected.

In the night shift on 5th October '82, MIC, methyl carbaryl chloride,
chloroform and hydrochloric acid started leaking from a joint in the pyroly-
zer transfer line. The joint did not have a gasket and it gave way when transfer of these materials was started. The gas plume moved toward the nearby
settlements and people ran away in panic. The gases also went into the residence of an inspector general who lodged a complaint with the police. The
police response was to go to the factory for an inquiry. However, the management told them that all was under control. The MIC supervisor V. N. Aggarwal who had tried to stop the leak suffered intensive chemical burns and
two workers were severely exposed to the gases. Mr. Aggarwal remained in
the hospital for two months and subsequently left the company.

Many leaks took place in the MIC plant in 1983 and 1984. MIC—chlorine, monomethylamine, phosgene, carbon tetrachloride. A mixture of all

these leaked with frightening regularity. Workers would scamper out of the factory during such leaks and work in the plants would stop for some time. Injuries suffered during these leaks were treated at the plant dispensary and usually the incidents were recorded as incidents of material loss and not as injuries suffered by workers.

Chouhan has persistently cross-checked Union Carbide. He also has cross-checked the government of India. What has been said has been matched with what has been done. By Chouhan's count, 3,000 died during the first few days after the gas leak. Approximately one person has died every day since—raising the death toll to over 10,000. And Carbide claimed that MIC was only a mild irritant—despite its own safety manual instructing that MIC is deadly even in very small doses. And the government of India, which had promised to be guardian of the people, concurred both in denying the death figures and in refusing to acknowledge long-term effects—working to limit Carbide's liability in order to hide its own.

Chouhan finds fault with both Carbide and the government of India. Their crime is duplicity. Carbide promised to bring science to India. The government of India promised to help make science India's own. Chouhan saw himself as an emissary, responsible for translating between different ways of knowing. Carbide sold him out, as did the government.

It is not that the technological promise was unwarranted. The problem was corruption. So Chouhan must work to recuperate Carbide's way of knowing, distinct from Carbide itself. Carbide needs to be punished so that modernity is exonerated.

Chouhan never thought that modernity could simply be installed in India. He knew that interpretive extensions and adaptations would be necessary. And he wanted to be the go-between. But the places between which he wants to move have to be cleaned up. Modernity must be disentangled from Union Carbide. The nation, somehow, must be disentangled from what the government of India has become.

(4.3) Sabotaging Sabotage

Workers from the Bhopal plant challenge Union Carbide's sabotage theory because it shifts liability away from management and diverts attention from the total context. In his book, Chouhan challenges the sabotage theory with detailed documentation of the safety lapses, maintenance failure, and general negligence in the plant since he was hired in 1975. These details attempt to breach the claim that the cause of the disaster was one nonmanagement person's activity in an interval of less than five minutes. Chouhan

FIGURE 4.2 KISSING CARBIDE/NEHRU   India, represented in a Nehru hat, star-crosses with Union Carbide Corporation, who wears both a company logo and Eveready batteries as earrings. Painting by T. R. Chouhan.

insists that the "whole story" has more of a history, both within the plant itself and within the broader processes of transnational capitalism.

Chouhan's arguments call Union Carbide on its own terms. He argues that MIC storage tank 610 was contaminated with water, but also with residue from the pipes through which the water flowed. Only due to these by-products could the trimerization have occurred with the rapidity it did. Had water been added through a clean hose, directly to the tank, trimerization would have occurred, but over a much more extended time period. At ambient temperature, pure water would have caused violent reaction only after twenty-three hours. According to Carbide's own schedule of events, complete trimerization occurred within three and half hours' time, between 10:00 P.M. on December 2 and 1:30 A.M. on December 3.

Chouhan argues that this "water-washing theory" is supported not only by accounts of a routine maintenance exercise that was carried out without proper safety precautions, but also by features of plant design that allowed water to pass through vent gas headers made of carbon steel and through a pressure gauge made of copper tubing. In the absence of a slip bind, back pressure inside the headers pushed water back into the storage tank, carrying with it pipe filings from vent header and pressure gauge walls. According to Carbide's own description of the properties and requirements of MIC, it "must be stored and handled in stainless steel. . . . Any other material may be unsuitable and possibly dangerous. Do not use iron or steel, zinc or galvanized iron, copper or tin" (Union Carbide Corp. 1976, 1). To cut costs, vent headers were nonetheless made of carbon steel; the storage tanks were pressurized with copper tubing. Chouhan argues that "the cause of the Bhopal disaster was not a saboteur or even one, indifferent management decision. The cause was in the plant design itself and in the corporate ethos of Union Carbide."

Chouhan situates Carbide's activity within a changing market for chemical agricultural inputs. When the MIC unit was set up in Bhopal in 1979, a plant design was adopted that allowed bulk storage. This was done in anticipation of selling MIC produced in excess of the plant's own formulation requirements to other suppliers of MIC-based pesticides. Clearly, this indicated expectation of a huge growth in demand, supported by claims that MIC-based pesticides were "environmentally sound." Carbide did not anticipate a future in which farmers would be disillusioned by the cost of chemical inputs or by increased crop vulnerability to new pests. Nor did Carbide anticipate famine years in which farmers were simply unable to produce the cash necessary for purchase of chemical inputs. And no one expected the spread of small pesticide producers who relied on cheaper components. By 1984, when the plant was manufacturing only a quarter of its licensed capacity, Union Carbide India Limited (UCIL) had been directed by Union Carbide–Hong Kong to close the plant and prepare it for sale. Negotiations to settle the shutdown were completed in November 1984.

Chouhan's book includes graphs showing the connection between the changing market context and the management decisions leading to the disaster, retelling the oft-told story of how corporate loss is externalized and the risk burden borne by workers and communities. Chouhan's documentation of intensified personnel reduction runs a parallel chronology to Carbide's plans to relocate the plant.

It was not, however, only after dramatic losses that conditions at the Bhopal plant were unacceptable. As early as 1978, workers joked that "anything could happen in this plant." The event that provoked this remark was a fire in the plant during which the company fire truck was sitting up on a jack with all tires removed. The company "managed" the fire by removing outspoken union leaders and entertaining government officials at posh hotels. Periodic safety audits were managed in a similar way. Years before the disaster, these audits pointed out major safety concerns and explicitly warned of the possibility of a runaway reaction in the

MIC storage tanks. They also recognized that there was no evacuation plan for the surrounding community. Bhopal managers told Danbury that to coordinate such a plan would overly publicize the dangers posed by the plant, which had to be avoided, since they had recently avoided relocation away from the city by having their production labeled "non-noxious" by zoning authorities.

Chouhan's story concludes with an argument for the urgency of a trial in which Union Carbide could be forced to disclose in-house research data on the properties and dispersal proclivities of MIC. Chouhan knows that the data may not exist, and even if they do, they may not be reliable. But he demands the data nonetheless as a matter of principle.

By challenging conventional configurations of what counts as "trade secrets," Chouhan articulates a broad critique of the way knowledge is produced and circulated in contemporary society. His strategy of double take on official proclamations, simultaneously demanding and questioning data from Carbide, itself produces a new form of knowledge, wary but respectful of the power of factual claims to truth.

(4.4) Systems Theories

When I came to Bhopal, I knew that I would need expertise that I did not have. But I never dreamed that I would need to learn to draw, much less understand, the piping configurations that comprise a chemical plant. Every flow path, linkage, and barrier had to be retraced. The difference between pipes made of stainless steel and those made of iron had to be marked. The precise timing of contamination and trimerization had to be charted.

It was in 1990 that this process of learning began. Chouhan arrived at my house with a pile of papers filled with technical drawings, chronologies of events, and endless lists of names—of officials from both Carbide and the government of India. He said that his book was almost finished, that he just needed me to do the typing. But there was little prose and no narrative structure. After a coherent introduction, the manuscript slipped into a tirade of stories told through documentation, rather than narration. The material was extraordinary. But it did seem to need explication. And Chouhan definitely had an interpretation.

When I first met Chouhan, the work had lain dormant for some time. But the desire to write a book continued to nag Chouhan. Because he thought the truth about Bhopal had not been sufficiently told. Because Union Carbide's sabotage theory specifically defamed workers. Because, somewhere in the story, he discovered his own voice.

Much of Chouhan's story had already been told. In the first months following the gas leak, Chouhan served as a primary witness for journalists and for India's Central Bureau of Investigation (similar to the U.S. Federal Bureau of Investigation). He provided details to sustain the water-washing theory, accounts of declining personnel training in the plant, and criticism of daily operating procedures. But he was not granted authorship. His name is not mentioned.

At some point, Chouhan was encouraged by middle-class activists to write his own first-person account, to be published in book form. Their support lasted long enough to write a few paragraphs of introduction and to spur Chouhan into a mammoth effort to secure supporting documents. He needed help, since he had decided that the book must be written in English. While his spoken English had become quite good, it couldn't translate onto the written page.

Thus began a long, often frustrating, but always very rewarding collaboration. Chouhan and I worked for hours each week for the next two years. He would tell me stories; I would find a place to write them in. His knowledge of plant design, production processes, and maintenance procedure was extraordinary. His memory of everyday details was sobering.

If at Chouhan's house, our work was done in the main room, which opened onto the courtyard. It was a narrow room with a high ceiling, encircled by Hindu icons, laminated pictures of foreign places, and boxes. High shelves along one wall overflowed, as did the one cupboard, which fit between a bed and a television, neatly covered with an embroidered towel. The walls were a bright blue, dulled somewhat by the absence of windows. The door could be closed at night by drawing together two rough planks of wood with an iron latch. During the day, the door remained open or draped by a curtain.

Outside four houses faced one another, sharing a courtyard, water tap, and bathroom. *Tulsi* (basil) plants grew by every doorstep. People—especially children—moved freely between the houses and across the roofs, which was where the children played. Women did laundry in the courtyard, but most of their work was inside in small kitchens meticulously organized for incessant cooking. Tangled aromas of chili, cumin, and ginger seemed ever present, as did the hiss of kerosene stoves, which, when not cooking a meal, were brewing tea to be served in the early morning, again around ten o'clock, again after lunch, again in the late afternoon, and perhaps in the evening.

Chouhan's wife was quiet, very kind, and very, very organized. When not cooking or washing clothes, she helped the children with homework, amidst occasional reminders from Chouhan that she performed better in school than he, earning particular recognition for her skill at writing and math. But she rarely did the shopping, and never alone. Chouhan handled everything outside the home through weekly trips to a sprawling market, where he would carefully select just the right amounts of each vegetable required for the week.

Chouhan's motorcycle was parked outside the house in a narrow alley that led to another, broader alley, which led to a street. It seemed quiet in the alleys, between walls that separated public from private lives. But the alleys had their own order, stylized in quiet opposition to the frenzy of the street. Chouhan knew everyone. First, the pan seller, whose small cart was always parked in the same place, across from a shop that ground flour and repaired bicycles. His glass case had rows and rows of cigarettes, tied in small, hand-rolled bundles or packaged in the blue and yellow Indian equivalent of Marlboros. The top of the case served as a workspace, lined with the jars of cloves, lime, and betel needed for a properly prepared pan. Chouhan didn't need to ask for what he wanted. One pan, a little hotter than most, and a bundle of bidis. Opposite Chouhan's house was Kamla Park, one of the largest parks in Bhopal, where the Old City flows into the new.

Chouhan sees the disaster through the prism of this fine-tuned organizational form. The space of family and household, shaped to be well-functioning systems. I realized the importance of this prism only as I watched it fall apart. Chouhan lived in his house for twenty years. Then he was forced to move so that the landlord could turn the space into a clinic. Chouhan moved to a house nearby, but without the enclosed protection of a courtyard or the open expanse of a roof. His children had to play in the alley. There was no place outside for his wife.

In the wake of disaster, the many systems through which Chouhan lived and understood the world began to break down. The disaster thus operates as a symbol for him, drawing out the connections between many different systems, revealing their corruption.

(4.5) Demonizing Technology

The epilogue of Chouhan's book outlines what "The Fight Against Toxics" will entail. Everyone has a role to play in a tightly coupled system. Hazardous technology is at the center, boxed in by layers of protective policies and people. Company owners must acknowledge the fundamental immorality of putting profits before people. Unions must take responsibility for more than worker welfare. Communities must organize themselves to keep watch over plants. Government regulators must actually implement existing policy on environmental and occupational safety. Journalists must acknowledge that Bhopal consists of many tragedies—health, environmental, legal, and also media. They must not tire of the Bhopal issue, since citizens have a right to know about the risks of hazardous technology.

Chouhan knows that it is not just a matter of improving technology. So he has proposed a toxics *yatra,* following the Indian tradition of pursuing great social or religious causes with mass marches from village to village and city to city—linking everyone and everything in a great protective net. Victims and workers of Bhopal would travel the country, and perhaps the world. They would share the story of Bhopal and tell people of the roles they must play to prevent disasters in the future.

People would learn that hazardous technology is a demon that cannot be banished, but can be caged. If not restrained, it will crush everyone in its path; spew poison from every orifice and stare, red-eyed, at the glory of its destruction. The rampage will cross the cities and the countryside, fouling water, air, and soil. Chouhan has painted a striking image of "The Unforgettable Night" to drive his message home. The MIC storage tank is the bowel of the beast. The production unit is the heart. Chemicals flow through every artery. People are processed and spit back out. The demon itself issues a warning: "Run for it!" Shouted to the slum dwellers of J. P. Nagar. To women bathing, or drowning in a poison lake. To a woman at a well.

Chouhan's own warning is also ominous: HAZARDOUS TECHNOLOGY IS A DEMON INVADING EVERY PART OF THE BODY AND SPEWING POISONS INTO THE ENVIRONMENT. UNION CARBIDE IS THE DEMON WHICH SPEWED DEATH AND DESTRUCTION OVER THE CITY OF BHOPAL ON DECEMBER 2, 3 1984.

(4.6) Be-laboring Environmentalism

It has been difficult for me to understand Chouhan's relationship to middle-class activists in India. At times, there seemed to be real camaraderie. Most often, however, middle-class activists seemed to be patronizing him. Sometimes they were just plain cruel. Chouhan would burst into a room, excited about a new idea for preventing future disaster. The response was sarcastic. We would sit covered in papers, struggling to make part of Chouhan's story hold together. Someone would walk by, throwing out comments suggesting that we shouldn't bother, that the story had already been told. The importance of "a worker's account" was always gestured at rhetorically. But social practice did not sustain it.

Industrial workers occupy an ambivalent place within the Indian environmental movement.[2] Much like their American counterparts, particularly prior to the 1980s, workers are often blamed for being part of the problem. In India, however, it is even more difficult to blame workers for "choosing" to work in the plants. Poverty lingers so close that begrudging people industrial employment is somewhat ridiculous—unless you are convinced of

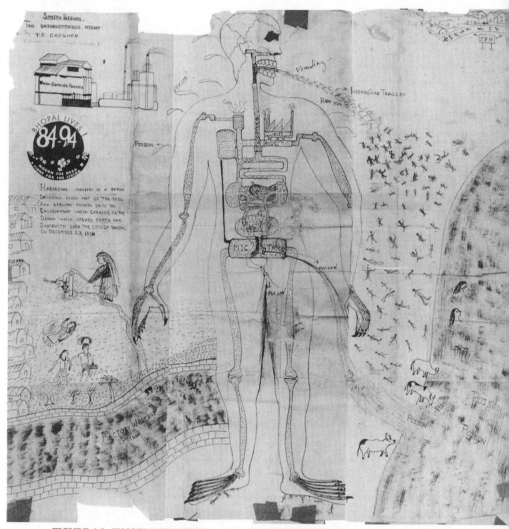

FIGURE 4.3 CHOUHAN'S DEMON   "The Unforgettable Night," representing technology as a demon that spews poison and people and crushes everything under its feet. The demon's body harbors Union Carbide's plant in Bhopal. The demon's bowels are the MIC storage tank. Painting by T. R. Chouhan.

the priority of rural struggle, convinced that the primary problem is extraction from the countryside to sustain urban modernity. Many in India's environmental justice movement hover around this line. It makes it difficult to incorporate Bhopal at all, much less Bhopal's industrial workers.

(4.7) Repetitions

The Bhopal disaster woke people to the news that "better living through chemistry" could be a deadly affair. They also learned that environmental risk is constituted within transnational processes that link together people and issues once considered separate. Residents of Middleport, New York— linked to residents of Bhopal. Indian progressives, dressed in the rumpled khadi of Gandhian alternatives—linked to slick corporate consultants claiming allegiance to the world, rather than to nation. Human rights issues—linked to environmental issues.

Reference to Bhopal has become a way to comment on many things. Within articulations of Bhopal, one hears how health is defined by those exposed to toxics, by those charged with rehabilitation, and by those who think themselves immune. One hears about the role of technology in socioeconomic development, changing perceptions of the nation-state, and emerging awareness of the power of multinational corporations. One hears about the many promises of democracy and its legal institutions and about the role the law should play in adjudicating different visions of the past and of the future. One hears and sees how political advocacy is imagined, enacted, and restrained.

Where among all these people and issues can the Bhopal disaster be located? At what time did the disaster occur?

There was, of course, a singular catastrophic event that killed many, injured many more, and thrust people around the world into new understandings of the way risk percolates through modern society. But there are also other times of great significance—times when responsibility was assumed or not; times when the disaster was set up, even if not determined. Chouhan and others working to understand the Bhopal disaster have retraced these times as a way to query what happened and why.

Some of the activists I worked with pushed the origins of disaster back to 1600, the year the East India Company was chartered in London, establishing how India would be related to the West—through the multinational corporate form. Some fixated on 1974, the year Union Carbide India was granted an industrial license by the government of India to manufacture as well as formulate pesticides in Bhopal—establishing the government's role in the disaster. Still others emphasized the early 1980s, when the Green

Revolution in India was faltering—revealing that the promise of modern science and technology was not as straightforward as once believed.

Chouhan and other workers tended toward more microscopic analysis, zooming in on developments in the Bhopal plant where they worked: 1978, when the alpha-naphthol plant went online—and didn't function like the training manuals promised; 1981, when maintenance worker Ashraf Lala died of phosgene inhalation; 1982, when resignations escalated and the number of trained personnel in the plant declined dramatically—alongside worker agitations, which were ignored; 1983, when a five-day week was introduced in the plant—as part of a rationalization scheme; November 1984, when plans were finalized to close down the Bhopal plant; around 8:30 P.M. on the night of December 2, 1984, when the water washing of pipes downstream from the MIC storage tank began—without insertion of slip binds to keep water from backing up into the storage area; around 11 P.M. on December 2, 1984, when MIC was first detected in the air surrounding the MIC tanks.

INTERVIEWS WITH FORMER WORKERS, UCIL BHOPAL COLLECTED BY T. R. CHOUHAN

V. N. Singh (Token No 4554, Age 34)
Originally hired for alpha-naphthol plant

I joined UCIL as a plant trainee 28 March 1977. I had six months of classroom training and six months on the job. I was confirmed on 27 September 1978 and began as a regular operator in the alpha-naphthol plant. When we were in training, we thought the alpha-naphthol plant was a very sophisticated, continuous manufacturing process. But when we actually started work in the plant, we came to know how badly the plant was run. The whole process was often modified, mainly conversions from a continuous process to batch process. Once they shut down the plant for long periods for modifications; even afterwards, the plant did not run successfully. We heard that the naphthol plant was declared a failure plant when it was shut down permanently. While working in the naphthol plant, I got exposed to various chemicals so many times. Once, I was burned by hot chemicals coming out from the edge runner, a machine for fusion reaction. After the shut down of the naphthol plant, we were asked to take MIC plant training without any official letter. They started training first in the classroom for two months without ever providing individual copies of the process and safety manuals. The training was just a formality, with no intention of really teaching us anything. Our teachers were MIC plant personnel. After two months, they transferred us to the MIC unit for on-the-job training. After fifteen days, management asked us to take independent charge of a regular plant operator's position. At that time, the MIC unit was facing a shortage of manpower. In the event of any problem or abnormality, I had to take guidance from my colleagues or supervisor. Without help, I had to stop the work. Most of the

original MIC operators had resigned from UCIL; those of us transferred
from the naphthol plant took charge without proper training. On 2nd December 1984, I was on second shift. I reached the plant at 10:45 P.M. and
took charge of the phosgene reaction unit from the previous shift operator,
Rehman Khan, who told me that water washing of the phosgene stripping
was still in process and that the safety valves downstream are on. After taking charge, I remained in the plant. Around 11:30, we felt the strong presence of MIC, causing eye irritation. Some of my colleagues informed the
supervisor of the MIC plant. In the meantime, we saw that water was dropping from the second level of the structure and that near that place there was
more evidence of MIC. After we saw the water, we again informed the supervisor, Shakil Quresi. By this time, it was 12:00. Quresi told us we would
see after tea break. I went back to my unit and stopped the water washing
around 12:15 A.M. When I stopped the water washing, all the water from the
lines came out through the open bleeders. Near these open bleeders, I also
detected MIC. About 12:30, the supervisor was trying to find the leak but
could not. All of a sudden, around 12:50, they saw that the tank pressure
gauge was reading above its range. At same time, we saw a heavy release
from the vent gas scrubber atmospheric line. Then I broke the alarm glass
to start the loud factory siren. This was to warn other workers and to call the
rescue squad. After a few minutes, the loud siren was turned into a muted
siren. Then the rescue squad came to the MIC plant and tried to arrest the
toxic release by putting large amounts of water spray through fire hydrants.
After some time, the leak was uncontrollable so everyone started to flee
from the MIC unit in the opposite wind direction. I also ran away from the
MIC plant. At that time, we were not thinking that the Bhopal people would
be badly affected by the leak. Only the next day did we come to know the
intensity of harm caused by the toxic release from our factory.

   Then I heard that UCC said that some disgruntled worker had put water
into the tank. They only identified this worker as someone who had recently
been transferred to the Sevin plant. The only worker to meet this description
was M. L. Verma. Verma and I reached the MIC unit together at 10:50 on
2nd December and remained together, along with other colleagues, all the
time up until when we fled the factory. It is impossible that M. L. Verma put
water into the MIC tank. Also, it is a highly dangerous task to open a valve
leading directly into the MIC tank; if M. L. Verma had opened such a valve,
he may have been the first victim of the gas.

Mukund Patharadkar (Age 35)
Originally hired for alpha-naphthol plant

I joined UCIL on 13 February 1978 and began one year of training. Then
I was confirmed as a regular operator for the alpha-naphthol plant. I also
worked in the formulation plant and Sevin plant when the alpha-naphthol
plant was shut down. After permanent closure of the alpha-naphthol plant,
I was transferred to the Sevin plant and worked there up until the disaster.
When I was working in various plants, I felt that both personal and plant
design safety was very good on paper but in practice not up to the mark.
Theoretically, full care was taken to use safety equipment but the safety in

operating the plant and during mishaps was very lax. Also, critical pollution levels in the work area were not regularly checked. Because of these inadequacies in safety procedure and plant design, conditions were often unbearable. Some times the leak of chemicals led to the shut down of the unit. It was the dedication of Carbide's hourly rated workers that enabled the successful operation of the plant for eight years. In my opinion, if the Bhopal plant had been in a professional industrial belt where workers could have compared their conditions with those in other factories, the plant would not have remained open for even one month. Only because UCIL management shrewdly kept the workers unaware of normal safety standards was operation continued. The government should have made sure that these standards were enforced. Justice in Bhopal must include government recognition of how Carbide continually victimized the workers. I resigned UCIL 25 May 1985, after the disaster and UCIL had declared closure of the factory. I joined Bombay Dyeing and Manufacturing Company's DMT plant at Patlganga, District Raighar, Maharashtra. After closure of UCIL, all the hourly rated workers got compensation under 25(0) of the Industrial Dispute Act and additional compensation from Union Carbide. However, those of us who anticipated the closure and sought other jobs before the declaration were left out. The government of Madhya Pradesh offered me a clerical job in the Treasury Dept but the salary was too low to maintain my family. Bombay Dyeing is better than UCIL: I got two promotions in just four years, facilities are better, safety conditions are better.

V. D. Patil (Age 36)
Originally hired for MIC plant

I joined UCIL on 6 November 1978 as part of the first group for the MIC unit. I had eight months of classroom training, one month on the job in the Sevin plant and then began work in precommissioning the plant. Out of 32 who joined with me, ten had resigned before the start up of the plant at end of 1979. In 1981, another ten of my colleagues resigned. In 1982, another ten resigned. Only two persons, R. K. Yadav, and myself remained. My colleagues joined UCIL believing in the name of Union Carbide but when they saw the working conditions, the facilities provided, the promotion policy and the behavior of the management they saw that they were not up to good standards. We also saw that senior operators from other units were no better off and were being treated like labor class workers. As my colleagues are all science graduates and diploma holders, they started planning to leave UCIL just after joining. As they got other opportunities, they joined other offices or factories. I did not get another opportunity until February 1984, when I got an opportunity in Bombay Dyeing as a foreman. I joined Bombay Dyeing and got two promotions in six years. Now I am a plant superintendent and also availing of other facilities I did not have at Union Carbide.

After I joined UCIL, other batches of trainees came and also quickly left. As a result, by 1982, the plant was mostly run by the trainees since many senior operators had resigned. In 1981, they stopped giving training for the MIC unit and transferred in operators from other units who did not have training in the MIC process. Up to 1982, the supervisory staff was well trained and did not change too often. However, after 1982, supervisory staff was also transferred in from other units without any additional training. The quality of the operating personnel became worse and worse. The condition of the plant was also deteriorating. This down slide began in 1982 and got worse and worse. When I resigned in February 1984, three of the most experienced people in the MIC plant resigned: MIC plant manager K. D. Ballal, Production Superintendent Satish Khanna and Production Assistant V. N. Agarwal; Safety Manager B. S. Rajpurohit and Plant Superintendent C. R. Iyer were transferred. No one in the plant had sufficient training and experience, leading to increasingly unsafe conditions. The disaster should have been no surprise to any one.

In 1980, the manpower for the MIC unit included 13 operators, all qualified and trained; 2 supervisors; 1 production superintendent, separate for the MIC unit; and one plant superintendent separate for the MIC unit. In 1981, operators were reduced to nine, including trainee operators. In 1982, operators were transferred from the Sevin to the MIC plant without proper training. In 1983, operators were reduced to six, mostly untrained and not qualified; there was only one supervisor common to the MIC unit and the CO unit; the plant superintendent was also made common for all units in Phase II; the production superintendent was made common to the CO and MIC units; the MIC manager was also made common to the MIC and Sevin units.

When I joined UCIL, I thought it was a reputed concern but with the duration of time my opinion changed. Operators were treated just like the labor class, just like bonded labor. The only difference was that we could leave the work site, so we did, as soon as other opportunities came our way. At the beginning, safety instruments and procedures for the MIC unit were relatively good; but as time passed, management became negligent, more and more untrained personnel were put on the job. The safety of the plant and workers became worse and worse.

At the start up of the plant, we were taking parameter readings every one-hour. In 1981, this was changed to every two hours. The phosgene system was provided with a lot of critical instrumentation but because of choking and corrosion, the instrument started frequent malfunctioning. Production was not stopped but continued, bypassing the instruments. There were many major and minor leakages during my time in the MIC plant due to corrosion of lines and equipment, malfunctioning of instruments and due to unsafe handling of process equipment. After the disaster, in 1986 I came to know that UCC had accused a disgruntled worker of connecting a water line to the MIC storage tank. This is not possible.

Dates and times are a way to query the disaster in concrete terms. They also reveal different understandings of historical development and of the relationship between past and future. People "tell time" by turning certain events into reference points—investing some things with meaning, while discounting others. Chouhan and Union Carbide can thus be differentiated. They tell the times of the disaster differently, setting up the past to configure very different futures.

FIVE

States of India

(5.1) Constitutional Contradictions

In the independence-era dreams of Jawaharlal Nehru, a pesticide plant in Bhopal would have been a temple of the new India, representing the dynamic synergism of science, industrialization, and socialism. Fifty years later Union Carbide's Bhopal plant continues to operate as an icon, but of a different kind. The contradictions of this chronology have provoked varied responses. For some, "Bhopal" is a simple aberration, a momentary incident of tragedy unconnected to either past or future decisions and goals. For others, "Bhopal" represents the unfortunate, but necessary, risks of progress toward national self-sufficiency and global competitiveness. Still others insist that Bhopal is emblematic of structural failure, providing a lens through which we must judge established institutions, their programs, and the programs relied on to challenge them.

The government of India has voiced all three of these responses, demonstrating the difficulties and contradictions Third World governments face within contemporary culture and political economy.[1] The paradoxes are harsh, particularly in India, where the contradictions are literally constitutive of the postcolonial state. The Indian constitution is bifocal. It guarantees political democracy, not unlike the U.S. Constitution. But it also has directive principles (Articles 36–51) that promise the pursuit of economic democracy.

The directive principles do not articulate rights. They are described as "fundamental in the governance of the country and it shall be the duty of the state to apply these principles in making laws" (Article 37) (Thakur 1995, 50). But they are not enforceable by the judiciary. There are overt contradictions. The directive that the ownership and control over material resources be redistributed to subserve the common good (Article 39), for example, cuts into property rights (Article 19). But there are also more subtle implications, as evident in the Bhopal case. The drive to deliver economic prosperity through industrialization meant that regulatory oversight of Union Carbide's plant in Bhopal was a double bind.[2]

(5.2) Science as Reason of State

In late February 1989, Dr. Satish Dhawan, former chairman of the Indian Space Research Institute, announced to the press that Bangalore's scientific community would be filing a petition before the Indian Supreme Court to voice collective protest against the out-of-court settlement of the Bhopal case.[3] This announcement followed a lecture at Bangalore's Indian Institute of Science by physicist Anil Sadgopal, who argued that the Supreme Court would find it difficult to reject a review petition filed by the scientific community and that the two primary reasons for the "Bhopal sell-out" were lack of human conscience and bad-quality science education.

According to Dr. Sadgopal, the settlement of the Bhopal case raised serious questions about the social character of organized science in India. Most basic, the established research organizations like the Indian Council of Medical Research had failed to recognize the critical interface between scientific and legal matters and thus failed to fully understand the significance of their own research findings. Furthermore, unwarranted secrecy and refusal to subject research data to open scientific scrutiny had resulted in the propagation of false theories and confusion about treatment protocols, which ultimately served the interests of Union Carbide.

Dr. Sadgopal's arguments referenced a world in which science had lost straightforward credibility as a mechanism for producing the conceptual and material forms necessary for sustaining a just society. In India, the reasons for this loss of credibility are highly visible in the debates over "development" within which the Bhopal disaster is often understood. Sadgopal's arguments therefore resonated strongly in the community of activists challenging the development paradigm that has guided economic planning in India since independence. The story of Bhopal has become part of this challenge, becoming permeated with images of opposition that challenge both government agendas and the scientistic credos that legitimate them.[4]

(5.3) Categorical Imperatives

While the Bhopal case was settled out of court in February 1989, the rationale for the settlement award figure was only announced three months later.
It relied on medical data provided by the state government of Madhya Pradesh, which placed only 5 percent of victims in compensatable categories.
The data were based on a scoring method of health assessment that hierarchically ranks different body systems and categorizes patients according to
a lettering system indicating degrees of injury and disability. The method
"wrote out" any possibility of multisystemic damage or long-term effect
and "wrote in" a system wherein victims would have to prove their injury
before local claims courts, with documentation from government hospitals
and morgues. In August 1994, as people around the world prepared to commemorate the tenth anniversary of the disaster, a total of 82,523 individual
cases had been cleared by claims courts in Bhopal. The award for each
death ranged from $4,000 to $12,000 (Alvares 1994; Jaising 1994).

VOICES OF BHOPAL, 1990
BHOPAL GROUP FOR INFORMATION AND ACTION

Interview with Suleman Khan, age 50, resident of Ashoka Garden
I have been working as a booking agent in the Madhya Pradesh State Road
Transport Corporation for the last 24 years. I was on duty right after the disaster and during Operation Faith when the Corporation's buses were used to
carry people who were running away from Bhopal. I started having serious
health problems about a month after the gas leak. In May '86 I was transferred to Piparia, 150 kilometers away from Bhopal. There were no facilities
for medical treatment in the Piparia hospital. So I had to absent myself from
my duty and come to Bhopal. They stopped my salary. I wrote many applications to get myself transferred back to Bhopal. More than 2 years later,
they transferred me back but I am still not getting my salary. I was admitted
in the MIC ward in June '87 and remained there for more than three years.
I was so breathless they had to put me on oxygen. The doctor in charge of
the MIC ward has written on my papers that my lungs are badly damaged.
I have to take 8 to 12 tablets in a day to be able to talk, move about or just to
breathe. I have written 17 letters to the officials of the Corporation, 7 letters
to the Chief Minister and 4 letters to the Prime Minister. I have requested
them to pay me my salary, give me some monetary relief and do something
for my medical treatment. These medicines don't seem to be working. In
August '87, I was called for medical examination by the Directorate of
Claims. I was in the MIC ward at that time. Then, after 2 years, when I was
still in the hospital, they sent me a notice that said that I have been put in
Category 'B'. My wife was also admitted in the MIC ward. She too has been
told by the Claims Directorate that she has suffered only temporary injury.
Like they said for me. I wrote a letter saying that there was something

terribly wrong in putting me in Category 'B'. The doctor in charge of the
MIC ward put in his recommendation in that letter. It has been almost a year
since and they haven't replied yet.

Interview with Abdul Zahoor, age 30, resident of Baug Umraodulha
I get swelling in my stomach. I become extremely uneasy and cry out in
pain. Sometimes this happens all through the night. I am tired of x-rays
done and the doctors say nothing about my disease. I have gone to all kinds
of doctors, the big ones too. I have been to Sajjad Nursing Home, the J. P.
Hospital and to the Hamidia Hospital. I have even gone to the Hakims and
Homeopathic doctors. But it has been like this. Pain, pain, all the time. I
became weak, had body aches and fever for a long time after the gas, but
earlier I didn't have this pain in my stomach. They have done my medical
examination but now they tell me I have been put in "B" category. I had
shown them my medical papers. Still. I haven't been able to work for the last
two years and have stayed in bed all the time. I depend on my brothers for
food and my treatment. Something has to be done for this pain in the stomach. I am getting the interim relief of Rs. 200 [$8], but it isn't enough for my
treatment. I have to spend 700 to 800 rupees on treatment every month. And
yet there is no relief.

Interview with Asad, age 14, resident of Ibrahimpura
I get breathless and often I am down with fever. Also, I cough a lot. I go to
school but I cannot study. I forget things easily and my eyes burn. I study
in a government school. Ever since the gas, I am always taking medicines.
Those doctors who were examining me, I told them I have breathing problems. But they have sent this notice that says I've been put in "B" category.

Interview with Shakila Bano, age 30, resident of J. P. Nagar
Right after the disaster, I was admitted to the Katju Hospital. Then I was
admitted to the Hamidia Hospital for a long time in the MIC ward. The doctors told me that my x-ray pictures showed that my lungs were badly damaged. I had filed my claims and was called for medical examination. The
medical examination, they said, was necessary to make my case strong for
compensation. They did all kinds of examinations. They did blood tests,
sputum tests, urine tests and also took x-ray pictures. I was once again admitted to the Jawaharlal Nehru Hospital after that. Now they sent me this
notice that says I have been put in category "B." It says that I only suffered
temporary injury due to the gas and I am all right now. Even now I cough so
badly all the time; I throw up blood sometimes. I have pain in my chest.
When I get admitted to the hospital, the doctors do not let me go home.
"You are still very sick," they say, and ask me to stay on at the hospital.

Reliance on the Madhya Pradesh data was a badly flawed strategy for
two reasons. First, it assumed a system of claims courts in which victims
could prove their health status—a system that cannot be said to exist in
India. Second, it negated the possibility of compensation distribution based
on geographical indicators of exposure—which would have placed far

more victims in compensatable categories and would have offset concern about creating new forms of social stratification in Bhopal via uneven distribution of rehabilitation resources.

There was alternative medical data, produced by an equally authoritative institution, the Indian Council of Medical Research (ICMR). These data were based on geographical indicators that marked areas as highly, moderately, or mildly affected. The results were staggeringly different than those that legitimated the settlement. The 1990 ICMR report acknowledges 521,262 individuals as exposed, with 97 percent (508,230 individuals) evidencing small-airway obstruction that could lead to future morbidity and impair the capacity to work. Of the exposed population, 32 percent were acknowledged to be overtly symptomatic, with that percentage increasing each year. Death rates among the affected population were more than double that of the unexposed population, and the data documented significantly high incidences of spontaneous abortion, stillbirths, and infant mortality. Data produced by the ICMR have never been fully released to external researchers or to medical practitioners in Bhopal. The "needs of the Court" have justified the secrecy, within the provisions of the Official Secrets Act.

(5.4) India, Inc.

The settlement of the Bhopal case was upheld in October 1991, shortly following an intensification of commitments to the International Monetary Fund (IMF). Soon afterward, a seminar was convened to address the changing investment climate in India following liberalization, with particular focus on the legal issues pursuant to foreign investment and to the possible problems created by stringent tort liability. The seminar was inaugurated by the Union Finance Minister. The opening speaker was India's Chief Justice Kania. Participants included the Union Minister for Law and the "estimable" F. S. Nariman, Union Carbide's Indian counsel for the Bhopal case. At the following press conference, reporters were reminded that the handling of the Bhopal case was evidence that India is an amiable site for foreign investment, symbolizing Indian commitment to the New World Order. The guarantee of a fair and reasonable approach to corporate liability was the out-of-court settlement between Union Carbide and the government of India, which had represented victims of the Bhopal disaster in a role of *parens patriae*.

Progressive activists have told me that this statement is evidence of an important shift in political culture. Since independence, fear of exploitation had shaped India's engagement with the international order. By the 1990s, the greatest fear was of *exclusion.* One effect is pervasive suggestion that *any* kind of participation in the international order is good. Critical discussion of different ways of participation—once discussed in terms of dependency theory—becomes irrelevant.

(5.5) Remaking India

One of the most resilient images of political activism in response to the Bhopal disaster is that of a woman medical doctor who responded to the out-of-court settlement of the case by taking a broom into the chambers of the Indian Supreme Court and proceeding to clean the floor. The dirt she sought to remove was of many sorts. In particular, she was responding to a settlement amount far short of projected rehabilitation costs, legitimated by medical data that categorized gas victims according to a hierarchical ranking of body systems that denied both multisystemic damage and reproductive disorder, which her own research had confirmed.[5] She was also marking the way that Left response to the proceedings leading to the settlement had tangled itself in the mire by accepting an ordering of the court around the opposition of Union Carbide and the government of India, which had granted itself sole right to speak on behalf of victims in the capacity of *parens patriae.*

Dr. Sathyamala's willingness to deal with the dirt was not authorized; despite her obvious expertise, her name rarely appeared in media accounts of the settlement. Instead, the *New York Times* interviewed Bud Holman, Union Carbide's legal counsel. He described work toward the settlement as "like walking up a pitch black, winding staircase, and you never know how much further you have to go in total darkness. Then, all of sudden, it's light and it's all over." Kamal Nath, Ministry of Forestry and Environment, also described the logic of the settlement in simple terms: "we could leave it open ended, or clinch it, at some point."

Images of Bud Holman being blinded by the light and Kamal Nath wanting a clinching conclusion are instructive, indexing the multidimensionality of Sathyamala's critique. Sathyamala did not expect the government of India to respond well to the Bhopal disaster. She had little faith in the independence-era promise of scientific socialism, to be achieved through the trusteeship of indigenous elites. On the other hand, Sathyamala did expect the government of India to respond well to the Bhopal disaster through legal action that acknowledged continuing disaster in Bhopal. For this to be achieved, the government of India would have to admit medical studies produced by the nongovernmental groups—studies that countered medical studies produced by the government itself. Sathyamala thus voiced a critique not only of the substance of actions taken by the government of India, but also of the way the Indian government constituted itself as a government.

Government actions were taken and justified without acknowledgment of knowledge produced outside itself. The government of India operated like a closed system, incorporating all that it encountered into its own logic, discounting what resisted explanation as noise. Sathyamala challenged this figuration. She sought to remove the dirt from the Bhopal case—by acknowledgment of dirt—matter that could not be dealt with through standard operations of the government as a system. Sathyamala called for a different kind of system—a system that operated effectively as a system, but not through exclusion of things that upset its own logic.

(5.6) From the Study of Women

Disaster harbors particular contradictions for women. Bhopal is no exception. In the wake of disaster, women have gained access to public space and, in some cases, to wage-earning work. But they have also borne an unequal share of the effects of disaster. Many have husbands who can't, or won't, earn. Many are responsible for the care of sick children, spouses, siblings, and parents. Many are sick themselves. Many have faced the social stigmas of reproductive disorder. All must deal with a medical establishment that systematically discounts the particular problems of women's health.

Stories of malformed babies still circulate in Bhopal. Statistical significance is unclear. But they index fear that more is wrong than what people already know; December 3, 1984, becomes a pivot point around which normalcy breaks down. Interpretation of stories told by Ajeeja Bi and Shahida Bi are suggestive. Ajeeja Bi lives in Kazi Camp, very near the Union Carbide plant. She had three children before the gas that were "normal." After the gas, she aborted three times. They were all born dead with skin the color of coal. The first was born at six months, the second at seven months, and the third at eight months. All were delivered in the hospital, but doctors have not told Ajeeja Bi what is wrong. Shahida Bi lives near Bharat Talkies, near the Railway Station. She had a son a few months after the gas leak that was all right. Her next child was born dead, with no eyes and no limbs. The next child seemed all right, but died soon after birth. Then a child was born with skin that Shahida Bi says looked scalded. And only half its head was formed; the other half looked like a membrane filled with water. Shahida Bi's body has remained covered by a rash.

Reference to reproductive disorder as a sign of systemic malfunction began soon after the gas leak. In the immediate aftermath, voluntary groups worked with the assumption that MIC and other gases released were potential teratogens. Debates over whether abortions should be recommended were stalled by the clampdown on information from government hospitals and by Union Carbide's unwillingness to offer any insight into the composition and effects of the gas released. Entrenched problems with the delivery of abortion services in government health care facilities complicated the debate and embedded the issues of Bhopal in long-standing critiques put forward by women's health activists.[6] Within a few months, the medical establishment in Bhopal did explicitly deny increased rates of spontaneous abortion and refused to consider recommending that pregnancies be terminated. Confidence in the establishment position was low.

In February 1985, two women doctors, Rani Bang and Mira Sadgopal, conducted a study that resulted in a paper titled "An Epidemic of Gynecological Disorder." This study found high rates of leucorrhea (94 percent), pelvic inflammatory disease (79 percent), excessive bleeding (46 percent), and lactation suppression. It also found an increase in stillbirths and spontaneous abortion. The establishment dismissed the study because it was based on a small sample of fifty-five women. The capacity of the voluntary sector to conduct studies sufficiently large to be considered methodologically valid has remained a problem in Bhopal. Efforts to respond to the particular health problems faced by women have been particularly affected.

An important effort to skirt the problem of research scope was led by Dr. Sathyamala in May 1985. The study was a project of Medico Friends Circle (MFC), with the support of many other voluntary organizations. Dr. Sathyamala's study was titled *Distorted Lives: Women's Reproductive Health and Bhopal Disaster* (Medico Friend Circle 1990). The three areas selected for the study were in Municipal Ward 13, one of the officially recognized "severely exposed" wards in Bhopal. Starting data on the three areas were drawn from a study carried out in January 1985 by the Centre for Social Medicine and Community Health, Jawaharlal Nehru University (Centre for Social Medicine and Community Health 1985). The total population in the three areas was between 31,000 and 35,000. Basic data included the following:

J. P. Nagar
  (1,998 households)
Mortality rate: 65.3/1,000
Morbidity rate: 66%

Kenchi Chola
  (1,300 households)
Mortality rate: 35.7/1,000
Morbidity rate: 91.9%

Kazi Camp
  (1,950 households)
Mortality rate: 46.7/1,000
Morbidity rate: 54.6%

Dr. Sathyamala's study confirmed indicators of reproductive disorder documented by Rani Bang and Mira Sadgopal. It also documented significantly increased incidence of spontaneous abortion and stillbirths. Before the gas leak, the fetal death ratio was 6.43; in September 1985, the fetal death ratio was 31.25. Spontaneous abortion was shown not to be confined to the immediate postgas period, but continued in high incidence even ten months after exposure. And high incidence of abortion was not confined to those pregnancies conceived before exposure; spontaneous abortions in conceptions after the leak were five times higher than before.[7]

Dr. Sathyamala's study was not published until 1990 because of controversy among researchers over methods. But the findings circulated, as indicators of possible chromosomal aberration, as confirmation that toxins were residing in the body and could pass the placental barrier, and as proof that gas exposure had multisystemic effects.

(5.7) And Yet It Moves

BHOPAL GROUP FOR INFORMATION AND ACTION
1 November 1990

Dear friends,

The sixth anniversary of the Bhopal gas disaster is drawing near. In the past few months, communal and casteist forces have caused widespread damage and dissension all over the country. The media and, through it, the attention of the general public, has remained dominated by issues such as Bofors, Mandal Commission and the Ram Janambhoomi-Babari Masjid controversy. While these are important issues, they are but manifestations of the deeper malaise that pervades all aspects of life in India. The media, as it happens to be, has engrossed itself with such manifestations and the deeper political/developmental/environmental issues have been relegated to secondary positions of concern. Narmada dams, Kaiga Nuclear Reactor, Balliapal Missile Range, increasing automation, rising foreign investment, work hazards, amniocentesis, displacement of tribal people and the impoverishment of peasants are hardly copy in these times. Neither is Bhopal.

Unbeknownst to the media, the most critical medical, rehabilitation and legal issues on Bhopal remain to be addressed in a meaningful manner. Death, disease and despair continue to stalk the shanties; the suffering continues in Bhopal. So does the struggle. The "valiant victims" continue their battle on the streets for rehabilitation and for punishment of the guilty officials of Union Carbide responsible for the genocide in Bhopal. It has been difficult for the courageous women of the Bhopal Gas Peedit Mahila Udhyog Sangathan (Bhopal Gas Affected Women Workers Organization), the largest organization of gas victims; many of them continue to suffer from gas related diseases. They have been harassed, beaten and arrested by the police;

the organization has run into huge debts and the press is more often inclined to ignore them. And yet they continue to struggle. Sometimes they win.

The provision of interim monetary relief of Rs200 per month per person by the Central government to all the residents of the 36 municipal wards declared to be gas affected is one such victory. Along with extra legal initiatives for monetary relief, the Sangathan had, in June 1988, filed a petition for interim relief in the Supreme Court of India. Consequently, on 13th March the Supreme Court ordered the Central government to provide for interim monetary relief to the gas affected people for three years. The relief disbursements started in June and, according to latest figures, 3,03,707 gas victims have so far received installments of this relief. Provision of such relief has made a difference in the lives of the gas victims; they are eating better, are not forced to do hard physical work and are slowly paying back the debts they had incurred in the last few years. The disbursement of interim relief as being done by the Madhya Pradesh government is, however, beset with serious problems. The pace of disbursement continues to be slow and close to half the gas affected population is yet to receive this relief. According to official figures, over 34% of the beneficiaries can not be traced, a situation brought about due to incorrect recording of names and addresses of the gas victims and the absence of an infrastructure for monitoring migration. Corruption is rampant and in the location of identification and dispersal centers, no attention has been paid to the conveniences of the gas victims. Remedial measures suggested by victims' organizations have largely been ignored by the State government.

While the National Front Government at the centre deserves commendation for its initiatives on interim relief, it has paid little attention to providing adequate medical care and economic rehabilitation to the Bhopal victims. Six years are about to pass and the medical treatment of the gas victims continues to be the same as in the immediate aftermath of the disaster; this is largely symptomatic treatment. While Union Carbide, through its refusal to provide information on the composition of the escaped gases, their effect on the human body and the means to deal with these effects has impeded the evolution of a proper line of treatment of the gas-affected people. The government hospitals and the research projects of the Indian Council for Medical Research (ICMR) have contributed little in this direction. Even today, over three hundred thousand of the people who were exposed to the toxic gases continue to suffer from symptoms like breathlessness, loss of acuity of vision, fatigue, bodyache, menstrual irregularities, loss of appetite, depression, anxiety, etc. The 1989 report of a long term epidemiological study conducted by the ICMR has indicated that death rates in the severely affected areas are more than double and that in the moderately affected areas are little less than double compared to the control area. According to a study sponsored by the Jawaharlal Nehru University, New Delhi in 1989, 70 to 80% of the gas affected population in the severely affected area and 40 to 50% in mildly affected areas suffer from medically diagnosed illnesses. Ongoing research being conducted here and abroad indicates widespread and permanent damage to the immunological system of the gas victims and the presence of pre-malignant conditions in many of them.

Assessment of injuries suffered by the gas victims due to their exposure to Union Carbide's gases is another critical area that has been neglected by the government. As is evident, a proper assessment of the personal damage wrought by Union Carbide is essential for extracting adequate compensation from the multi-national. And yet what has been done so far in this regard by the Directorate of Claims of the state government has led to gross underassessment of injuries. According to the latest figures, of the 3,57,485 claimants whose injuries have been assessed, 1,72,776 people have been found to be only temporarily injured and 1,54,813 have been considered to suffer no injuries at all, and only 40 persons have been placed in the most serious category of injury. It follows then that if compensation from Union Carbide is based on these figures furnished by the Directorate of Claims, over 90% of the gas victims will either receive minimal compensation or none at all. Over 1,00,000 gas victims, a majority of them children, have yet to register their claims with the Directorate and 2,47,000 of those who have registered their claims remain to be medically examined. Adequate compensation is not only a lawful right of the gas victims; it is essential for their survival as well. Unless drastic steps are taken for a proper evaluation of the injuries, it is quite possible that an overwhelming majority of the gas-affected people will be denied compensation.

The Mahila Sangathan has called for the setting up of a National Commission (comprised of representatives of the State and Central government as well as non-governmental individuals and organizations) that would re-organize and reorient medical research so as to evolve a proper line of treatment of the gas victims. The proposed commission would also review the injury assessment done so far and implement measures for a proper evaluation of the damages inflicted by Union Carbide. Interim monetary relief being only a short term measure of providing sustenance, it is imperative that the National Commission design and implement adequate economic rehabilitation measures so that the gas affected people are provided with jobs that are conducive to their health conditions and which enable them to make a decent and dignified living.

The high morbidity among the gas affected people and particularly the damage done to their immunological system calls for healthy living conditions. Apart from token gestures, the government has done little in this regard. Most of the gas victims live in shanties that are damp and have poor ventilation and the people in the most severely affected slums do not have access to clean drinking water. Early this year, the Citizens Environmental Laboratory analyzed samples of water and soil collected by us from the vicinity of Union Carbide's now closed factory. The reports of this analysis show that drinking water wells in the communities are contaminated with di- and tri- chlorobenzenes, chemicals that damage the liver, kidneys and respiratory system. Soil samples from the waste impoundment area contain polynuclear aromatic hydrocarbons, which are known carcinogens. Such a situation calls for large-scale investigation and consequent clean-up action. Despite repeated representations the government is yet to initiate either of these.

Meanwhile, the case against Union Carbide is yet to cross the preliminary stages of litigation. Until September this year, there were hearings in the Indian Supreme Court on the petitions seeking review of the settlement reached in February 1989 between the previous government and Union Carbide. The death of the Chief Justice in October has necessitated a rehearing of the review petitions, which is likely to begin in February next year. It will take several months before a judgment on the validity of the settlement is awarded. The case for compensation from Union Carbide and criminal prosecution of officials of the Corporation can only proceed in the Indian courts after the settlement has been invalidated. It certainly will be a long haul and members of the Sangathan, which is one of the review petitioners, are resolute to carry on with the litigation until adequate compensation is extracted and Union Carbide officials are criminally convicted. Meanwhile, the prospects of pursuing the case in the courts in the United States have brightened with a favorable ruling in the state courts of Texas. Subsequent to the admission of a petition filed by a group of injured Costa Rican farmers against two U.S. based multinationals, it is likely that "foreign" litigants will not be barred from moving courts in Texas on grounds of "forum non conveniens." By all accounts, the struggle for justice in Bhopal will be a long one.

Your support in this struggle is essential. The solidarity expressed by individuals and organizations from different parts of the country and the world has significantly strengthened the resolve of the victims to continue with the struggle. Bhopal needs also to be remembered and acted upon so that we are more aware of the "silent Bhopals" happening around us, more equipped to pre-empt Bhopals likely to happen in the future. Through this letter, we seek your support of the demands of the gas victims, as follows:

1) Union Carbide Corporation should be held liable for the death and destruction caused in Bhopal and it should pay compensation that is adequate to take care of the medical care and rehabilitation needs of the gas affected people.
2) Officials of Union Carbide responsible for making the decisions that led to the disaster should be criminally prosecuted.
3) The illegal, unconstitutional and immoral settlement reached between the Government of India and Union Carbide in February 1989 should be overturned.
4) Union Carbide should pay interim relief to the Bhopal victims as ordered by two Indian courts.
5) Union Carbide's assets in India should be handed over to the workers and there should be a ban on its future business operations in India. Particularly, the now closed pesticide plant in Bhopal should be completely dismantled and Union Carbide should pay for environmental clean up of the grounds.
6) The Government of India must set up a National Commission on Bhopal that meaningfully addresses the issues of medical care, injury assessment, economic and social rehabilitation of the gas victims with the urgency they deserve.

With these rallying points, the Bhopal victims will be observing the sixth anniversary of the gas disaster on 3rd December 1990. On this day, the sorrow and anger of the people of Bhopal would collectively be demonstrated—rallies will be organized, marches will take place, effigies of Carbide officials will be burnt and oaths will be taken. Through this letter, we invite you to come to Bhopal on that day and participate in the observance

of the sixth anniversary. For those who cannot come to Bhopal, may we suggest the following solidarity actions:

1) Organize public demonstrations, meetings, exhibitions, burning of effigies, die-ins,
   etc. before Union Carbide plants and offices of other public places.
2) Organize press conferences around the issues of Bhopal and demands of the gas
   victims.
3) Organize letter campaigns supporting the demands with letters being sent to the
   Prime Minister of India (Prime Minister, North Block, New Delhi) and the Chairman of Union Carbide Corporation (Robert Kennedy, Union Carbide Corporation,
   39 Old Ridgebury Road, Danbury, Connecticut 06017-0001).

If you send us reports of your protest actions or wish to send messages of
solidarity, we would be happy to include them in the news sheet that we
publish and distribute to gas affected people.
> Remember Bhopal!
> In solidarity,
> Dr. Rajeev Lochan Sharma
> Satinath Sarangi

The first year I was involved in Bhopal activism on behalf of gas victims
focused on the settlement of the Bhopal case, on the medical categorization
scheme that supported the settlement, and on the problems and importance
of interim relief. The Bhopal Gas Peedit Mahila Udhyog Sangathan—the
Women's Union—had filed a formal plea for interim relief in 1988. The
Indian Supreme Court first ordered the state of Madhya Pradesh to begin
distributing interim relief in April 1989, in the midst of highly visible protest against the settlement. The first order, on April 21, 1991, directed payment of Rs. 750 per month ($30) to the family of each adult who had died.
At the same time, the court asked the Madhya Pradesh government to provide data on other categories of victims in dire need of cash relief. By
August, an order directed payment of Rs. 1,000 ($40) to 7,687 categorized
as having permanent injuries. The Women's Union protested this mode of
relief distribution, arguing that the categorization of injury was arbitrary
and veiled the magnitude of need in Bhopal. The Women's Union also argued that relief distribution based on medical categorization would stratify
the community.

The Women's Union's demands for distribution of interim relief according to geographic indicators were finally sanctioned in January 1990. The
first plan was to distribute "one time" interim relief payments to almost
600,000 people living in the thirty-six gas-affected wards. By March, the
plan had been revised again. Instead, Rs. 200 per month ($8) would be
given to all identified gas victims for three years. Also, Rs. 500 per month
($20) for three years was promised to those categorized as permanently

disabled. Meanwhile, the lump sum payments to heirs of the dead and to those having permanent injuries were to continue.

The problems with plans for distribution of interim relief were practical—and political. The government of Madhya Pradesh was responsible for distributing the relief. State officials said that it would require at least eight months to carry out preliminary identification of victims. Meanwhile, local banks insisted that they did not have the infrastructure to administer more than 500,000 new accounts and the disbursal of Rs. 10 crore every month.

The process of identifying victims was macabre; victims had nightmares about their inability to prove their identity; acquiring a name on official lists became an obsession. There were thirty-six identification centers that issued notices to appear before them, based on three documents: the Tata Institute of Social Sciences (TISS) survey, a computer list of claimants, and electoral rolls. Once called, the victims were to bring the notice to appear, their own copy of the TISS survey form, their ration card, and three photographs (for which they were reimbursed).

If all the documents were complete, the victim was then sent to an assistant beneficiary identification officer (BIO), who confirmed that all documents were indeed complete and issued a stamped receipt. Next the victim saw a banking officer, who was to open an account in the victim's name. Money for children was deposited with the mother.

Withdrawals were made every month at banks around the city. Often one family would need to visit many different banks, often distant from each other. This cost of travel to collect interim relief became a new burden, since little effort was made to coordinate bank locations with where victims lived. Collection of the money usually took an entire day due to long queues.

The Women's Union traveled to Delhi to protest problems with the distribution of interim relief in June. Almost 2,000 women made the trip, traveling without rail tickets—as is allowed in India when organized citizen groups seek an audience with their government leaders. The women sat throughout the day in front of the prime minister's house. The memorandum delivered to the prime minister complained that many people had been left off the relief lists, often due to trivial mistakes in the spelling of names or listing of age. It also detailed the practical problems faced because of the location of the identification centers and banks. And it demanded that a list of victims be made public so that victims themselves could watch out for corruption.

Efficient distribution of interim relief in Bhopal has never been realized.

Distribution of final compensation through community claims courts has been far worse. With time, however, the significance of interim relief became clear. For some, it truly made the difference between life and death. Many began taking prescribed medicines more regularly. They could buy a bit more food and were able to pay off some debt—particularly to landlords. The potential role of the government of India could finally, even if briefly, be imagined.

Six

Situational Particularities

(6.1) Matter Out of Place

Bhopal is in the very center, the navel of the country, a mark on the geographical body of India. The region was once known as Malwa; now it is the state of Madhya Pradesh. Bhopal is the capital.

Most of the state is a high plateau, inhabited primarily by Indo-Aryan peoples, but also by Gond and Bhil tribals. Ashoka's Mauryan Empire once reached here, stretching from what is now Afghanistan across most of India. The stupas Ashoka built at Sanchi aren't far from Bhopal, as a crow flies. The Buddhist message is increasingly remote.

The city of Bhopal is on the site of the eleventh-century city of Bhojapal, founded by the legendary Raja Bhoj, who commissioned the lakes around which the city is built. Legend teaches that Raja Bhoj was afflicted with an incurable disease. A sage advised him to excavate a lake fed by 365 springs so that he could bathe daily in its waters. 359 springs were located near the headwaters of the Betwa River, 32 kilometers from Bhopal. Six more streams were located by one Kaliya Gond. Taking advantage of the ring of hills around the headwaters of the Betwa, Raja Bhoj's engineers built two enormous dams. The waters of the six streams located by Kaliya Gond were diverted into the catchment area the two dams created. The Upper Lake was, in those times, said to be the largest and most beautiful lake that adorned the surface of India. Even today the dams that created the lakes are regarded as exemplary specimens of hydraulic engineering skill (Shrivastav and Guru 1989, 35).

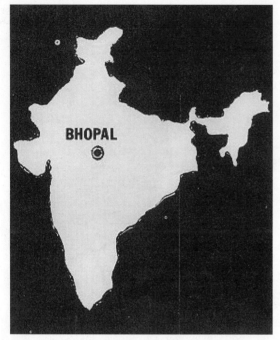

FIGURE 6.1 INDIA MAP
WITH BHOPAL AT CENTER
The country of India, with the
city of Bhopal at its center, depicted in the pamphlet titled *No
Place to Run: Local Realities
and Global Issues of the Bhopal Disaster,* published jointly
by the Highlander Research
and Education Center (United
States), the Center for Science
and the Environment (India),
and the Society for Participatory Research in Asia (India)
in 1985.

The city of Bhopal that stands today was established by Afghan chief
Dost Mohammed Khan during Aurangzeb's reign (1658–1707). Taking advantage of the confusion caused by Aurangzeb's death, Dost Mohammed
carved out his own small kingdom. Muslim control over the city was sustained until independence, through collaboration with the British. During
World War II, the locality of Bairagarh was a prisoner-of-war camp. After
independence, the Bairagarh camp was used to house Hindu refugees from
Pakistan. Muslims from across India also immigrated to Bhopal in large
numbers, seeking the security of the largest Muslim state after Hyderabad.
In 1961, half of the residents of Bhopal were immigrants.

Bhopal has never been a pastoral locale. It has drawn people into itself
out of violent currents. It has been a place of migrancy, of continual up-
heaval and of reinscription. What hasn't changed is the way the poor of
Bhopal are swept into grand narrations, in scripted roles.

It is this inscription that is pointed to when advocates for gas victims
insist that "Bhopal is no isolated misery." The slogan is repeated at protest
rallies, when gas victims gather to hear about the progress of legal proceedings, and in pamphlets grasping for ways to renew the significance of Bhopal in public memory. Some say the slogan exposes how the Green Revolution, which the Carbide plant served, was a major cause of agricultural

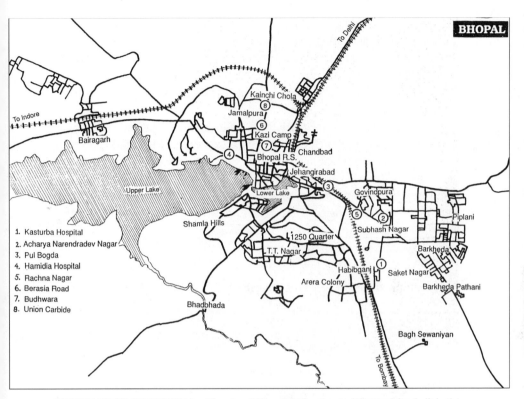

FIGURE 6.2 MAP OF BHOPAL   The city of Bhopal, laid out around the two lakes built in the 11th century by the legendary Raja Bhoj. The Union Carbide plant is at the northern edge of the old city, north of the lakes. This map was included in the pamphlet titled *Bhopal: A People's View of Death, Their Right to Know and Live,* published by Eklavya in 1985.

labor displacement and migration to the cities. In this account, many of those living near the Carbide plant had already been victimized by the same processes that culminated in their 1984 exposure.

The slogan is also used to index the victimization process that led to certain communities living in such close proximity to a hazardous industrial facility. The most affected colonies are among the poorest in Bhopal. They are situated just adjacent to the plant, so residents were aware that "poison" was being produced because of the frequency of small leaks and routine emissions that caused nausea and other effects. They were never in a position to protest because they had illegally set up their houses without land ownership or government permission. They continually feared that the slums, not the plant, would be relocated.

Even when union workers tried to mobilize their support in pressing for greater safety standards, residents felt the security of their homes as the greatest risk. They saw the workers as an elite group trying to protect itself.

Ironically, in the elections just preceding the disaster, these residents were given "pattas" (land rights) by a politician trying to pull their votes. Later, trying to justify these grants, the politician insisted that he never saw them as a "final solution." As "dirt," residents of the slums adjacent to the Carbide plant should have been displaced. As possible vote banks, they had to be consolidated.

Most gas victims do not participate in these interpretations. They may recall why they came to Bhopal or why they were unwilling to challenge the operations of the plant in their midst. Some even know the story of how the lakes of Bhopal were built. Or the story of Shah Jahanabad, whose name identifies one of the colonies west of Berasia Road. But they haven't been given speaking parts. They lack idioms of explanation because the lines do not belong to them. They are sucked into the narrations of others, as bit players. They are in their own story, as foreigners.

(6.2) From Confusion to Categorization

VOICES OF BHOPAL, 1990
BHOPAL GROUP FOR INFORMATION AND ACTION

Interview with Bano Bi, age 35, resident of Chhawni Mangalwara
The night the gas leaked, I was sewing clothes sitting next to the door. It was around midnight. The children's father had just returned from a poetry concert. He came in and asked me, "What are you burning that makes me choke?" And then it became quite unbearable. The children sleeping inside began to cough. I spread a mat outside and made the children sit on it. Outside, we started coughing even more violently and became breathless. Then our landlord and my husband went out to see what was happening. They found out that some gas had leaked. Outside there were people shouting "Run, run, run for your lives."

We left our door open and began to run. We reached the Bharat Talkies crossing where my husband jumped into a truck full of people going to Raisen and I jumped into one going towards Obaidullahganj. It was early morning when we reached Obaidullahganj. The calls for the morning prayers were on. As we got down, there were people asking us to get medicines put on our eyes and to get injections. Some people came and said they had made tea for us and we could have tea and need not pay any money.

Meanwhile, some doctors came there. They said the people who are seriously ill had to be taken to the hospital. Two doctors came to me and said that I had to be taken to the hospital. I told my children to come with me to the hospital and bade them to stay at the hospital gate till I came out of the hospital. I was kept inside for a long time and the children were getting worried. Then Bhairon Singh, a Hindu who used to work with my husband, spotted the children. He, too, had run away with his family and had come to the hospital for treatment. The children told him that I was in the hospital since morning and described to him the kind of clothes I was wearing.

Bhairon Singh went in to the hospital and found me among the piles of the dead. He then put me on a bench and ran around to get me oxygen. The doctors would put the oxygen mask on me for two minutes and then pass it on to someone else who was in as much agony as I was. The oxygen made me feel a little better. The children were crying for their father so Bhairon told them that he was admitted to a hospital in Raisen. When I was being brought back to Bhopal on a truck, we heard people saying that the gas tank had burst again. So we came back and went beyond Obaidullahganj to Budhni, where I was in the hospital for three days.

I did not even have a five paisa coin on me. Bhairon Singh spent his money on our food. He even hired a taxi to take me back to Bhopal to my brother's place. My husband had come back by then. He was in terrible condition. His body would get stiff and he had difficulty breathing. At times, we would give up hope of his survival. My brother took him to a hospital. I said that I would stay at the hospital to look after my husband. I still had a bandage over my eyes. When the doctors at the hospital saw me, they said, "Why don't you get admitted yourself, you are in such a bad state?" I told them that I was all right. I was so absorbed with the suffering of my children and my husband that I wasn't aware of my own condition. But the doctors got me admitted and since there were no empty beds, I shared the same bed with my husband in the hospital. We were in that hospital for one and a half months.

After coming home from the hospital, my husband was in such a state that he would rarely stay at home for more than two days. He used to be in the Jawaharlal Nehru Hospital most of the time. Apart from all the medicines that he used to take at the hospital, he got medicines like Deriphylline and Decadron from the store. He remained in that condition after the gas disaster. I used to take him to the hospital and when I went for the Sangathan meetings, the children took him to the hospital. He was later admitted to the MIC ward and he never came back. He died in the MIC ward.

My husband used to carry sacks of grain at the warehouse. He used to load and unload railway wagons. After the gas, he could not do any work. Sometimes, his friends used to take him with them and he used to just sit there. His friends gave him 5–10 rupee notes and we survived on that.

We were in a helpless situation. I had no job and the children were too young to work. We survived on help from our neighbors and other people in the community. My husband had severe breathing problems and he used to get into bouts of coughing. When he became weak, he had fever all the time. He was always treated for gas related problems. He was never treated for tuberculosis. And yet, in the post-mortem report, they mentioned that he died due to tuberculosis. He was medically examined for compensation but they never told us in which category he was put. And now they tell me that his death was not due to gas exposure, that I cannot get the relief of Rs. 10,000 [$400] that is given to relatives of the dead.

I have pain in my chest and I get breathless when I walk. The doctors told me that I need to be operated on for ulcers in my stomach. They told me it would cost Rs. 10,000 [$400]. I do not have so much money. All the jewelry that I had has been sold. I have not paid the landlord for the last six years and he harasses me. How can I go for the operation? Also, I am afraid that if I die during the operation, there would be no one to look after my children.

I believe that even if we have to starve, we must get the guilty officials of Union Carbide punished. They have killed someone's brother, someone's husband, someone's mother, someone's sister; how many tears can Union Carbide wipe? We will get Union Carbide punished. Till my last breath, I will not leave them.

The evening of December 2 seemed no different. It was a Sunday. Men went to films or hung around tea stalls—talking, still, of assassinations, riots, and Rajiv Gandhi, about events and worlds from which they were thrice removed. Delhi was the center of these worlds; Bhopal was a stop on the rail line that took you there. Some knew the distance precisely: 705 km between Bhopal and Delhi; 12 hours by rail, at a cost of approximately $4. The night train was the best way to go. Board around 11:00 P.M.; reach Delhi by mid-morning. All those living in the railway colony knew the rhythm, as did the rickshaw drivers and all those who ran small shops and kiosks alongside the tracks.

Children played cricket in the alleys and flew kites from the roofs. Squares of blue, green, and red dotted the sky. Falls all the way to the ground were rare. The kites would be caught on rooftops or between buildings. Children dashed in and out to retrieve them.

Women continued the cooking and cleaning. Most wore extra shawls or sweaters to buffer the cold. Throughout the day, Hindi pop music blared from tea stall loudspeakers, interrupted only by the call for prayers. By evening, the streets became quiet. Workers came and went from the ten o'clock shift at the Union Carbide plant.

No alarm announced the presence of deadly gas in the homes and streets of Bhopal. People awoke thinking neighbors were burning chili peppers. When it became impossible to breathe, they fled into dark streets. With no information on what was happening, they simply ran with the crowd, catching rides on passing vehicles whenever possible. Families were separated, old people and children abandoned. Many ended up on the outskirts of the city or beyond, without money and only the company of strangers.

But it was only when the sun rose that the magnitude of the disaster became apparent. Dead bodies of people and animals blocked the roads. Vegetation was yellowed and shriveled. A smell like burning chili lingered in the air. Meanwhile, Carbide officials had informed hospitals that methyl isocyanate (MIC) was like tear gas and could be treated with oxygen, antacids, and water washing of eyes. Police were making rounds of the city, informing people that the danger was past and asking them to return home. But no one believed them. Already there were rumors that the police had started dumping bodies in the river in a conspiratorial effort to hide the devastation.

Rajiv Gandhi announced that it was safe for victims to return home before an official assessment was made. Meanwhile, it became known that food was being brought in from outside for officials and other elites. A hard, green crust had appeared on the surface of stored food, but people were told that there was no "scientific reason" for concern that it was contaminated. A helpless dread spread throughout the city, with people very conscious of their lack of control over the continued risks.

When victims approached the Carbide factory for medical aid and information, desperate managers afraid of rioting suggested that another leak might soon occur. This set in motion a second mass flight and intensification of the uncertainty. People continued to die, but there was no information on possible antidotes or remedies. Clearly, the effects of the gas were not only temporary, and there was increasing fear of long-term consequences. If they lived, would their sight be restored? Would they be able to breathe easily? How were they to earn a living if disabled? What about the effects on unborn children?

These questions were met with silence. Voluntary workers had flocked to the city to render any aid possible, but they were barred from official information sources. The clampdown reached from the highest city authorities to the ward nurses at government hospitals. One doctor, when queried about the rate of spontaneous abortion, said that officially he couldn't even tell his own name.

Exclusion from information and resources necessary for medical relief has characterized victim experience of the Bhopal disaster. So has exclusion from the law and other forms of political authority. Victims have been constructed as juridically incompetent and denied the right to opt out of representations made in their name. They have been told to prove their suffering. They have been asked to be forgotten, to make way for an "increasingly normal Bhopal."

(6.3) Statutory Violence

Beware.
That Night is Back
Once again murderous December is upon us
The Wind turns cold and yet we sweat,
We sweat with fear as the wind turns cold

And the city starts to wear a familiar look,
The air of that terrible night, the night of poison
When the death merchants rifled our guiltless people.
Coughing, shrieking, wailing, screams,
A city sobbing, drowning in noise,
Some imploring, others blind with tears
Blinded to one another's plight,
Stepping over each other's bodies in their panic
To escape alive from the claws of death,
Poor helpless and rich, all shared one fate.
The weak lay like logged trees along the roads
And mothers nursed dead babies:
Oh what those cruel gases did in one night
Sparing neither human nor animal,
Trees, plants, flowers, leaves—all withered, all died—
The killer brand burned whatever it touched.
Still from silent burning grounds the dead
Whisper loud warnings:
"No-one should have to die such painful deaths."
"Beware the devotees of Mammon,
Watch out for those who thrive on war,"
Come voices out of the silent graveyards.
Once again murderous December is upon us
We sweat with fear as the wind turns cold.
     Rafatul Hussaini (poet and gas survivor) wrote this "Ghazal"
          for the 10th anniversary of the disaster.[1]

Outside the gates of Union Carbide's Bhopal plant there is a statue of a woman running with her children. Her face is uncovered. Her fear and grief are marked in stone. Victims say they like the statue because it holds things still and allows them to remember. The woman is separated out from the crowd, recognizable as an individual, even if she doesn't have a name. She marks a refusal of privatization, a challenge to the state's insistence that sorrow be borne alone, managed through individualized claims that offer "cash for one's children, and purdah [veils] for truth."

These sentiments are voiced with a deep cynicism and with a savvy awareness of dislocation. As one victim explained, "[T]hey come to our door, fill out their papers, say that we are being counted . . . but they speak behind the veil, knowing that we can not see their faces or recognize their words." Another victim stated the problem more bluntly: "[T]hey count us, give us identification cards and medical categories, but tell us nothing we don't already know . . . we are a problem—that they will fix in their big offices with numbers and papers."

Bhopal has become a city of paper. Faded, watermarked forms are tick-

FIGURE 6.3 STA-
TUE   The concrete
statue of a woman
running with her chil-
dren that stands out-
side the gates to the
Union Carbide plant
in Bhopal. Photo by
Mike Fortun, 1996.

ets to any possibility of compensation. Gas victims carry these papers with
them, laced together with string, clutched to their chests as they board
crowded buses, moving between the Collectorate, the hospitals, the claims
courts, their homes. Traffic in hope.

Most cannot read the papers they protect. But most can explain their
meaning. When you meet them, they want you to look—to see for yourself—the prescriptions, the proof of a hospital stay, the ultimate categorization. "Read this to me." Perhaps the most repeated phrase in Bhopal, a
response to a rehabilitation scheme that has traded hard copy for cash compensation, banking on literacy as a scarce resource.

(6.4) Depression and Paralysis

Many mourn the fact that gas victims have been denied the opportunity to
mourn. Those who died in the immediate aftermath were dumped into mass

graves or the Narmada River, or burned in mass crematory fires. Many bodies were never identified. Families of the deceased never knew what happened or where bodies were left to rest. There was no punctuation, no ritual to recognize entry into a new scheme of things.

Lack of punctuation continues to characterize everyday life in Bhopal. Because the dead were never properly buried or burned. But also because of the ways toxic chemicals inhabit the body. There is no closure on toxic exposure. It can kill quickly, with recognizable signs of tragedy. Vomiting. Choking. Respiration foreclosed by water, which has filled the lungs in a desperate attempt to protect sensitive membranes.

Most, however, die slowly. Debilitation becomes a way of keeping time. So the paralysis now visible in Bhopal seems significant on many counts. I heard about it again and again when I returned to Bhopal in 1996. It evoked a deadening stasis and depression—an inability to move, operating on both figurative and physical registers.

The paralysis develops progressively, often starting with tingling in one leg or a hand. With time, the whole left side of the body or everything from the waist down is numb. I was shown the certificates, where this paralysis had no register. I was told of other people whose bodies now showed small signs that they, too, were on a downward spin. Limbs that fall asleep if left idle for even a moment, burning muscles, twitches suggesting that the body was operating according to a logic all its own.

I don't know if paralysis of the bodies of gas victims is of statistical significance. I saw it everywhere. Perhaps because of the ways victims who served as my guides directed my eyes. Perhaps because of the resonance with which paralysis seemed to embody all of Bhopal, operating allegorically to signal more than what I could take down in my notebook or hold within my imagination.

(6.5) Refusals

Many victims of the Bhopal disaster call for revenge. The children of Bhopal are particularly merciless. An eye for an eye. A tooth for a tooth. Their stories compounded the violence of their slogans. Sunil, age eight in 1984, lost nine members of his family in the disaster—his parents and seven siblings. Another brother and a tiny sister were left in his care. Rajiv tells of a game he and his friends play. All run about desperately, in all directions, with their eyes tightly shut. Then someone calls "gas," and everyone falls down dead.

The elders of Bhopal also focus on the need for punishment. Their graffiti peppers the external walls of the Carbide plant, demanding that Warren

FIGURE 6.4 CHOUHAN'S HANGING PEOPLE   T. R. Chouhan's interpretation of what science and technology, and Union Carbide's Bhopal plant in particular, have offered the people of India. Painting by T. R. Chouhan.

Anderson, CEO at the time of the disaster, be hanged. For the annual commemoration of the disaster, participants first visit the cremation grounds
and then the Muslim burial site. Last, they convene at the plant gates. An
enormous effigy of Anderson is brought out for display and immolated. It
is a great honor to be chosen to light the fire.

Some observers criticize the bald violence of these words and actions.
Others insist that the calls to violence are symbolic, providing a way to
express sorrow and suffering in a context wherein no other language is
available. Still others insist that the calls to violence should be heard literally, as due in a world in which corporate misconduct has proceeded without accountability for far too long. I hear the calls to violence as a response
to suffocation, as a desperate effort to find words that could produce room
to breathe. Graffiti is one of the more permanent interventions victims can
make. Ironically, it inscribes their desire *not* to be represented. In these
words and gestures are critiques of all available modes of describing disaster, whether official or oppositional. Victims now tell me that they don't
want their own stories to be told. They don't want photographs, and they
don't want their names mentioned. They demand that we find some other
way to remember Bhopal.

(6.6) Maelstroms

There were communal riots in Bhopal in December 1992, just following
the eighth anniversary of the disaster.[2] By official accounts, 139 people
were killed, at least 50 in police firings. Interpretations of the cause of
the riots in Bhopal vary. The crudest provocation was the destruction of
the Babri Masjid at Ayodhya by *kar sevaks*—Hindu fundamentalists who
claimed that the mosque was built on the site of the birthplace of the Hindu
god Ram. Controversy over the site had brewed throughout the 1980s. On
December 6, 1992, a mob of 300,000 demolished the mosque, provoking
riots across India. *India Today* reported that casualties included 1,700 dead
and 5,500 injured (Thakur 1995, 17). In broader terms, the riots were provoked by a general communalization of politics, fanned by the Bharatiya
Janata Party (BJP). A BJP government was established in Madhya Pradesh
in 1990.

Few expected communal violence to reach Bhopal, least of all government administrators—who called an emergency meeting to discuss law
and order issues related to the Babri Masjid, only to discuss other cities,
such as Indore and Jabalpur.[3] Bhopal is known for its history of communal
harmony, reaching back to the Nawab period. Even in the recent past, communal violence had never reached Bhopal. While I was still in Bhopal BJP

rallies were common, rife with anti-Muslim slogans and speeches detailing how Muslims covertly sympathized with Pakistan, threatened the demographic balance by having too many children, and refused to integrate with Indian culture—especially the culture of religious toleration, best represented by Hinduism. I was always told that these displays of fundamentalism were brought to Bhopal by outsiders and wouldn't take. The Bhopal Gas Affected Working Women's Union was especially proud of its diverse membership, occasionally suggesting that the organization should provide a role model for others. But members of the Women's Union did not expect communal hostility to gain momentum in Bhopal itself.

The riots started on the morning of December 7, 1992, shortly following cable television coverage of the destruction at Adyodhya. The first violent incidents were in Old Bhopal, deep within the gas-affected areas. Muslim youth attempted to enforce a *bandh* (shop closure) and then proceeded to attack shops, vehicles, and government property. Their weapons were country-made bombs, rifles, swords, and simple rods. In Jahangirabad, near the home of Jabbar Khan, a police constable was killed, and police opened fire. Rumor fueled the momentum. Word circulated that arms had been stockpiled in Tajul Masjid (the largest mosque in Bhopal) and that truckloads more were en route from Indore, along with a tanker of gunpowder. Pilgrims lodged at the Masjid, awaiting the annual Ijtema rituals, were identified as Pakistani agents. Later these pilgrims were traced to the Philippines. Doctors confirmed that none of the injuries they treated was carried out by sophisticated arms.

Curfew was imposed around 10 A.M. on December 7, stranding many in schools and places of work. The separation of families increased apprehension, turning rumor into an ever more volatile force. By the evening of December 7, newspapers were confirming these rumors, running headlines that escalated death tolls in Bhopal as well as the number killed in police firings in Adyodhya. These accounts were authenticated in ways familiar to Bhopalis: "According to special correspondent Mark Tully," of the British Broadcasting Service (BBC) (People's Union for Democratic Rights and Sanskritik Morcha 1993, 4). Tully has visited Bhopal often and has produced commendable reportage on the continuing disaster for gas victims, based on interviews with affected people. His programs have kept the Bhopal story on the air in Britain and circulated as classroom films in the United States.

Despite the curfew, December 7 turned into a horrible day. Some claim that enforcement of the curfew was purposefully ignored to allow Muslims time to perpetrate enough violence to justify Hindu retaliations. In any case, the violence continued to escalate, particularly along Berasia Road—

which extends north of the gates of Old Bhopal toward Union Carbide. Shops along the main road were the first targets, then homes and neighbors. Incidents at Bafna Colony had a particularly tragic, ironic twist. The colony had grown in recent years to a population of nearly a thousand; some attribute the growth to the colony's classification as "gas affected," thus offering some promise of compensation. The entry to the colony, just behind a complex of shops facing Berasia Road, was lined with homes of relatively affluent traders and salaried workers. As one moved deeper into the colony, homes became smaller, darker, and more densely arranged; the homes of Hindus abutted those of Muslims. On the morning of December 7, Muslims gangs coming from Kazi Camp and Kabad Khana entered the area armed with petrol bombs, swords, and knives. Their target was property, which they burned or looted. Hindu women took refuge in the homes of Muslim neighbors while their houses sizzled. Some Muslims' houses also burned. Most argue that the Muslim houses were damaged simply due to proximity to the homes of Hindus. Others claim *kar sevaks* who came into the colony later were responsible. Still others claim that Muslims set fire to their houses themselves in order to receive compensation.

Search operations began in the early morning hours of December 8. Police went house to house confiscating weapons, arresting men, and, by many accounts, "misbehaving" with the women—many of whom were alone, since fathers, husbands, and sons had fled to avoid arrest. In many cases, women remained alone for almost a week, or up to a month if the men of the house were in jail. Intense claustrophobia mingled with an extraordinary desire for safe places, cut off from the world outside. Police were as feared as the roving gangs. Rape was the biggest threat. Looting of precious household items, such as cooking vessels, sewing machines, and bicycles, was also feared. Even simple maintenance of daily life became traumatic; if there had been any milk to start with, it was all gone by the end of the first day of curfew; food stores quickly diminished; trips to communal water pumps became hazardous, if possible at all.

In the first two days, 24 people died, 150 cases of arson were registered in 13 places, and 350 people were arrested. Of the 24 dead, 17 were killed in police firings.

By the morning of December 9, press reports were even more provocative, with headlines proclaiming an "Attack on girl's hostel." By evening, the lead claimed that "Hostel girls flee to save their honor." The story began at Marharani Laxmibai College in Budhwara, a Muslim locality on the north bank of the Lower Lake. When the curfew was imposed, many of the 3,000 students, all women, had reached the campus. Almost a third had come from the BHEL industrial area across the city.

Some of the students lived nearby and were able to reach home; others were transported across the lake to the college hostel in Professor's Colony, in New Bhopal. These students traveled on a boat operated regularly by the college and thus did not have to cross the Old City to reach their destination. Rumors about their fate escalated nonetheless. The details were lurid: eighty girls had been raped, their breasts cut off, mass molestation. News coverage of the story in the Bhopal city edition of *Nav Bharat* circulated to over 16,000 households, though not in Old Bhopal itself because of the curfew. Later analysis of this coverage by civil liberties groups acknowledges its import, describing how, particularly during communal riots, "what is believed to have happened . . . becomes the cause of what actually happened" (People's Union for Democratic Rights and Sanskritik Morcha 1993, 1).

The molestations of college girls described by newspapers did not happen. Members of the civil liberties investigation team confirmed this with a senior official at Marharani Laxmibai College. Yet the story continued to operate with horrendous force. The same official also reported a chance meeting with a man in a police lockup, charged with burning seven members of a Muslim family to death. The man's refrain: "Let me out for just one day, I will burn another hundred Muslims. They have dishonored our sisters" (People's Union for Democratic Rights and Sanskritik Morcha 1993, 32).

By December 9, much of the violence had shifted away from Old Bhopal to newer parts of the city. In Chandbad, a sixteen-year-old girl was stabbed repeatedly. In Acharya Narendradev Nagag, a pregnant woman was stabbed. In Piplani, two women were burned to death after having been stripped and threatened with rape. Eleven women died in the Bhopal riots; in three cases, postmortem reports confirmed rape (People's Union for Democratic Rights and Sanskritik Morcha 1993, 32).

Sometime during all this, the Jumerati Darwaza—one of the gates marking the traditional perimeters of Old Bhopal—was completely burned. The Darwaza became a protected monument in 1956, but never received the attention of historical preservation. Nor did it acquire a communal identity—until the early 1990s. By 1992, the Darwaza was associated with Muslim rule by both Hindu and Muslim communities. Shops in the bazaar that surrounds it were attacked by Muslim youth on the morning of December 7; Sindhi shops were especially targeted. Later, Hindu mobs retaliated, accessing the top of the Darwaza from the upper stories of a nearby house. Police involvement in the destruction was clear.

Many blame the BJP for the politicization of the Darwaza, and for the riots more generally. Some specifically refer to the measures taken by the

Ministry of Urban Housing and Welfare in the name of city beautification. The Women's Union links the politicization of monuments in Bhopal to these slum clearances of 1990, directed at Muslim slum dwellers most known to vote for the Congress Party.

The operations of the relief apparatus in Bhopal cannot fully explain the riots, but they are an important part of the backdrop. The riots were, in part, an effect of ubiquitous cynicism about the role of law in society. This cynicism has escalated dramatically since the Indian Supreme Court upheld the settlement of the Bhopal case in October 1991. The settlement order reinitiated criminal proceedings, but a trial had not commenced. And it was widely known that the court-ordered extradition of Union Carbide CEO Warren Anderson had not been pursued by the Indian government. The October 1991 order also led to the establishment of local claims courts under the control of the state government, and hence the BJP.

Bhopalis, again, have been cast in scripted roles—as foreigners, caught up in plots that are not their own. Plots in which some people benefit much more than others. Plots in which police and government officials play major roles. Plots that revolve around brutal categorization schemes. The enduring effect is a tense, recognizably lawless Bhopal that is even harder to live in than before.

(6.7) *Flagrante Delicto*

I learned many things in Bhopal, many of them unexpected. I learned how to draw the piping configurations that lace a chemical plant together and how much it means when storage tanks are made of iron, rather than stainless steel. I learned to remember numbers and dates, aware that there is a politics to how we count and how we pass time. I learned to write on demand, knowing there was no time for habits of perfection, or procrastination. I learned how powerfully language inflects the world and how much it matters when we opt for one mode of description over another.

I also learned of the perilous work of crossing divides of class and culture. I learned that political organizations, no matter how well intended, are not immune from pressures that sustain the very social processes they seek to challenge. I learned how difficult, and important, it is to acknowledge complicity with much of what one critiques. I learned that questions regarding effective strategy are endless, and ever capable of imploding any plan of work. Should our work be in community organizing? At street protests? Within the regimes of law? In literary pursuit?

In all, I learned that the social space of contemporary activism is shaped by brutal double bind. It is a negotiated space wherein conventional models of ethical conduct fail. Conventionally, ethical conduct demands integrity, an adherence to moral principles that mime those of mathematics. Integrity roots itself in the integer, not the fractional. An integer is an undivided whole untouched by symptomatic indications of systemic failure. The integer, as root of conventional ethics, postulates logical coherence and a linear equation of the true and the real. To have integrity, one must believe in formalism, discounting all that undercuts established concepts and laws, universalizing the truth of one way of knowing.

My activism in response to Bhopal could not be launched with integrity. Our purpose in Bhopal was to mark dysfunction, forever insisting that the "whole story" offered by official calculations was fractionated from within. Our method was irregular and insisted on the extraneous. We argued for more, better science, while we collected stories and published life histories. We worked for local remedy that seemed at odds with fundamental and preventative reform. We relied on organizational structures continually touched by practices that exhausted our theories and divided our words from our deeds.

In all, activism in response to Bhopal has refused elegance and confounds rationale. Evaluation of work in response to Bhopal is therefore difficult. Others offer analyses, often within the rhetoric of risk management and tort law. Nothing much we did in Bhopal counts. The legal proceedings have generated scant resources for victims and have led to a plan of distribution that fosters schism within local communities. No liability has been established, and the structure of the case could set an insidious precedent for handling mass torts, particularly in the Third World. Attempts to build alternative social forms have been abandoned, or exhausted of critical momentum. In sum, most of what we advocated has not materialized. Like disaster, our advocacy can be read as an externality, without register in recognized accounting practice.

I refuse this accounting, as a matter of habit as well as of method. Victims insist that I "Remember Bhopal." Persistent refusal of conventional accounting practices and their strategies of exclusion is key. We must refigure what "counts," granting relevance to indicators that disrupt glossy promises of a harmonized global order. We must write in recollections of a past that shape a future more conversant with harsh opposition. The challenge is to translate between science and social progress, between law and justice, between democratic ideals and everyday pluralism. There is work in both scholarship and direct political engagement. Neither is sufficient; both are crucial.

blitical engagement must speak in a language sensible within the
utions being challenged. Opposition can be strategized, but is
nstrained by logics and rhetorics deeply imprinted with the mas-
of a powerful present. Scholarly language pursues different tactics, insisting that what isn't sensible within these master codes nonetheless
"counts." The logic and rhetoric of critical scholarship are rarely amenable
to the demands of court proceedings or the agenda setting of bureaucracies. Instead of engaging the present, this scholarship sets up the future—
differently.

Acknowledgment of my own everyday exclusions seems imperative.
While in Bhopal, I persistently wrote off dissent among victims and ignored the mismatch among agendas of anarchism, collectivism, and politics. As I continue writing about Bhopal as an ethnographer, other exclusions replace or reinforce those of the field. Some of these exclusions are
practical; some are theoretical. As I extend my research to focus on how
"Bhopal" has affected risk flow and resistance in the United States, I get
out of touch with daily struggles in India. As I begin to study how corporations construct their worst-case scenarios and other indicators of control,
I forget to read newsletters sent from the grassroots. As I teach young engineering students, I foreground possibilities for technological and managerial reform, relying on conventional faith in science as a social resource.

Awareness of these exclusions is humbling and at times paralyzing. Heroic images of scholars as activists without double binds madden as much
as they lure. Instead, I try to work an image of entanglement, with no hope
of exoneration. *Flagrante delicto.* A legal descriptor of being caught in the
very act of committing the offense. A useful descriptor of my experience
in activism. A useful reminder that my own theorizations should cause disease, reminding, in turn, that the Bhopal disaster is, in fact, a disaster of
thought, language, and sociality as well as of broken bodies and a contaminated environment.

SEVEN

Opposing India

(7.1) In the Middling

By April 1985, American lawyer John Coale had determined that families of every person killed by the gas leak in Bhopal could receive anywhere from Rs. 300,000 to Rs. 800,000—between $25,000 and $67,000. Coale placed at least 2,000 families in this category. People with catastrophic illness would receive anywhere from $8,000 to $33,000. Coale argued that this category included as many as 25,000 people. These figures were published in a Hindi-language handbill titled "Compensation Settlement for Gas Victims: Truth and Reality." Authorization came from Coale's claim that he had already negotiated a settlement in American courts. According to the handbill, the settlement was "the largest ever given in a personal injury case." If victims accepted the settlement, payment was promised "in a few months instead of years" (Jenkins 1991 [1989], 92–93).

The government of Madhya Pradesh responded to Coale's handbill with a handbill of its own, warning against the "temptation offered by suspect middlemen." These "selfish self-appointed middlemen" sought to wrest the welfare of gas victims out of the hands of the government by talking of a settlement of the Bhopal case, which was not possible because the Indian Parliament had passed a law that gave the government the exclusive right to represent victims. Victims must rest assured that their welfare "is secure in the hands of the Government," remembering that the middlemen in Bhopal had not been endorsed by the government of India, nor ever would be (Jenkins 1991 [1989], 92–93).

GOVERNMENT NOTICE
GAS VICTIMS—BEWARE OF THE TEMPTATION
OFFERED BY SUSPECT MIDDLEMEN

THE STATE GOVERNMENT'S ATTENTION HAS BEEN DRAWN TO-
WARDS THE ACTIVITIES IN BHOPAL OF SUSPECT MIDDLEMEN
WHO ARE CONNIVING TO GET A MEAGER COMPENSATION FOR
GAS VICTIMS—INSTEAD OF GETTING THEM A SUITABLE LEGAL
SETTLEMENT. THE GOVERNMENT HAS ALSO LEARNED THAT
THE GAS VICTIMS ARE BEING MISINFORMED BY THESE MIDDLE-
MEN THAT IF THEY DO NOT ACCEPT THEIR PROPOSITIONS, THE
CASES OF THE VICTIMS WILL BE DISMISSED FROM AMERICAN
COURTS. THIS MISLEADING STATEMENT IS BEING MADE TO
MAXIMIZE THE SHARE THAT MIDDLEMEN RECEIVE FROM ANY
SETTLEMENT.

THE GOVERNMENT OF INDIA HAS ALREADY FILED SUIT
AGAINST UNION CARBIDE IN AN AMERICAN DISTRICT COURT
AND FULL ATTENTION IS BEING PAID TOWARD IT. PARLIAMENT
HAS ALSO PASSED A LAW THAT GIVES THE RIGHT TO REPRE-
SENT GAS VICTIMS EXCLUSIVELY TO THE GOVERNMENT OF
INDIA. THESE SELFISH SELF-APPOINTED MIDDLEMEN WHO ARE
TALKING ABOUT A SETTLEMENT HAVE NOT BEEN ENDORSED
BY THE GOVERNMENT OF INDIA. NEITHER WILL THEY EVER
BE ENDORSED—BECAUSE THIS WOULD BE COMPLETELY INAP-
PROPRIATE. THE GOVERNMENT WILL ONLY ACCEPT A SETTLE-
MENT IN WHICH ALL GAS VICTIMS WILL BE FULLY AND APPRO-
PRIATELY COMPENSATED. THE WELFARE OF THE GAS VICTIM
IS SAFE IN THE HANDS OF THE GOVERNMENT. GAS VICTIMS
SHOULD NOT BE MISLED BY THE TEMPTATIONS AND ASSUR-
ANCES OF MIDDLEMEN. FOR OBVIOUS REASONS, UNION CAR-
BIDE WILL MAKE EVERY EFFORT TO REACH THE SMALLEST
POSSIBLE SETTLEMENT. FOR THIS REASON ALL PEOPLE AS-
SOCIATED WITH THIS SHOULD WATCH OUT FOR MIDDLE-
MEN WHO ARE TEMPTING THEM TO REACH SETTLEMENTS
BY WHICH THEY MAY LOSE THEIR JUST AND APPROPRIATE
COMPENSATION.

THE WELFARE OF THE GAS VICTIMS IS SECURE IN THE HANDS
OF THE GOVERNMENT. (Jenkins 1991 [1989], 92–93)

By the end of June 1985, the government, with the assistance of doctors, had identified only about 1,000 "severely affected persons" and had given each an average of $118. Approximately 14,000 identified as "moderately affected persons" received payments of approximately $16. By the end of

October, families of 1,499 of the dead had received about $830 in compensation (Shrivastava 1987, 93).[1]

Meanwhile, disaster had occurred elsewhere. On June 23, 1985, an Air India plane crashed off the coast of Ireland. Relatives of the approximately 300 dead received compensation of nearly one million rupees each, while the government of India mounted a multimillion dollar expedition to recover parts of the plane from the bottom of the sea (Alvares 1994, 116).

The voluntary sector in Bhopal was not immune to these disorders. Already fissures were surfacing, hinting at the structural faults upon which extended engagement with Bhopal would be built. Activists in the voluntary sector had to work within, even if against, the many schemes for categorizing Bhopal. Some worked to implement official schemes, providing medical relief or running job training programs to sections of victims that may have been excluded from benefits simply as a matter of scope. Responding to the limits of the government's organizational efficiency, they worked to extend the logic of official schemes beyond their practical reach. Others spent their time opposing, or circling, official schemes by producing data that highlighted their inadequacy, by developing "people's plans" for alternative forms of rehabilitation, and by mobilizing public support, both within Bhopal and beyond. But participation in the identification and typification of gas victims was unavoidable, even if one's argument highlighted a rate of spontaneous abortion left out of official calculations or the way ration cards benefited some people much more than others.

Like the American lawyers in Bhopal, voluntary sector activists were "middlemen." They worked between officially recognized parties, on a sliding spectrum surrounded by acrimonious debate. Their task was to facilitate flows of resources, including information, across different interests and perspectives, working at crossroads unmarked by signs indicating exactly where middle-class activists should stand. Some worked as extensions of the state, arguing that disaster was no time for political critique. Others worked more adjacently, critiquing the state because it was not playing enough of a role in Bhopal, denying the magnitude of injury as a way to deny responsibility. Still others argued that the disaster threw the fundamental illegitimacy of the state into high relief, demanding community governance of rehabilitation schemes, free of all government support.

Historically, there has been a recognizable role for India's middle class in helping the poor. "Upliftment" of the poor has been a national goal. The middle-class social worker had a legitimate place in various programs, within broad goals of development and national autonomy. Even if within "oppositional" political parties, the place of the middle-class activist was recognizable. Camaraderie within postcolonialism seemed both necessary and possible.

Middle roles are always difficult to define, but particularly so when once stable points of reference begin to move.[2] By the 1980s, such reference points for India's middle class were indeed fluid. The Delhi riots following Indira Gandhi's assassination in 1984 brought the atrocities of the Congress system to the capital city itself. Police stood by as Sikhs were dismembered or set on fire. Many people credit voluntary associations for providing the most humane response (Omvedt 1993; Kothari 1993). Talk of "a crisis of governance" in far-flung villages had been percolating to the surface of political commentary for some time. The Delhi riots were proof that the crisis was not only on the periphery.

Talk of "internal colonialism" also began to circulate with greater force, particularly with reference to the Punjab and Assam, where state repression gained extraordinary momentum. Meanwhile, identity politics defined the limits of the political spectrum, in anticaste movements as in right-wing efforts to promote "Hindu Raj." But the Congress Party continued to emphasize "national integration" even while moving away from socialist rhetoric to accept a large loan from the International Monetary Fund in 1981, with corollary mandates to move away from a state-directed economy toward the promises of the free market. Evidence of an "enclave economy" was already visible in the consumerism of the urban, largely upper caste middle class. Production of refrigerators, motor scooters, and televisions soared. The Maruti car, once scorned as a product of Sanjay Gandhi's dealings abroad, became a prized status symbol.[3]

Foreign money and technology secured a new place on the Indian landscape during the 1980s, as did the expertise needed to develop and consume it. Rajiv Gandhi, "Mr. Clean," brought his "computer boys" to power with him, along with plans to privatize the Indian economy and to promote the role of multinational corporations. Unity remained a key theme with Gandhi's emphasis on "akhandata," unbrokenness. Meanwhile, drought ravaged the Indian countryside, particularly in Gujurat, Maharashtra, Karnataka, and Tamil Nadu (Omvedt 1993, 187). Protests were blamed on "the foreign hand" of neoimperialism, often identified with the CIA.

It was in this context that the Bhopal disaster brought many middle-class progressives together, to work with gas victims, but also to work out new identities and strategies for the Indian Left. Most dispersed to other parts of India, often to other progressive projects, after a few weeks or months. Many stayed involved in the issues of Bhopal from afar, often keeping Bhopal as an important reference point, especially in work responding to negotiations over the General Agreement on Tariffs and Trade, and liberalization more generally.

Over the years, I have met many activists who worked in Bhopal during that first tragic year. Few that I met had worked in the more "establishment" efforts of the Lions Club or even the Self Employed Women's Association. Most had been involved in what came to be known as the "Morcha," a coalition of gas victims and activists from outside Bhopal, credited with both "unproductive disruption" and the creation of a new form of voluntarism, particularly suited for India's troubled times in the late twentieth century. Thus, even modes of opposition in Bhopal became understood as models—as categorical statements of progressivism in an era in which established blueprints were considered obsolete if not complicit with all that they sought to overcome.

For many, involvement with the Morcha was a troubling experience. Memories of their time in Bhopal seem marked by a doubled ugliness. They remember the suffering of victims and the inadequacy of all that they could offer them. They also remember ugliness within the collective middle-class effort to respond to Bhopal. This ugliness was more than an effect of inadequacy. It seemed to have something to do with entanglement, complicity, and the problem of being middle class in a country undergoing dramatic change, which could not be expected to benefit over 90 percent of the population.

These activists could speak frankly and even proudly about "the crisis of alternative politics" in India during the 1980s. They had a place within this crisis as bearers of responsibility to create "new political idioms" and "new organizational models." Bhopal somehow exceeded these mandates. Some would say that the problem lay in the disjuncture between Indian ways of knowing and the mechanistic logic of industrialization, opposing Carbide's Bhopal to a tribal Bhopal. This articulation worked well, but still couldn't grasp the dismay that so many voiced in their remembrance of Bhopal. Many simply shook their heads and admitted that they didn't know what had happened in Bhopal that first year, but that they would never go back. Others tried explanations based on particular personalities, bruised egos, or ambitions to be "the new Gandhi." Still others claimed that Bhopal became a backstage for practicing sexual liberation, with all expected melodrama. One woman insisted that Bhopal caught activists in "the death throes of Leninism." Few I spoke with thought any of these explanations were sufficient in themselves.

Like environmental politics elsewhere, Bhopal entangled the middle class in particularly personalized ways. Many were carriers of the expertise that had brought about and legitimated what was now labeled "destructive development." They also were the beneficiaries of such development, even while able to turn their expertise into a tool of opposition. Some activists admitted the compromise of their position, acknowledging their resemblance to those they challenged. They spoke the same language. They were experts in the very forums that they blamed for creating exploitation. The challenge was to use their resemblance as a means of resistance, subverting from within, forging an appropriate technology of expertise. Many insisted it was a dubious battle, even while engaging it. Others insisted that resistance through partial resemblance had a great history in India, reminding me of colonial fear of hybridization so evident in Kipling's Eurasian figures. The children of English officials and Indian women posed a real threat, undercutting the distinction between ruler and ruled that justified the Raj—even though the lovers themselves always suffered terribly for their transgression (Kipling 1929 [1891]).

(7.2) To Be an Environmentalist

What is it to be an environmentalist when the label is shared by George Bush and a tribal from the hills of India? Is the mineworker who cooks his rice with firewood from ancestral lands a "forest thief," as conservationists contend? Where do gas victims of Bhopal fit within an ecological imaginary?

These questions were asked in Bhopal and elsewhere in India during the late 1980s, as part of a broad effort toward indigenous environmentalism. They index a struggle over territory and language that has realigned relationships among the state, "social action groups," and "the people." They suggest how environmental politics provoke questioning of traditional Leftist strategies and their vocabularies of opposition.

The case of Sitaram Sonvani is suggestive. Throughout the 1970s and 1980s, Sitaram organized tribal communities in eastern Madhya Pradesh to protest the abuses of the Forest Department. Saving the forests from the devastation of commercial timbering was their only chance of survival, so he advised them to take the "necessary steps" to reclaim it. The tribals of Madanpur and Rampur responded. A reserved forest area that they had cultivated for twenty-five years was "sold" to a contractor by villagers imagined and invented solely for the purchase. The contractor moved in with teak planting. The tribals removed the plants. On the morning of August 4, 1990, seventy-five young saplings were carefully dug up and set aside to be replanted on the borders of the property. Then the police came and the tribals were arrested, beaten, put in their place.

PEOPLE'S UNION FOR CIVIL LIBERTIES
Raipur, Madhya Pradesh/September, 25, 1990
The Report of a Fact-Finding Team
ARREST OF AN ENVIRONMENTALIST AS A "FOREST THIEF"???

Introduction

Sitaram Sonvani, a social worker from Gariaband Tahsil of Raipur District in Madhya Pradesh was taken for interrogation by the Police Inspector S. R. Sahu and Forest Ranger N. P. Upadhyay on September 10, 1990 around 3 P.M. from village Piparcheri, where Sonvani had been living for the past five years. His colleagues were assured by these officials that they were taking Sonvani only for discussions; and that he would be brought back by them on time for participation in the celebrations for Archya Vinoba Bhave Jayanti on 11th September—for which preparations were being made by village-folks and social workers of the area.

It was only when Sonvani did not turn up that an extensive search began for him. The obvious place to inquire about him was the Police Station. Arjun Singh Sidar—a co-worker of Sonvani—had to swim across the flooded river to reach Gariaband; only to be told by Thanedar Sahu that "the police was not the keeper of Sonvani" and that "he would be put behind bars if he did not stop asking embarrassing questions to the police."

Left with no other option, Arjun Singh Sidar came to Raipur and contacted other social activists, including the Organizing Secretary of the National People's Union for Civil Liberties (PUCL). Representatives of about 14 non-governmental organizations met and submitted a Memorandum to the district officials at Raipur. Suspecting foul-play by the forest and police officials of Gariaband, a search party of social activists was also dispatched to Gariaband.

It was, however, with the co-operation of the high officials that Sonvani could be traced to the Raipur Central Jail. It was also informed that Sonvani was arrested under IPC Sections 447 (criminal trespass) and 37 (theft). A team of social activists then met Sonvani in Jail, and collected background information.

A group of about 25 social action groups and trade unions then met on September 15 at Dalli-Rajhara (Durg District) to take stock of the situation arising out of Sonvani's arrest, and chalk out a strategy for future action. . . .

In the meantime, Sonvani, in the true tradition of Gandhi Ji—whose philosophy has been instrumental in motivating him for social action—decided to undertake fast-unto-death in Jail from September 20. . . .

Background

Garjai Pani Reserved Forest Area is located at the borders of Patwari Halka Numbers 17 & 18 between Madanpur and Rampur villages of Gariaband Tahsil. This reserved forest area has been without any plants and trees for the past 25 years. The tribal-peasants (belonging mostly to Gond & Bhunjia tribes of the area) have been cultivating the land for the past 25 years or

so. Even today, the boundaries of the farm-plots are prominently standing
4½ feet high. But, no tribal-peasant has been provided with any legal rights
to cultivate this land, as yet.

Sr. B. D. Sharma, Chairman, Scheduled Castes & Scheduled Tribes
Commission of the Government of India had visited the land in question
in 1988–89. There are photographs available showing B. D. Sharma, along
with the local officials of Gariaband and Sitaram Sonvani with the tribals
cultivating this land. The officials agree that this was a strong case of allot-
ting the land to these tribal-peasants. Several Petitions, Memorandums and
Prayers were submitted in the past five years under the leadership of Sitaram
Sonvani from local to district to state level.

In spite of it, the Forest Ranger Sri N. P. Upadhyay of Chura Forest Range
used the 100 acres of land for Teak Plantation in 1989, covering this piece of
land being cultivated by tribal-peasants for the last past 25 years.

This infuriated the tribal-peasants, who made several representations to
the officials and politicians. But, to no avail. . . .

The tribal-peasants of Madanpur & Rampur went to Sitaram Sonvani for
further advice, as he was instrumental in awakening them about their rights
on the forest. As an environmentalist, Sonvani re-iterated his earlier position
about the harmful effect of commercialization of forest over the neglect of
tribal life. Sonvani also clarified his position on plantation of teak, eucalyp-
tus, and pine. He advised them to take necessary steps, even the path of agitation, to reclaim their ancestral land for the only means of livelihood they
were left with.

Coupled with this advise against the backdrop of the Chief Minister's
announcement, a group of about fifteen tribal-peasants went to Garjai Pani
area, and started clearing of the teak plants—about a year old. In fact, they
had removed about 75 plants carefully and kept these aside with the intention to plant these later on the boundary of the land. The plants kept now in
the Police Station at Gariaband stand witness to this fact that the tribal-
peasants had no intentions to destroy the plants.

It is at this time that the forest officials under the leadership of Forest
Ranger P. N. Upadhyay reached the spot and caught every one of them. The
agricultural implements and 75 teak plants were also confiscated. The tribal-
peasants (of whom five were women) were dumped in one jeep and taken to
the Forest Depot at village Parsuli.

The tribal-peasants were booked under IPC 447 & 379, and produced
before the Magistrate at Gariaband. Fourteen out of these 16 were released
on bail on 6th August, and the remaining two on August 31, after they lan-
guished in Raipur Central Jail for more than 25 days.

The forest officials kept the arrested at village Parsuli inside the Forest
Depot for almost two days and two nights. This time was utilized to torture
them, harass them and conspire to implicate Sitaram Sonvani and Sumitra
Bai, the two social workers of repute.

The forest officials chose one Bhagi, alias Bhage (in fact Bhagirathi of
Madanpur village) to falsely implicate Sitaram Sonvani. Bhage was beaten
up badly. According to eyewitnesses, his head was banged against the mahua
tree inside the Forest Depot. Bhage was so badly shaken up that he ran away

under the cover of the night on the pretext of easing himself out. Bhage is missing since that day. His wife and social activists have been holding the forest officials responsible for his disappearance! According to one version, police and forest officials have purposely kept him in hiding because they do not want the tales of their tortures to be carried far and away. If he is shown to be absconding, the police have a case to delay the production of chalan in the court after 90 days. Thus, gaining ground to falsify the case.

Similarly, Sumitra Bai continues to live in village Madanpur as usual. But, the police are recording her to be absconding. . . .

What Are Sonvani's Crimes?

Sitaram Sonvani is an undisputed follower of Gandhian principles. He is being considered an environmentalist amongst the people and social action groups. Born in village Gurur of Durg district of Chattisgarh, Sitaram belongs to a family of landless. During the past five years, Sitaram Sonvani has been working in Gariaband Tahsil among the kamars and bhunjia tribes. The kamars—bamboo workers—are at the verge of extinction because of the loss of their traditional trade of bamboo products depending on forest produce—bamboo, which is rarely available to them in spite of a special Government Project called Kaman Project.

About 200 families of kamars have been engaged in self-employment schemes in decentralized units to enable them to gain self-reliance. Sitaram Sonvani started and supervised this project. Under the guidance of Sitaram Sonvani, 6 primary and one middle schools are being run through village level committees. Literacy campaigns in 65 villages, collective farming and fisheries are other development efforts being promoted in the area under the able guidance of Sitaram.

Sitaram is also responsible for initiating jangal bachao-manav movement on the pattern of "Chipko movement" of Sunder Lal Bahuguna. It is the women of Maragaon village who led the Chipko movement in 1989 in Gariaband Tahsil. . . .

An investigation into the life and work of Sitaram Sonvani will bring to the fore many more examples of conflicts and confrontations with the local officials and business interests, who are out and out to destroy the good work of social and economic upliftment of the poor and the oppressed.

Conspiracy

A criminal combine of forest and police have conspired to falsely implicate Sitaram Sonvani and his close associates in criminal cases. . . . (People's Union for Civil Liberties 1990)

Sitaram saw the Mandanpur tribals being taken to jail as he sipped his tea at a small kiosk a kilometer from the village. While they had been uprooting teak saplings, he had been 16 kilometers away, at a typing center, preparing a report of a meeting held the night before. Eyewitnesses confirmed his whereabouts. Sitaram was arrested, nonetheless. He was released six weeks later, after fasting for twenty-five days. The police became afraid he would die in custody. Then thousands of kamars demonstrated in front of the District Tribal Welfare Office. Only then was Sitaram released. No valid reason was ever given for his detention. Official guardians of the forest were not reprimanded.

Sitaram was arrested because he was a catalyst. His "crimes" were many. In 1975, he formed a cooperative of tribals displaced by the Seekaser Dam. The tribals received no benefit from the dam, yet were forced to pay the irrigation tax. The cooperative took control of a small yellow earth mine once leased out to an influential contractor.

Then Sitaram helped village Mahyabhata build a grain bank to store surplus so that drought did not cause dependence on moneylenders. Next he helped kamar tribals with self-employment schemes to circumvent their loss of traditional work—after conservationist legislation had cut off their access to the bamboo products of the forests.

The list continues. Sitaram became a nemesis of environmentalism. But, with time, he also insisted that he was an environmentalist himself. Tribal rights, practices, and knowledges became a counter to conservationism. India's environmental justice movement gained momentum.

Many activists say the movement first really came together at Harsud, one of the villages slated to be submerged by the Narmada Dam, where there was a mass rally in September 1989. The protest was against "destructive development" generally. Participants came together again in Bhopal for the fifth anniversary of the disaster. A loosely organized coalition called Jan Vikas Andolan (People's Development Movement) was formed, and a series of meetings was organized to work out an "alternative vision." These meetings garnered high hopes during the years I was in Bhopal. They were always riveted by debate, but the sense that a new form of environmentalism could be built from the grassroots was strong. The goal was to reorient the Indian Left away from the distribution of benefits and around the distributions of costs—of resource degradation in particular. Bhopal fit within the emerging imaginary uneasily, even if stubbornly. Bhopal was the sign of what should not have been. It was difficult to turn Bhopal into a site of possibility. Rural and particularly tribal ways of life harbored the ideal.

On the ground, however, "the tribal" perspective was complex and open-ended. Chattisgarh Mukti Morcha (CMM) provided an inspiring example. CMM carries a red-green flag to signify worker-peasant unity. It is based among low-caste and tribal miners and peasants in eastern Madhya Pradesh—the same area where Sitaram works. CMM was built into a powerful organization under the leadership of Shankar Guha Niogi, who had an unblemished reputation and savvy organizational sensibilities. Niogi was assassinated in September 1991 as the CMM was preparing for a strike to contest the treatment of contract laborers in the mines.

The militancy of CMM throughout the 1980s led to concrete gains. It also led to brutal repression. But CMM went beyond agitation to become a "social movement union." The broadest commitment was to connect the issues confronted by industrial workers with the issues of the surrounding countryside. Struggles to bring about land reform and to maintain prices for rice and forest produce were linked to opposition to mechanization of the mines, to industry monopolization of water supplies, and to promotion of a general need for health and educational services. The oppression of women was explicitly articulated. So was the need for "alternative development."

The CMM borrowed from different cultural and ideological traditions to respond to complexly interconnected problems. Historically entrenched ways of organizing issues and people were retrofitted. Issues and people once left separate were brought together. It was not a harmonic model. It pulled environmentalism into the contradictions, needs, and possibilities of the grassroots.

Members of CMM came to Bhopal fairly frequently while I was there in the early 1990s. A large contingent always came for major rallies and for the anniversary. There was a commitment to build regional solidarity. There also was a sharing of possibilities. CMM had "given a new meaning to green." It also had figured out how to build organizational strength without homogenizing its members. And it had built new institutions to materialize its commitments. Miners in Chattisgarh operate a hospital. It serves people poorer than the miners themselves. And it involves middle-class professionals in ways that rely on, but rearrange, their expertise.[4] Bhopal still reaches for something similar.

(7.3) Social Problems, Technical Solutions

The state of Madhya Pradesh is not known for its radical traditions, as are states like Maharashtra, Bengal, and Gujarat. Nonetheless, voluntary efforts to combat the suffering and uncertainty in Bhopal came quickly. Initially, organization was completely ad hoc. Despite their exhaustion and sickness, victims themselves showed an extraordinary resourcefulness and compassion. They were joined by activists from around the country. A particularly large contingent came from Hoshangabad, a few hours south of Bhopal, where activists had run a rural education project since 1972.

Within days, the Zahreeli Gas Kand Sangharsh Morcha was formed to coordinate their efforts—quickly becoming referred to simply as the Morcha—the collective.

The Morcha brought together activists from many different backgrounds. Local people worked alongside professionals, academics, trade unionists, Gandhians, and others in a mammoth effort to respond to the urgency of the situation.[5] The initial cohesiveness of the Morcha was not sustainable. Within a year, participating groups had splintered into diverse, often divisive, groups oriented by very different perspectives on how advocacy for gas victims should proceed. The work of the Morcha nonetheless set the stage for grassroots response to the Bhopal disaster. Most important, the Morcha did not simply prioritize relief and relegate critical perspectives to secondary importance. From the outset, the Morcha was as concerned with means as with ends. There was a continual attempt to evolve a democratic organizational structure, with particular attention to the role of local people and women.

The Morcha also took a very critical stand on "Bhopal" as an issue in the ongoing debate about the role of technology in the development of Indian society. The Morcha represented the Bhopal disaster as the outcome of economic planning that prioritized high-speed growth and imagined bureaucracy as the appropriate response to unequal distribution of its benefits. Perhaps most significant, it questioned the ability of technology to solve problems of its own creation, working on the line that there can never be true rehabilitation or justice in Bhopal because there are "no technical solutions to social problems."

The Morcha's critique of the state and its programs was not limited to a critique of planning orientations and other macrolevel operations. The Morcha blamed local officials for allowing Union Carbide to be operated so negligently and so near residential colonies. They also exposed both the inefficiency and corruption of official relief efforts and the government's attempt to avoid welfare responsibilities through overall denial of the magnitude of injury and need. But the Morcha recognized that the voluntary sector could never provide more than a modicum of the required services. So it worked toward the greatest possible government accountability. Ravi Rajan provides an overview, from the perspective of Sujit Das, a Morcha supporter:

> The Morcha contends that the gas victims have the legal right to free and comprehensive medical relief as well as economic rehabilitation and compensation; that Union Carbide (UC) and the government are not only morally and legally obliged to provide such relief speedily but they alone possess adequate resources to do so; that the moral obligation of the Indian

citizen is to see that this is done. The priority task to this end is to engage all effort to persuade and pressurize the U.C. and the government to fulfill their obligation instead of frittering away time and energy in philanthropic operations sustained by funding from dubious sources. In any case, there will be no dearth of relief activity. But the Bhopal disaster is not identical with run-of-the-mill natural calamities. Here, the politics of secrecy, the politics of multinationals, legal liability, the absence of a democratic polity, federal culture in a state machinery, etc. have all contributed to build up a complex web of mystery, conspiracy, conflict and apathy. Hence, the overriding priority of agitation and mass action. (Rajan 1988, 27)

Rajan explains how this perspective turned into a four-pronged strategy: "to mobilize victims to fight for their basic needs; to create alternative data to counter the governmental efforts at erasure; to present, in some areas, 'people's plans' as alternatives to government programs; and to establish a network of various organizations to debate and act on the various 'larger issues' raised by Bhopal" (Rajan 1988, 27).

The Morcha would continue to emphasize that Bhopal was not a natural disaster. Two things were accomplished with this move. First, it unsettled tropes of destiny—what one activist described as "god did it" explanations. By denaturalizing Bhopal, the Morcha wrested the disaster out of philanthropic gestures of kindness that equated Bhopal with earthquakes, floods, and other phenomena where human decision-making could not be queried. Denaturalizing Bhopal turned the disaster into a resolutely social problem. The "cause" of the disaster was spoken of in terms of collusion—between a bureaucratic, increasingly authoritarian state and a multinational corporation.

Denaturalizing Bhopal also unsettled tropes of unity, expressed particularly in the Indian government's claim to be the guardian of gas victims. The Morcha demanded support from the government, but would not participate in suggestions that "India" was the unified opposite of Union Carbide. Nationalist cohesion was broken down. The Morcha positioned itself against the state, as the defender of civil society. This articulation allowed Bhopal to become a key reference point in emerging debates over economic reforms in India. Bhopal became a place where what comes after postcolonialism began to be worked out.

(7.4) Imagining Disaster

An essay about the Bhopal disaster, written eight months after the gas leak, begins by reminding readers that the history of the twentieth century has been marked by violence. "Auschwitz and Treblinka, Hiroshima and Guernica, My Lai and Pol Pot's Kampuchea. To this glossary we must add a new

name—Bhopal" (Visvanathan and Kothari 1985, 48). The essay then details how difficult such an addition would be.

The essay was written by Shiv Visvanathan and Rajni Kothari, two
prominent members of the Lokayan group—a group formed in 1979 as
a research-action project for the Centre for the Developing Societies in
Delhi.[6] By 1985, Lokayan's annual report was offering bold criticism as
part of an effort to invent new idioms and social forms to carry progressive
social change. Highlighting increasing political centralization and authoritarianism, and "the emergence of violence as the principal instrument of
the State to settle disputes," the report claims that

> [t]he State has lost its élan. The bulk of people are caught in the grim battle
> for survival. The elite has given up its leadership role and the counter-elite is
> nowhere in the picture. The many protests and struggles in sight are com-
> pelled to be limited to micro-spaces—and often turned against each other.
> In this scenario, how does one build a feasible overall strategy for change?
> (Omvedt 1993, 190, quoting Lokayan Group 1985, 5)

Lokayan's response to its own question about an "overall strategy for
change" was multifaceted and based on a number of sweeping critiques—
of the state, of nongovernmental development agencies, and of established
communist parties. The practices, artifacts, and knowledges associated
with modernity—particularly science and technology—were argued to be
part of the problem, and what linked the state, nongovernmental organizations (NGOs), and the established Left together.

Funding was another key issue. At the outset, Lokayan had been funded
through West Germany. By the mid-1980s, it had given up foreign funding
in order to materialize its analysis of the way such funding could distract
from efforts to build "Indian alternatives." Lokayan did not argue that foreign funding determined what could be done in India, and thus should be
unilaterally scorned. Instead, it used debates over funding as an opportunity to elaborate on what it meant to pursue indigenous solutions. Its work
was oriented by the idea that indigenous solutions could be brought to the
surface, through a "specific methodology . . . of learning through linking
issues and constituencies and actively participating in such linkage. It is
a nonvanguard, nonterritorial, nonparty political process" (Omvedt 1993,
190, quoting Lokayan Group 1985, 7).[7]

The linkage the Lokayan group advocated was timely, synchronizing
with the increasing prominence of "grassroots" efforts around the world.
Thus, the idioms and social forms Lokayan tried to build were simultaneously local and global, drawing on critiques and strategies with transnational salience, as well those "homegrown" in India itself. Lokayan was

also able to build on the opportunities created by changing attitudes in funding organizations that led to financial support for overtly dissident activities—often referred to as "action" initiatives, as opposed to development initiatives. By the mid-1980s, this meant that there was a new tier of organizations in India that were "non-party" radical organizations, often staffed by full-time, paid activists-employees. These new organizations had the freedom to pursue options discounted by more established organizations. They also, however, created opportunities for what many pejoratively referred to as "activist entrepreneurialism." Lokayan's critique of the role that Third World elites had played in postcolonial development initiatives was thus complicated by new modes of elite activity—intended to do something different, but certainly not immune to predictable elite impulses.[8] Lokayan's own decision to decline foreign funding was, in part, an acknowledgment of how complicated doing something different could be. Many other middle-class groups in the 1980s developed similar logics, finding ways of working without outside funds, even when it significantly curtailed the possible scope of their activities. The implicit message was that money mattered, both symbolically and materially.

Lokayan also prioritized other strategies for circumventing the paradoxes of middle-class advocacy. The first was an emphasis on "grassroots" grounding. The second was an emphasis on the restructuring of middle-class groups—to disrupt internal hierarchies and to make them more able to work collaboratively both with other middle-class groups and with non-elite groups. Breaking "the syndrome of collective inability" was the overall goal. A third strategy involved conceptual critique—to ready activists for the "peculiar inversion" of issues and strategies that made fixing problems far from straightforward. In engagement with issues of communalism, for example, Lokayan emphasized how communalism was once considered a revivalist tendency that provided "protection to cultural and ethnic minorities such that they could build themselves up and join the mainstream." By the 1980s, communal conflicts were argued to be a "direct by-product of our modernization and secularization." There was, then, no easy answer. Communalism had clearly become a problem, but secularism could not "fix" it—because of its insistence on homogenization (Lokayan Group, 1985).

Bhopal posed similar difficulties. In the essay Visvanathan and Kothari published in the *Lokayan Bulletin* in the summer of 1985, they were clear: Bhopal was difficult to think about, much less engage strategically. In Visvanathan and Kothari's reading, the Bhopal disaster had two distinct episodes. The first episode was the spectacle of the gas leak itself—the international media event that oscillated between "a statistics, *Guinness Book*

of Records style and a teratological episode in Ripley's *Believe It or Not.*"
Between the two extremes lies the second episode of the disaster, in the
banality of the aftermath, marked by silence and erasure. Visvanathan and
Kothari explain that

> it is into this space that the bureaucratic discourse has thrust itself forcing
> everyone to speak "bureaucratise." Any debate soon appears like war be-
> tween two clerical factions. The latent consequences have been devastating.
> It has rendered the victims tired, inarticulate and speechless, and voluntary
> activists shrill. The pain of their protest sounds more and more like noise,
> discordant to the picture of "an increasingly normalizing Bhopal" (Visvana-
> than and Kothari 1985, 52).

To understand what has happened in Bhopal, they call upon us to grasp
what Susan Sontag has called "the imagination of disaster," the "collective
representations, the myths, the symbols, the allegories, the images,—secu-
lar and religious—available for understanding of catastrophe." What they
want to show is "how the very structure of modern industrialism encodes
the understanding of the catastrophe. It involves, in the case of Bhopal, the
bureaucratization of the catastrophe and links with wider images of modernization" (Visvanathan and Kothari 1985, 49). What Visvanathan and
Kothari wanted to query was how a different imagination could be drawn
out and how people, both dominant and oppressed, could work to increase
or limit the availability of such an imagination.

To do this, they explain how "the catastrophe was localized and controlled into a series of humdrum acts." And,

> as one doctor put it, "the victim had to be processed, carved out and milked
> and every one of the little bureaucrats, the government doctors, the police-
> man, the tehsildar, the ration shop owner, social worker and political goon
> had to have his cut. It was the remnants that went to the victims." The mind
> itself was on the assembly line. (Visvanathan and Kothari 1985, 50)

Bureaucracy thus needed to be understood as institutionalized objectifica-
tion and vivisection, to allow control and domestication. The disaster became a banality—a banality not unlike that described by Hannah Arendt
in trying to account for Eichmann's violence (Arendt 1977). Banality is an
effect of split vision, and it is "not only in the assembly line. It acquires
a more lethal form in the everyday perception of bureaucrats" (Visvanathan and Kothari 1985, 49). Problems are encased in modules, routinized,
and serialized to accommodate the office file, ready to be discussed ad
infinitum.

Visvanathan and Kothari insisted that it was not "the corruption of the
clerk" they were talking about, but "the ritual procedure of coping with
disaster embedded in a bureaucratic grid. In fact, one could even contend

that all the little corruption domesticated and marginalized the sheer enormity of the disaster. To counter this, one must attempt to see it all through the eyes of the victims" (Visvanathan and Kothari 1985, 50). Yet one could not be a victim unless the bureaucracy said so.

A victim could not be a victim without a ration card—or if excluded from bureaucratic censuses and surveys. Visvanathan and Kothari emphasize the enormous power of these instruments, explaining how "absence from them can be warrants of death; inclusion, virtual certificates of survival" (1985, 48). They also discuss the arbitrary ways distribution of relief and compensation was determined and the way science had been used to legitimate exclusion.

Science, in their view, had become captive of the state. And Bhopal "revealed the bankruptcy of the professions." A return to the perspective of victims thus seemed important, but also was far from straightforward. Many of the victims had already been sucked into a machinic logic—by the relief effort, but also by the media, for whom they played starring roles. There was a sense, Visvanathan and Kothari said, that "the government is auctioning off pieces of the catastrophe to the nearest kin with Rs. 10000 as the price of forgetfulness" (1985, 48).

Visvanathan and Kothari try to insert what they have seen and heard in Bhopal into a binary logic. Myths of manufacture are opposed to creation myths. The vivisection and serialization enshrined by Descartes' method, Adam Smith's division of labor, and Hobbes's social contract are opposed to the integration of the calendrical cycle. The gaze of the modern hospital is opposed to diagnosis of the whole person. Prestructured catalogs of systems are opposed to victims' own perspectives. Carbide's Bhopal is opposed to a tribal Bhopal.

These oppositions don't really work. Visvanathan and Kothari suggest this themselves:

> There is something about man-made disaster that human memory finds difficult to retain and comprehend. A flood or an earthquake is easily internalized into the calendrical cycle. The language of myth and religion is easily available to conceptualize it. Man made catastrophe seems unmetaphorical and aridly secular. It is subjected to quick erasure, to a kind of amnesia that not only denies the victim basic compensation but even the right to remember. . . . There is something deeper and inarticulate here, expressing something latent in all of us. Something about Bhopal is different. It is not that we are unused to catastrophes. We have had the Bombay plague, the Bengal famine; floods are part of the ritual cycle of our imagination. Yet, Bhopal, industrial, robotics, does not fit our universe of perceptions. It is not the wrath of our Gods or even our local viruses and demons. It is alien to our norms, like a huge piece from some other jigsaw. The scale, the sheer alienness of the size of compensation sets it apart, as something that can only be

expressed in the language of the American stock market. Indian lawyers
somehow feel that the Carbide plant is not a machine in Vishvakarma's
world. It is so American, let us enact it out in America. Our legal texts lack
precedent for this. Yet Indira Jaising's warning that we might have to ready
ourselves for such battles is not something we can ignore. (Visvanathan and
Kothari 1985, 54, 58)

Visvanathan and Kothari's essay provides a detailed analysis of the mi-
cropolitics of disaster in its first months. It also articulates themes that have
recurred in the middle-class progressive response to the disaster in later
years. Their recognition of the difficulty of conceptualizing Bhopal may be
more overt than usual. But many share their sense of a deep failure in all
reports of Bhopal, oral and written. Such reports, in their view,

[e]ither . . . reified Bhopal, in all its specificity or left it nebulous. What
they lack is a clear metaphor, one iconic image around which the conflicting
imaginations can be brought into play. Bhopal is still a catastrophe in search
of a metaphor, a vision that is both within and beyond Bhopal. . . . In fact,
Bhopal as a disaster has yet to graduate into a tragedy. A disaster is an objective event, something out there. It becomes a tragedy when human beings respond and there are flaws in the response. (Visvanathan and Kothari
1985, 60–61)

This proposal for what still must come in the aftermath of the Bhopal disaster is hardly straightforward. It suggests, however, a sentiment within
which many middle-class activists I have known have worked out a response to the disaster. Tentatively. Reaching for something that can't yet
be named. Pursuing new linkages, as a way around available—and obsolete—idioms and social forms.

(7.5) Encroaching on Civil Rights

The government of Madhya Pradesh carried out slum demolitions in Bhopal in the summer of 1991. A report on the demolitions was put together
by the People's Union for Civil Liberties (PUCL). The PUCL report provides a snapshot of disaster on the ground. It also illustrates an important
mode of advocacy.

The civil liberties report has been an important means for circulating
the advocacy of middle-class progressives in India, particularly since the
1970s, when the Emergency spurred the formation of the People's Union
for Civil Liberties and Democratic Rights (PUCLDR). Many describe the
PUCLDR as providing an important opportunity for collaboration among
"liberals," "humanists," and "radicals." The terror of the Emergency is
said to have help solidify the collaboration. PUCL was formed in 1980,
after a split in the movement following Indira Gandhi's return to power.

By the 1980s, the discourse of civil liberties and human rights faced
further complications, partly in response to critiques of "state-centered solutions" to social problems—alongside attempts to consider the dominant
development model in terms of justice. An introduction to a collection of
essays published in 1989 to encourage *Rethinking Human Rights* pinpoints
a key paradox: "If 'formal law' is its final referent, then the human rights
movement ends up primarily appealing to the state to efficaciously implement its own laws, and such an activity can end up creating serious problems in situations where the state and the legal system enjoy limited or no
legitimacy" (Sethi 1989, 9). Most activists I knew who worked within this
critique acknowledged that recourse to law and appeals to the state would
remain important. But they also recognized the importance of developing
a *politics* of human rights that could widen notions of citizenship beyond
legalistic definitions. Bhopal provided important examples.

The Bhopal disaster clearly makes demands on the law, but it also calls
for something more—something that is difficult to articulate, yet that can
be sensed in the dense detail of a traditional civil liberties report.

A civil liberties report is replete with detail. The documentary technique
is careful and has an objective tone. The dominant trope is metonym—
things proceed by contiguity and sequentiality. There is expansiveness to
the account, however. While the focus is local, many links are established
to position the story within broader narratives. Bhopal becomes part of the
history of violence against the urban poor, suggesting links to social movements working to redefine ownership, the city, and pollution. And Bhopal
is linked to institutionalized communal bias.

Something important is accomplished through these reports. Tragic
particularity is pulled into a discourse that distributes significance. Local
events become part of a bigger story. A space is created where mundane
details have meaning.

PEOPLE'S UNION FOR CIVIL LIBERTIES
Madhya Pradesh/June 1990
**Report of an investigation into the "anti-encroachment drive" in
the gas affected slums of Bhopal**

ENCROACHMENT ON CIVIL RIGHTS

A large part of the population in cities lives in slums. In the eyes of the law,
most slum areas are considered illegal settlements, but for the people who
live in these slums there are no options. An estimated 40 million city dwellers live in slums, which is 20% of the urban population. Almost half of the
population of Bhopal, like in many other cities, resides in slums. A majority

of these slums happen to be affected by the world's worst industrial disaster—the toxic gas leak from Union Carbide's factory in December 1984.

Since last year, the State administration has been carrying out an "anti-encroachment drive" as a part of its "city beautification plan." This drive has led to a situation of terror in the slums.

During this drive, there has been large-scale demolition of slums situated along the Upper Lake of the city. A large number of people have been evicted from their settlements. The People's Union for Civil Liberties has carried out an investigation on this matter. This is the report of the investigation team that included Mr. Om Prakash Rawal (Eminent social-thinker), Mr. Lajja Shankar Hardenia (Journalist, Economic Times), Dr. Arun Kumar Singh (Geologist), Ms. Tultul Biswas (Student, Barakatullah University) and Dr. Rajiv Lochan Sharma (Doctor).

The city of Bhopal, besides being the capital of Madhya Pradesh, is popularly known as the cultural capital of India. The city, situated in the plateau of Malwa in Central India, can be divided into three distinct areas. The old city, which was established during the reign of the Nawabs, where most people are dependent on wage labor and petty trade for their livelihood. The New city, where government offices and staff quarters were built after Bhopal was made the Capital of the state, is to the south of the Old City. Most of the residents of this part of the city are government officers and other employees. The industrial township of Bharat Heavy Electricals Limited (B.H.E.L.) is the third distinct segment of the city of Bhopal.

The total population of Bhopal is about 10 lakhs [1,000,000], out of which approximately 6 lakh people reside in the old city. About half the population of the old city is settled in slums. According to the records of the government, close to 50,000 people in Bhopal live in 160 "illegal" slums, most of which are in the old city. The city has many large and small lakes, the Upper Lake and the Lower being the two largest. Shakir Nagar, Fatehgarh, Sajda Nagar, Sidhi Ghat are some of the densely populated slums along the Upper lake. About 15,000 people had been living in these slums in small dwellings made of wooden planks, stones or mud. The land along the lake is owned by different individuals and institutions. Some of the land belongs to the Madhya Pradesh Wakf Board, some owned by the individuals, and rest is mostly occupied by the people who hold entitlements (pattas) given to them by the government. Some of the residents do not have pattas.

Most of the men in these slums are either auto rickshaw/tempo/minibus drivers, handcart pushers or daily wage laborers. The women work as wage laborers, housemaids or betel nut cutters. People also roll bidis at home to make a living.

A survey conducted by the Indian Council of Medical Research (I.C.M.R.) after the 1984 gas disaster indicated that 55% of the population in the old city is Hindu and 43% is Muslim. Though the two communities are close in size, it is the Muslim community that has suffered due to the demolitions. Our investigation team found that among 800 families evicted, only four were Hindus. Mr. Babu Lal Gaur, the State Minister for local self-government told the investigation team that 30% of the population in demolished slums is Hindu. However, the Superintendent of Police informed the team that 99% of the affected people were Muslims. Most of the residents of these slums were known to have traditionally voted for the Congress Party. According to Fazalludin, a retired army man and resident of Sidhi Ghat, during the recent elections, a majority of the names of the residents had disappeared from the voters list. According to the local English newspaper, the people in these slums who were Muslims and supporters of the Congress Party were shifted out from one assembly constituency to another so as to benefit the ruling party in the state known to have Hindu fundamentalist inclinations. Both the minister for local self-government and the Superintendent of Police held that such charges were without any basis.

The Revictimization of Victims

The residents of these slums had all been affected by the 1984 gas disaster. The toxic gases that leaked from Union Carbide's factory caused more damage to people living in hutments. The doors and windows of the pucca [hard wall] houses were closed that night and the gases could not enter them as easily as they did into the hutments through cracks in the walls and roofs. People in these slums were thus heavily exposed to toxic gases and continue to suffer serious illnesses.

Due to the exposure to toxic gases, gas victims suffer from multi-systemic illnesses. The impairment caused to the lungs and muscles of the affected people had led to a reduction of their capacity to do work. According to the I.C.M.R. [Indian Council of Medical Research] the gases have caused lung damage in 97% of the people. All the ousted slum dwellers interviewed by the investigation team complained of weakness, breathlessness on walking, and impairment of vision due to gas exposure. Despite this, most of them earned a living through hard physical labour. I.C.M.R. reports state that in the gas affected area, 74% of the people do not go for any work and that on 36% of the days in a year the people can not go to work due to illness. Due to the inadequacies of the Government's medical care system the people have to either sell their household goods or take loans at high interest rates to pay for their medical treatment. As the ousted people have been relocated at a considerable distance from the city they will have to spend money on traveling to their work place. While the government has not been able to provide jobs to the gas affected people it has rendered a large number of them jobless by moving them far away from the city.

Since last year the central government has been providing monetary relief of Rs. 200 per month per person among the gas victims. Most of this money is being spent on medical treatment or in paying back past debts. The investigation team had a view of the household belongings of the evicted people and these consisted of a few cooking vessels, couple of cans, bedclothes and occasionally a cycle or a sewing machine.

The exposure to the toxic gases has resulted in reduced working capacity and a range of illnesses among the people. The eviction of the people has resulted in increased misery.

The Demolition of Houses

As one goes across the slums along the Upper Lake, the first one is at Retghat and is called Lal Imli Wali Masjid Ki Basti. The demolition of the houses started here on the 26th of May. Early that morning the city authorities accompanied by about 200 policemen, 10 with arms, descended on the settlement and demolished 75 houses in one day. Some of the houses that were demolished in this slum were more than hundred years old, including Nazar Mahal, which was a building of archaeological importance. Most of the residents here had ownership rights over their pieces of land. Aftab Ahmad, who lived in this slum, reported that he had with him his land registration papers but his house was demolished after issuing three notifications to him in twenty days. He was not paid any compensation for his house. The minister for local self-government, however, informed the investigation team that all people having ownership rights have been compensated for their houses and that they have been provided with plots of land to build their houses. According to Aftab Ahmad, there are many like him who have not been given any compensation.

The demolition continued until 30th May. Every morning the employees of the municipal corporation came with their bulldozers and a large posse of armed policemen and demolished 100 to 150 houses. The policemen and their guns were too intimidating for the people and they watched on silently as their houses were turned to rubble. A few who tried to oppose this destruction were assaulted by the police, abused and threatened. The investigation team was told by Liyakat, Mansoor, Shahban and Munni Bai that the demolition of the houses was carried out in this manner till 30th May.

Terror and Protest

Meanwhile, terrorized by the thought of their houses being demolished, two women—Musarrat (30) and Pan Bai (36)—from Ammu Khalifa Ki Bagiya committed suicide by drinking kerosene oil. The other residents of the settlement protested against the destruction of their houses and threatened to commit mass self-immolation. They also submitted a list of demands to the authorities in which they asked for stoppage of demolition during the rains and payment of compensation for their houses. The slum was built on land belonging to the Wakf Board and there was an order from the High Court staying demolition of houses built on this land. The municipal administrator refused to accept the order and the demolition of the slum was stopped only after people threatened to immolate themselves en masse.

The next slum to be demolished was Fatehgarh. In the morning of 31st May when the demolition squad reached Fatehgarh, about 50 women stopped them in the way holding kerosene cans threatening to immolate themselves. Ms. Shahbana who was leading the group said that she had dowsed herself in kerosene and that had scared the officials from going ahead and they went back. After this hundreds of women sat on a dharna (sit-in) in the slum. The dharna continued for three days and demolition of the houses was discontinued by the officials. According to the superintendent of police most of the people participating in the dharna were outsiders but Ms. Munni Bai, Nafeesa Bi and Rabia, who were arrested while on dharna, informed the investigation team that all the women who sat on dharna belonged to that locality.

On 3rd June at five in the morning Fatehgarh was encircled by policemen, armed men from Special Armed Force (S.A.F.) and Central Reserve Police Force (C.R.P.F.) and mounted police. All access to the slum was blocked. Three of the women sitting on dharna, Ms. Munni Bi, Nafisa Bi and Rabia Bi, said their slum was surrounded by hundreds of policeman. After the encirclement a policeman came to the women on dharna and told them that some policewomen wanted to talk to them and asked them to come to the main street. All the women sitting on the dharna went to the main street, where they were arrested by the police and told that their houses will not be demolished and no one would be harassed. Sixty women were arrested in this dramatic manner causing a major setback to the growing opposition to the demolition. It is possible that some of the local leaders connived with the police and the city administration and allowed such arrests to happen and were later rewarded with land plots in the city itself.

After the arrest when all opposition was silenced demolition of the houses was carried out under heavy armed police presence. Masood, a resident of Shakir Nagar, reported that there were two to three policemen at every house. The whole slum was surrounded by hundreds of policemen and no one was allowed to go across the police cordon.

In this manner 130 houses of Fatehgarh were destroyed on 3rd June and the demolition drive was restarted. Houses were demolished by bulldozers and in the presence of a massive police force. By 6th June the whole of Fatehgarh was razed to the ground and several houses were demolished in Sazda Nagar on 6th and 7th June.

Why Were These Slums Chosen for Demolition?

Among the 160 "illegal settlements" in Bhopal why were the slums situated along the Upper Lake chosen for demolition? While these slums were demolished as illegal settlements no action was taken on other prominent illegal constructions like a hotel at a busy crossing near the railway station or that on illegally occupied land by a minister in the state government near Jawahar Chowk. The government and the city administration were giving varying responses to this question. At first it was said that the demolition of these slums was necessary to construct a road along the lake. But government buildings in this area were not demolished and in fact a few government buildings are being constructed there.

Later the government stated that people along the lake had to be evicted as they were polluting the Upper Lake and thus posing a hazard to the city's drinking water source. Both the minister for local self-government and the superintendent of the police offered this as the reason to the investigation team. The investigation team during its visit to the demolished slums did not see any sewage pipes going to the lake from any of the slums. The team was informed by local residents that night soil from these slums used to be carried off in trolleys and never discharged into the lake. The investigation team saw several open drains along the lake. The Public Health Engineering Department, which is supposed to protect the lake water from pollution, was part of the demolition drive. The executive engineer of this department refused to give any information when the investigation team asked him about the sources of pollution of the lake. The minister for local self-government stated that all sources of pollution of the lake would be stopped within three years. Prominent sources of pollution of the lake are the Research Laboratory of Union Carbide, Bharat Bhavan [an arts complex], the Chief Minister's Residence, the posh Kohe-Fiza colony and Hamidia Hospital, all of which discharge sewage into the upper lake. If pollution control had indeed been the priority, certainly stoppage of these sources of pollution would have taken precedence over demolition of the slums. All the evicted people whom the investigation team met said that the anti-Muslim sentiments of the ruling party were the sole reason behind the demolition drive.

Legal Procedure Not Followed

The investigation team was informed by the evicted people that none of the residents in the slums of Sidhi Ghat, Fatehgarh, Sazda Nagar and Shakir Nagar had been issued notification about the demolition. The minister of local self-government said that they had discussed with all the slum people and the people had agreed to move away. The minister told the team, "We believed in negotiation not paper work. Paper-work comes in the way of our work." The investigation team did not come across any slum resident with whom the minister had any discussion. Rather, the people informed the team that they were warned just one hour before the demolition and were told to remove their household goods. According to the police superintendent, people were given notice several days prior to the demolition. The statements of the evicted slum dwellers and the different responses of the minister and the superintendent of police on the matter of issuance of notice leads the team to believe that no notice was served before demolition.

The minister stated that after the rains more "illegal slums" will be removed and then it will not be possible for the government to give land or compensation to the people who will be evicted. The people evicted during the current demolition have been compensated by the Public Health Engineering Department from its funds for the upkeep of the Upper Lake. The minister also informed the investigation team that in the next five years the residents of all the 160 "illegal settlements" would be moved away from the city.

Given that several residents of the demolished slums held legal "pattas," the government has clearly violated due process of law by evicting them without prior notification.

Supreme Court Order

While the demolition were going on in Bhopal, the Bhopal Gas Peedit Mahila Udhyog Sangathan [Bhopal Gas Affected Working Women's Union], an organization active among the gas victims obtained an order from the Supreme Court staying all demolitions. The order was passed in the afternoon of 6th June. This was reported in some newspapers on 7th. Despite the order, demolitions were carried out in the night of 6th June and from 7 in the morning till 11 in the night on 7th June. Chand Mian, Batul Bi and Ishwar, residents of Fatehgarh, informed the investigation team that their homes were demolished at 10 P.M. on 7th June, when it was raining.

The minister for local self-government who is also the state law minister said that demolition work was stopped right after they received the order of the Supreme Court. He also pointed out that the slums mentioned in the order had already been demolished and hence the government could not be restrained from carrying out demolitions in other areas. An active member of the petitioner organization reported to the team that the Supreme Court order had been received by the state government on the evening of 6th June and demolitions were carried out on 7th June regardless.

Rehabilitation

The people evicted from the slums along the Upper Lake were settled in three areas. 28 families were settled in Bag Farhat Afza in the city; 228 families were moved to Badwai and 400 to Gandhi Nagar—11 and 13 kilometers from the city. Hundreds of families have not been rehabilitated and they now have taken shelter in rented rooms or with their relatives in the city. About 800 acres is available in the city and it would have been quite possible to relocate the evicted people within the city instead of driving them away.

The investigation team visited the areas where evicted people have been relocated and found people living in sub-human conditions. It had been raining for four days and they had no shelter. There was no place where people could sleep and no means to cook food. The most basic facilities had not been provided. In Gandhi Nagar and Badwai only one hand pump was provided for the entire community. There were not shops where people could buy provisions. A government ration shop functioned for two hours in a day but one could get only small quantities. After it rained for four days on the shelterless people the government distributed 15 bamboos, 6 poles, 5 mats and 10 metre polythene sheets to each family so that they could build some shelter. Rupees one thousand [$40] were also given to each family as relief.

The officials have issued "kabzanamas" to each family thus giving them occupancy rights over 15x20 plots. The "kabzanamas" however bear no stamp or seal of the issuing authority and their legal validity is doubtful. A government doctor visited the area on two days. There were problems even with the meager amount of relief provided. People complained that to build a hutment 9 poles were required while they have been provided only with six. That the poles were not straight, the polythene sheets provided were not large enough and were so thin that they got torn by the wind. People also said that the doctor visits the locality for a very short period and does not carry enough medicines.

Transport is one of the major problems faced by the people. They have to come early in the morning for their jobs in the city and there are almost no means of transport. The government run bus service just makes two trips in a day. People have to spend four to six rupees every day on transport, which is a substantial expenditure for majority of the people who earn only twenty rupees [$0.80] in a day. In Gandhi Nagar, Ms. Sazda Bi said it was difficult to go to work leaving children at home. Her little son Javed had been bitten by a scorpion the previous day.

On the issue of rehabilitation the superintendent of police said that the plans for rehabilitation were all right but had people been settled after constructing roads and providing water taps, the work of relocation could have been done better and at half the cost. According to the government office engaged in the rehabilitation, rupees one crore [$400,000] was spent in the resettlement exercise and an estimated 50 lakhs [$200,000] were wastefully expended.

Conclusions

Housing is an important part of life and every human being has a right to decent housing. In the slums of Bhopal, people are in a state of constant terror due to apprehensions of being evicted from their homesteads.

The demolition of the gas affected slums without prior notification is a clear violation of due process of law. The eviction of people without making proper arrangement for their resettling and thus forcing them to be shelterless under the scorching sun and pouring rain is blatantly inhuman. Through inadequate medical facilities, denial of essential facilities and by not providing means of sustenance a great injustice has been inflicted on the gas affected people. (People's Union for Civil Liberties 1990)

The PUCL report articulates the materiality of hardship in Bhopal. Mundane things—like the number and alignment of poles distributed to build hutments, the number of buses plying between two points, and the bite of a scorpion—accrue significance. The stubborn resistance of those victimized is also told about, and those who resisted are given names. The report also gives names to the bureaucrats that delivered the violence. The source of the violence is located in party politics and religious bias, given institutional authority through the Bhopal Beautification Plan—a key component of the official plan to rehabilitate gas victims.

The demographic data provided by the report are important. The segmentation of the city of Bhopal. The high percentage of Muslims. The dependence of many gas victims on jobs requiring physical labor. Stratification *within* Old Bhopal, evident in different types of housing. Revictimization of communities hit hard by the gas leak. Apprehension of terror as a daily phenomenon.

Official insistence on negotiation—not paperwork—is particularly ironic. Paperwork can mean the difference between life and death in Bhopal. Loss of proof of residence can mean loss of all hope of economic relief. Official fixation on "illegal settlements" is, of course, too ironic for words.

The report hints at the complex problems that would emerge in any attempt to build public health infrastructure in Bhopal. But the complexities are difficult to appreciate when government officials exhibit such gross disregard for both rights and welfare. This is how the "second disaster" in Bhopal has been configured. Abuses of power have been so blatant that there has been little space for fine-grained analysis or honest engagement with the truly extraordinary task of figuring out what it would mean to rehabilitate Bhopal.

(7.6) *In Search of Famine*

In Search of Famine. A film I have never seen, despite looking for it again and again, in India and the United States. A film about trying to make a film about the Great Famine of 1943. A film about idealism, objectivity, and parallels between the making of films and the making of the progressive advocate. A film completed in 1980, yet never really finished.[9]

Mrinal Sen, the director, had made films about "radicals" before. They were cast critically, as figures beset by arrogance and an inability to understand the people they thought they were fighting for. In *In Search of Famine,* Sen took on another level of complexity, casting the radical as a filmmaker convinced that he could capture the reality of famine, represent it authentically, and thereby change the world.

Sen's film was based on a story by Amalendu Chakraborty. But Sen's film does not follow Chakraborty's story exactly. Afterward, Sen recounted how his film only partially followed the script written out earlier. The rest was a record of what happened at the shooting. In other words, he let the film slip, becoming a film about the difficulty of making a film about making film, and about the need to confess our incapacities.

The official Famine Inquiry Commission reported one and a half million dead in the Famine of 1943. Unlike elsewhere, the dead in Bengal were not counted among the honored. Soldiers in Europe ate Indian rice and were given medallions of silver and gold. Bengalis died quietly, along the roads to Calcutta. It was said there was food in Calcutta, and perhaps there was. But distribution failed and the dying continued, without ceremony, without honor.

Sen wanted to show the contradictions within the continuity of history. He also wanted to show that "there is a gap or a void between the physicality of reality and the artists' redemption of it, of which he may or may not be aware" (Sen 1985 [1983], vii). The goal, in my reading, was to query what progressive commitment could mean, once complexity had to be acknowledged and entrenched standards of the Good fell away—or at least became so recursive as to lose their referential force.

The script begins with the arrival of the film crew in a Bengali village called Hatui. They are there to reenact and document a tragedy assumed to be over. The crew leaves the village with the film unfinished. They could not figure out how to relate to the villagers. Nor did they succeed at representing what famine was, and is—again and again. They tried to contain the famine within the space of the film. But there was continual slippage between what was in the film and the lived reality of the village in which the filming was done—a slippage that the director may or may not have been aware of. The film crew was even held blameworthy. One old peasant put it bluntly: "They came to take picture of a famine and sparked off another one themselves" (Sen 1985 [1983], 47). The film crew's presence in the village had caused prices in the market to rise.

The crew set up "camp" in an old pleasure palace built by Shah Mahmud II 250 years before. The fountains no longer sprayed jets of water fragranced with rose. Persian damsels no longer splashed snow-white feet, singing and playing the sitar. But the elegance continued. Columns of marble. Empty expanses of space. Twenty-six rooms, with seventeen owners—all but one of which lived elsewhere—across India, and even in America. A woman called the Zamindar's Wife remained. Desolate in her loneliness. One of the film crew tried to befriend her.

The object of concern to the film crew fractured. Famine remained a concern. But they enacted something more. A story about famine, but also about being middle class and trying to understand suffering foreign to their own experience. A story about failures of representation, but also about the fecundity of any single image. Between shots, the actors played a guessing game, trying to identify which tragedy was represented by a particular photograph pulled from their archive. Was it 1943? Before? After? The photographs documented both specific historical moments and unchanged circumstances, so they were difficult to identify. Emaciated bodies recur. Recognizing specific times and places required peripheral vision. Their deliberations are built into the film itself—a film about trying to make a film about famine. Radical claims to objectivity are focused on, caught by different angles and exposed as inauthentic.

In the end, the film seems to be about limits. It is about famine, and thus about limits of political will and infrastructure. It is about ideology and its limits, and about art and its limits. And it is also about different castes, classes, and cultures—and about the limits of all attempts to move between them. One of their actresses had quit, unable to accept village ways. When the film crew tried to recruit a local replacement, they offended the villagers. The role to be filled was that of a prostitute, driven by hunger to affairs with timber contractors, rich with city ways. The villagers could not disassociate the filmic and the real. The film crew thought that they could.

The figure of a village woman named Durga harbors many of the contradictions. She has a child, and a husband who cannot work. They are poor to the point of hunger, so Durga earns what she can doing small jobs for the film crew. Then the director thinks of Durga for the role of the prostitute—a role cast off by the actress hired for the job, and by other villagers. Durga does not have to accept the part to be implicated. By the end of the film, her husband no longer trusts her. She has become the woman in the film, whether in the film or not.

The script ends as the crew pulls away from the village to return to Calcutta. There are several dissolving shots of Durga, wearing the same blank expression. A mid-shot, a mid-long shot, a long shot, then an extreme long shot that fades into a dot. The last words are off-screen commentary: "A few days later Durga's son died. Her husband is missing. Durga is all alone." The script says that the voice sounds lifeless. And that as the dot fades out, the title of the film reappears: IN SEARCH OF FAMINE.

It is not clear whether the director of the film lost his faith that famine could be captured and represented. He leaves the village silently, and without apology. The tragedy thus doubles. There has been no apparent movement on either side of the class divide. Durga and the other villagers remain stuck, as do the director and most of his crew. Communication has failed completely. Sen, writing about making the film, reminded his readers that the manuscript on which the film was based ended with a heart-rending question: "Where did you run off to Babu? Haven't you come to make a film on a famine?" (Sen 1985 [1983], vii).

(7.7) Catalytics

In December 1996, I returned to Delhi and Bhopal to update material I had collected earlier, to develop comparisons between India and the United States, and to try—once again—to understand how progressive advocacy works. The trip was both encouraging and frustrating. New hospitals had been built in Bhopal, but were not functioning. Basic medicines were not available at government dispensaries. Private clinics were flourishing, absorbing the dribble of money distributed in compensation courts. The operation of the compensation courts was inefficient and widely perceived as unfair. Many people said that lawfulness had broken down throughout Bhopal. Clearly, the disaster continued.

Controversy among progressives also continued. But so did progressive activism itself—in both Delhi and Bhopal—at the grassroots as well as in the courts, at press conferences, and in various other forums where collective action was crucial. In Delhi, I was able to meet with groups associated with the Bhopal Gas Victims Solidarity Organization (BGPSSS) to learn of ways they had remained involved in the Bhopal case, alongside their other work. The Delhi Science Forum, for example, had published the *Alternative Economic Survey* annually since 1992–93 to provide independent, expert opinion on the effects and implications of economic liberalization. The Bhopal case was referenced to illustrate problems associated with foreign investment, technology transfer, and inadequate regulatory oversight.

BGPSSS also remained involved in the particularities of Bhopal itself. Publicizing the results of new medical data was a high priority, as was involvement in the criminal proceedings of the Bhopal case—which had been reopened when the out-of-court settlement was upheld in October 1991. New medical data had been produced by the International Medical Commission on Bhopal (IMCB), a group of fifteen medical professionals from eleven countries who came to Bhopal in January 1994 to conduct an independent analysis of the health status of gas victims.[10] The IMCB's findings were significant, pointing both to continuing morbidity in Bhopal and to the inadequacy of existing and planned rehabilitation structures. At a press conference I was able to attend, members of IMCB reported that at least 50,000 people in Bhopal continued to suffer from total or partial disability because of their 1984 exposure to toxic gas. They also confirmed incidence of multisystemic injury and reported significant incidence of neurotoxicological disorder and posttraumatic stress disorder. Their critique of government rehabilitation schemes focused on the inadequacies of the centralized, hospital-based system of health care available in Bhopal and on the way patients continued to be treated in a manner appropriate for acute emergency care, rather than in a manner appropriate for chronic disease and disability. They advocated the establishment of community-based health units intended to serve no more than 5,000 people each.

One such community clinic had already been established in Bhopal, even though without the support of the IMCB and without the support of many gas victims—who had been alienated by controversies surrounding the fund-raising campaign for the clinic. The clinic had been funded through contributions from individuals who responded to appeals for aid in British newspapers—appeals that some people considered vulgar and even dishonest. Others considered the appeals worthy of awards.[11]

Newspapers in Bhopal covered the controversy, accusing those involved with the clinic of blackening the name of the government of India and of serving vested interests, rather than gas victims (Hasan 1996). Gas victims I spoke with were more concerned about how their stories had been used. The photographs used in the funding appeals particularly angered them.

One version of the appeal used a photograph of a dead child whose face was being buried. A hand hovers above the child's face, about to cover it completely. It is probably the most reproduced image of Bhopal. It has been on the cover of books, popular magazines, and the pamphlets produced by voluntary associations. It is a grueling image. But I had never before heard gas victims speak of it as vulgar. Later versions of the fund-raising advertisement used a photograph of a weeping woman. Some Bhopalis found it offensive as well.

The extraordinary acrimony among and toward progressives responding to the Bhopal disaster has always been difficult to understand. And difficult to avoid participating in. The acrimony surrounding the clinic was no different. But I didn't think it fair to argue, as some I spoke with did, that the middle-class activists associated with the clinic were "no different than other middle-men in Bhopal." Nor did I appreciate newspaper reports that "cheats are active not only in the country but in far off places like Canada to exploit the name of the hapless victims of the tragedy and raise funds in their names" (Hasan 1996). There may have been extraordinary disagreement among progressives, but I saw no scams.

I don't doubt that much of the controversy among progressives working on the Bhopal disaster has been counterproductive. I also know that such controversy is not unique to Bhopal, even though activists have insisted that Bhopal is "particularly ugly." But neither do I doubt the importance of middle-class activists as catalysts of social change—particularly in India. Their effect is often indirect. And their connection to "the people" is often unclear. But I do think they make a difference and have to be admired for their persistence.

When I was in India in 1996, I spoke to many people about what progressive activism is and about how middle-class activists contribute to progressive change. Some people shared my sense of the importance of middle-class activism. Others were more cynical. The best conversations mingled both, helping me understand the often strange conjunctions that bring middle-class activists together.

In one conversation, Surajit Sarkar told my husband, Mike Fortun, and me how he became involved with middle-class activism and where it has taken him. From elite schooling to banking, from Delhi to small villages, from rural education projects to work in educational television and film-making. The conversation did not address Bhopal directly, though it was in Bhopal that I first met Sarkar, learned about his work, and began to rely on him as a translator—not of languages in the traditional sense, but of modes of speaking and acting across caste, class, and ideological divides. Sarkar's extraordinary ability to negotiate and reflect on these divides made him an invaluable teacher and friend. It also helped him become a prize-winning television producer and filmmaker.

In some ways, Sarkar's career path as a middle-class progressive may be far from representative. But his stories do provide a sense of the cultural and social forms that produce—and that are produced by—middle-class progressive advocacy.

Interview Transcript/Delhi, December 1996

M. FORTUN: What about that Red school you went to, Suro?
SUBRAMANIAM [Sarkar's wife]: Red school?
S. SARKAR: I showed him that article on St. Stephen's College, that had a picture of the college and some computer-generated hammer-and-sickle stuff all over it. Some 1996 notions of how this elite English-speaking college is actually a hotbed of radicalism and Naxalites and stuff. It's hardly a Red school, yeah? In retrospect they would like it to be, and like us to believe that it was. When you go to St. Stephen's College, like when you go to any one of these better Delhi University colleges, classes are held on time, there's a fairly high pass percentage, stuff like that. One of the major ambitions that students have is to make it to a government job. So there is a whole set of competitive examinations that you sit for. Over a million students will be sitting every year for these exams, which are for the banks, civil services, forest service—not for the military, though. Even the big corporate houses—Tata, Birla, and such—they have campus interviews at these colleges, where fresh young graduates are ripe for the picking. So, being, at some point in time, a fresh young graduate . . .
M. FORTUN: What year was that?
S. SARKAR: 1980. And, like everyone else, I decided that college was not bad: you get to read a lot of books, you don't have to go home, you're staying in the hostel; it's a world without any rules and regulations. So, if I could extend it by another couple of years, why shouldn't I? Which basically means making a deal with your parents: look here, I'll do my Master's, and I'll apply for all these competitive exams. The other option is to go abroad, which a lot of people from St. Stephen's College did. But I wasn't much interested in that. So, like everybody else, I applied for these exams, and through them I got a job with the State Bank of India. For two years I was under training, which meant you were rotated to different desks in different branches.

. . . Eventually I discovered there was very little rush for vacancies to be filled up in small towns and villages. So I chose places which were small towns and villages in areas where I would have liked to go anyway, just on holiday or something.
K. FORTUN: You were romantic about village life back then?
S. SARKAR: Not romantic, but very curious. On an everyday level, people spoke a different language; nobody spoke English. To be continually in a place where nobody spoke English for days and weeks on end was definitely

a new experience for me. I had lived outside of metropolitan areas. But, here, nobody spoke my mother tongue [Bengali], nor did they speak English. And finding myself quite fascinated by it, I moved from one to another. I had options—you had to go through four branches in two years. I chose ones that were in places I wanted to visit: a village in Rajasthan, one in Agra—the industrial estate branch, where many of the units served by the bank have now been shut down by an order of the Supreme Court, because their pollution is ruining the Taj Mahal.

So I learned how small towns and villages operated. . . . One day I just gave them my resignation paper and went to this place called Kishor Bharati for three years.

K. FORTUN: How did you know about Kishor Bharati?

S. SARKAR: I kept coming to Delhi and meeting people and telling them how totally frustrated I was getting. I read some newspapers and talked to people, heard about some of these other-kind-of-organizations. Definitely better than working in the bank. But I had no better ideas than that. I mean, I liked hanging around with kids, but that's separate . . .

K. FORTUN: Did Kishor Bharati pay you?

S. SARKAR: Yeah, they paid me one-fifth of the salary I was getting in the bank.

K. FORTUN: Was that a worry for you, or did an eighty percent pay cut not matter?

S. SARKAR: It mattered when I left Kishor Bharati, because I'd run up debts with lots of people. That's when I came back to Delhi, after being in Bhopal. Part of the reason to come back to Delhi and find work here was because it was the only place where I could earn at a rate to pay back debts.

K. FORTUN: At the outset, did you see your move into the voluntary sector as final? As a no-going-back kinda thing? A one-way street? Lots of people I've known here seem to have taken that detour. Or, did you think in terms of working a few years, going into debt, then getting out, so you could pay the debts back, then begin the cycle all over again?

S. SARKAR: No, no. In fact, I wasn't even very sure whether I wanted to pay the money back. The people I'd taken it from wouldn't have said anything if I'd never paid it back. But if they got it back, it would be good for them.

Most of the people I met when I was at Kishor Bharati either had family income or property income to go back to. Or they were people who had gotten into it and adjusted to a much lower scale of living. Which means also cutting down on visits to meet family, and so on and so forth. So there were all kinds. They were all English speaking, or people who would learn English over time.

K. FORTUN: What percentage went to elite colleges like St. Stephen's?

S. SARKAR: Very few. Later I discovered there were lots of students from St. Stephen's College who were in voluntary groups like this across the country. But it was people who had graduated in 1980 or 1982 who were the last of the lot. After that, St. Stephen's College changed—along with the whole tone of things in India.

I wasn't much into poetry until I went to Kishor Bharati. One of the funny

things I discovered there was that a lot of poetry was quoted or used somehow. The point is that it was something very new to me. For a long time I
couldn't get used to it. Then I realized that I didn't like some of it. But some
of it I definitely did like. But it indicates the kind of people who were there
at Kishor Bharati: the kind of people who quoted poetry.

    Kishor Bharati was an experiment in rural education, headed by Anil Sad-
gopal, a physicist from the Tata Institute of Fundamental Research in Bombay. All places like Kishor Bharati, Eklavya—the education crowd in the
voluntary sector—have got a high percentage of Master's and Ph.D's.

K. FORTUN: And a lot of scientists, especially physicists. Wasn't Anil a
theoretical physicist?

S. SARKAR: I have no idea. I never discussed any physics with him. He
quit his job at the Tata Institute, and got this land on lease from the government of Madhya Pradesh, in Hoshangabad, to start some experiment in rural
education. Generally, educating the Indian masses has been a big thing since
1947. I didn't know all this then, but as I got into it I discovered that rural
education is the goose that laid the golden egg, as far as the NGO sector is
concerned. So there was a lot of funding, and a lot of people coming into the
field; some of these good minds would come in and get disappointed or become cynical or whatever, and move on.

K. FORTUN: How was education defined?

S. SARKAR: When Kishor Bharati started, they started with what is called
the Basic Education Program. This was a modification of a Gandhian educational tradition, in which you take a group of kids from a village and you
teach them the alphabet and the numbers and everything else, but you also
teach them how to plow the land using a tractor, how to repair machines,
how to make sure that all the kids around you have got their shots for diph-
theria, and whatever.

K. FORTUN: And did Surajit Sarkar the banker know how to do all these
things?

S. SARKAR: No, but that was about fifteen years before I landed up there.
Kishor Bharati started in about 1972 or 1973. In those fourteen years, they
had tried this Basic Education thing, which fell flat on its face because the
villagers said, "bug off."

K. FORTUN: Why?

S. SARKAR: They said, what do you mean, plow this land? Any stupid
child will tell you this is sandy soil and you can't grow anything here. And
they said, "no, we can improve it, we can add this fertilizer, we can do this,
we can do that." And the village fellows said "do what you will, this soil
won't grown anything, but if you go twenty feet on that side, beyond that
changed color of the ground, there it will work." And it was, actually, a situation like this—it sounds very comical but it happened. This was in the Narmada valley so, like any river valley, there is a distinction between the allu-
vial part and the rest of the ground, so soil qualities differ. The villagers
knew this. It was for reasons like this that the agricultural part of the Basic
Education Program fell flat on its face.

K. FORTUN: Why were the Kishor Bharati people so naive? Were they just
convinced they could make something grow in any soil?

S. SARKAR: In 1972 you were just five years away from the Green Revolution, chemical fertilizers, it's-possible-to-feed-India rhetoric. All of them had come from the U.S., or at least learned ideas from Ohio State or Iowa State—all from textbooks—which forgot the basics: first ask somebody who lives there about where things grow and where they don't! They didn't do that.

K. FORTUN: So by the time you landed up there it was clear that the basic education project was a failure?

S. SARKAR: They had lots of funny incidents. They tried to shift to crops that would grow in the kind of soil that Kishor Bharati had, but there was no merchant in the nearest market center who wanted to pick it up—because he didn't want to have a new crop on hand. Where was he going to get rid of it? He has to have his own distribution system, to pick it up and pass it down the line. So you can't just have an agricultural experiment in isolation. So they moved into milk, because dairy development was another big thing. They got these bulls—imported Holstein, premium kind of stuff—to service the local breeds, so you'd get hybrids that would give more milk and so on and so on. [laughs] I'm telling you stories that have been told to me ten, twelve years later, by the villagers. So they have this milk that doesn't taste like milk anymore, because it was so watery. So people get their own cows for milk to drink at home, and kept these cows to produce milk for the market—because they figured it would sell in town. But what happened? They put the Holstein premium milk, which, according to them is very watery, into big cans, which they take on cycles all the way to the nearest railway station. Then onto the train to Itarsi, which is the changing point for trains to Bhopal and Nagpur, to sell it to agents for distribution. And they found that this pain of carrying it from here to there was too much effort and it wasn't paying well. Before, men just went to town on their cycles to sell whatever they wanted to sell, or buy, with just half-hour cycling there and back. With the Holstein milk they ended up spending most of the day just to sell milk. So the dairy experiment also ended in disaster. People just gave it up. A few people continued to keep the cows, and they did catch on later, when a market for the milk developed. Holsteins sort of became the pride of that place about ten years after they first came into the area; the Kishor Bharati experiment was an experiment ahead of its time.

K. FORTUN: How was it that the cows came of worth ten years later?

S. SARKAR: It was ten years later when the dairy boom happened in the country, and networkers—dairy middlemen—came into existence. At that time, people who had maintained their cows, which were largely the richer farmers, found that they had a gold mine. Suddenly these cows meant money. For ten years they were basically a thing that didn't work and they couldn't get rid of. Then, suddenly, they became big. Basically, once you set up a system to carry the milk, it'll work.

K. FORTUN: Was this government money that was buying all these bulls?

S. SARKAR: Partly government, partly corporate, partly NGO.

K. FORTUN: Why wasn't KB in the picture when things finally worked out?

S. SARKAR: In the intervening ten years Kishor Bharati had stopped direct village educational intervention and moved into school education—

textbooks, curriculum development, training. They had forgotten about working with this village or any other village around. When I landed up there in 1987, Kishor Bharati was finally talking again about projects working in those same villages—in schools, but this time primary schools, and formal and non-formal work. So I was there as a teacher, and associate coordinator of the program.

K. FORTUN: How large was the staff of Kishor Bharati?

S. SARKAR: In the children's activity program, the number of people varied from 5 to 16. In all, KB had say 25 at its max—while I was there. Before my time, I don't know.

K. FORTUN: Did they think of themselves as radicals or Gandhians or Marxists or what?

S. SARKAR: Definitely Marxists. I never talked to them about it, but I'm sure, because there were a lot of Marxist books in the library.

The Kishor Bharati story is a story in itself, and it's a very big, fascinating story. I won't talk in detail about it, because it would lead this whole thing into another line altogether. In my mind it showed that people from different backgrounds and classes, middle-class types working with the rural poor— they can work together and a fairly good balance can be achieved. On the other hand, it also showed that isolated experiments really don't work. I mean, you can't just do good in one place, and hope it will sustain. But, at the same time, it doesn't mean you don't do good in one place. Even today, the solidarity is strong amongst landless, agricultural labor and tribal people in the region. The networking has stood well after Kishor Bharati was gone. And now it's continuing without any outside, middle class support at all.

K. FORTUN: And why couldn't it have happened without a middle class catalyst?

S. SARKAR: Some people have said it very simply: Village people don't go to some places. Their days are so occupied; they only need certain predictable things, meet certain predictable people. To get out of the rut, you need somebody out of the ordinary. Kishor Bharati happened to be the out of the ordinary thing there. But it doesn't have to be an institution every time. Even with individuals, it happens, where a new group comes together. Anything could have been the catalyst. Kishor Bharati happened to be the catalyst there.

K. FORTUN: But what needs to happen is new social interfaces?

S. SARKAR: Sparks. Sparks need to play.

K. FORTUN: Sparks that do what—bring together people that otherwise wouldn't meet?

S. SARKAR: Bring together people who share the same ideas, but live in geographical locations that, though near to each other, don't overlap.

K. FORTUN: Were the shared ideas there before the networking happened, or were they created out of the network? Was the culture of solidarity just waiting for a venue, or did it have to be crafted?

S. SARKAR: The single event that catalyzed the organization of landless peasants and small farmers was the gherao [picketing] of the silk farm and its manager. It led to police action that set the Sangathan's [movement's] political agenda for many years. It all started because the government decided

to stop local recruitment to a silk farm, which was a major source of employment for villagers—especially the tribals, because the tribals reared
silkworms in their houses, then sold the cocoons to the silk farm. And the
silk farmers would take the cocoons, take the silk out of it, and send it out
for processing. Some government scheme came along that said that this sub-
contracting was to be stopped, and all rearing to be done in the silk farm by
people to be hired by the government. And this had major employment consequences for the village. So they were very angry about this decision by the
government. Kishor Bharati started hassling the manager, and meanwhile
going up the line of authority to the Hoshangabad collector's office and the
Bhopal secretary's office. So they worked at both levels: pressuring the silk
farm itself, which the villagers were already planning to do, and linking it
together with this other network, to see if anything could be done.

K. FORTUN: Did it cause anxiety when Kishor Bharati moved into this type
of advocacy, or was it just an obvious extension of their educational work?

S. SARKAR: There must have been lots of debates. Kishor Bharati was the
kind of place where you debated everything—even whether someone was
going against the interests of the group if she took six chapatis instead of
four chapatis from the kitchen every day to feed her three dogs. I mean, it
was a paranoid place, man.

K. FORTUN: Sounds familiar.

M. FORTUN: Back up. How did you convince them you were qualified?
Here's this banker boy who had no experience in education . . .

S. SARKAR: In 1987 there was a dearth of people going into the voluntary
sector. There were lots of jobs outside the voluntary sector.

K. FORTUN: Does this support the argument that one of the reasons there
was so much activism in the seventies was because there were no jobs for
university graduates?

S. SARKAR: Absolutely. That was a major factor.

M. FORTUN: But, again, what were your qualifications?

S. SARKAR: Before I went to Kishor Bharati, I was told that they were going to try to run some schools. And they asked, "do you like kids?" "Yeah."
"Then you'll enjoy working with them." I think I said, "I guess so." "OK,
come." I'd already given in my resignation letter to the bank. I'd done that
without even seeing the place. So three months later, I just landed up.

   In a sense it was a very liberating kind of atmosphere. To begin with,
Kishor Bharati was an oasis of English-speaking people in a completely non-
English-speaking landscape. And people were extremely bright, intelligent,
thinking. So I did a lot of reading and talking with people there. By and by
work came along. We started by telling stories to kids, and hearing stories
from them in return. And then asking them if they would mind writing it
down, and then writing them down, cyclostyling it—you know, a cyclostyling machine?—stapling it together like a magazine, and bringing it out every week. And selling it at 25 paisa each in the villages, in the market.

   Then we decided that a group of kids who used to come to this place
where we met every night would form the nucleus of the evening school.
Since the project that was paying me was a Government of India project,
we had government permission to try the same thing in regular government

schools. There were five schools where we were trying the same experiment. The whole idea was to get kids to be able to read and write. Kids entering middle school could not even construct a simple sentence in writing, and could barely read the textbooks.

K. FORTUN: Why did you need to have an evening school? [phone rings]

S. SARKAR: Evening schools because sixty per cent of the children in most Indian villages will never be able to attend the regular government schools. For the simple reason that they have to work, at times that clash with school timing. So if the school is to be taken seriously, then the school timing has to be changed to accommodate kids, not the other way around.

I did this for about two and half years as a teacher, and then as an absentee teacher and a "picking up the pieces man" for about eight or nine months.

M. FORTUN: What's a "picking up the pieces man"?

S. SARKAR: "Picking up the pieces man" means that work of this kind, invariably, if it follows the interests of the community—which is a community of landless or small farmers—ends up having interests in land and property—in work that does not match that of the voluntary institution's official mission. There was a clash in Kishor Bharati over whether teachers in the children's activity program could stand for panchayat [local] elections. KB was being hassled by local elites with warnings that outsiders should stay outside of politics. And these people had political connections that could sabotage the next year's funding for this and other programs that KB ran. This became a bone of contention with members of our team who were standing for election in one of the villages. Because they believed it would give them a certain political voice—a voice that was not easy to get and was coming their way after a long struggle. Most of the teachers were village people. Tejy and I (we ran the project together) were the only regular non-locals. [phone rings]

S. SARKAR: So it was pointed out, in the middle of an argument, that in the constitution of Kishor Bharati there is a clause that says that office bearers of any project of Kishor Bharati cannot stand for election into the village panchayat. So Kishor Bharati, being the institution that controlled the purse strings, declared one fine day: shut your work. So we were forced to move out.

But we didn't want our work to stop. So what was decided was that Tejy and I (we ran the project together) would shift office to Bhopal, get the money through Eklavya and continue work in the village, until the official project timetable ended. Once the official project ended we would not have further responsibility for people's economic well-being. Many members of the team had become dependent on the income the project provided, so we couldn't just shut it down at the whim of Kishor Bharati.

In September 1990, it was all over. But the story doesn't have such a bad ending, because today there are something like ten schools running in ten different villages, all run by either members of the team or students who passed through those schools.

M. FORTUN: Then? . . . .

S. SARKAR: After Bhopal I moved to Delhi, looking for work. . . . Then, through a series of coincidences, a friend in Delhi University—Professor

Krishna Kumar—mentioned how somebody was looking for somebody to work with him in a television company. So I found out his name, address, phone number, and landed up at his office. He was the general manager of a production house, owned by the Times of India group of newspapers, which made programs for broadcast television. He got very excited that I worked at Kishor Bharati—"you mean you know science?" So I said "yes I know science, I know everything, what do you want?" But he had heard about Kishor Bharati, he had heard about Eklavya, he had heard about the Hoshangabad science teaching program.

. . . The main idea was to make a half-hour program on science, to be telecast once a month, in the beginning. But if there was a good audience, it would go to once a week. The stories were to be about science and technology all across the country. It was from an idea that the Ministry of Science and Technology had given to Doordarshan, that there should be such a show. The pilot had been lying somewhere for a long time, and once the competing channels started, Doordarshan decided to improve its image. So a lot of money was given to the program. It became a prestige program for the first two or three years that it was running: it had good time slots in prime time and good viewership—one of the highest for non-entertainment programs, second only to a program called "World This Week." About 36% of the Doordarshan audience, which was about 500 or 600 million people. So the first year was monthly programs, starting in October 1991, and then in September 1992 it became fortnightly for two or three months, and then weekly, from December 1992 until March 1993. Then it stopped for about 6 or 8 months. Then I went back on a freelance basis until September 1995, when I stopped for good.

M. FORTUN: Was there anything going on in the television industry at this time, that made new jobs open up?

S. SARKAR: No, when I joined it wasn't like that. The Gulf War hadn't happened. It was two or three months before that—I think I met him in November 1990. And the Gulf War was significant, because CNN came in here, BBC came in here. You got pictures of the war, and you somehow got to understand why your car was forced to stand for one hour before getting its fuel. But those two months put me in a different league altogether. Because suddenly, at the end of 1991, I discovered that my work was in major demand. Everybody's asking me to join them. You wonder why. And then you discover that all these different television channels that have started appearing all need people to run the software for them, and make the software for them. But there are very few people experienced in it. And I had experience by then. So, suddenly, I was in big demand. . . .

. . . [A]nd then the Wheel of Fortune and Lady Luck smiled upon this poor young man, and he made a killing in the Big City!

K. FORTUN: Is it proper to call you cosmopolitan? Where are you in the social structure?

S. SARKAR: I haven't a clue. I don't know how I would describe myself.

K. FORTUN: Well, we couldn't do it for ourselves, either. So just tell me yea or nay: Cosmopolitan?

S. SARKAR: I don't know what it means.

K. FORTUN: You are. Progressive? Yuppie?

S. SARKAR: I'm not a yuppie, but I have some number of yuppie friends.

M. FORTUN: And you're not a civil servant . . .

S. SARKAR: I'm not a civil servant.

M. FORTUN: . . . and you're not a banker.

S. SARKAR: I'm not a banker. I earn my living as a freelance film or TV producer.

K. FORTUN: Are you an intellectual? Well, you are, but . . .

S. SARKAR: Then why are you asking? [laughs] See, there is a way I can do it, which is by describing what I've heard others call me. People have called me a diggajh, which is basically a chappie—an organic intellectual, is one way of describing it. Others have called me a bloody bastard—actually a cynic. Another that I particularly like, actually my favorite one: they said, basically you figure out what makes Delhi tick, and you figure out how to use it for your own good—to get your life going. But this has been from people who are living in villages; no townspeople have come up with this one. Others think you're a social worker, and are very confused when they discover you don't make a living from social work. I don't know what social work means, but this is what they say.

K. FORTUN: How do you feel about the term activist?

S. SARKAR: [phone rings] Unclear. Hallo? [tape off] (Fortun and Fortun 2000)

EIGHT

Women's Movements

[handwritten: Find one example of performance advocacy? one form of advocacy. Think about victims voices]

(8.1) Beyond Their Means

In many accounts, 1986 was a year in which gas victims played few roles in significant events. Union Carbide struggled against hostile takeover by GAF Corporation. Right-to-know legislation was passed in the United States. In May, Judge Keenan dismissed the Bhopal case from U.S. courts on grounds of *forum non conveniens*. Keenan's dismissal led the government of India to file a suit against Union Carbide in the Bhopal District Court, claiming damages in excess of $3 billion. Soon after, the district court granted a temporary injunction restraining Carbide from selling any more assets—responding to Carbide's divestment during the GAF takeover bid, which many interpreted as a "put-on battle" to shift assists out of reach of the Bhopal litigation. Union Carbide filed a countersuit holding the government of India and the state government of Madhya Pradesh responsible for the disaster.

Meanwhile the Bhopal case reached the Indian Supreme Court for the first time, via a petition filed by two gas victims and their physician, Dr. Nishith Vora. Vora had run a clinic that administered sodium thiosophate—an antidote to cyanide poisoning. The clinic had been shut down by the Bhopal police in response to Carbide's insistence that cyanide could not possibly have been a factor in the disaster, despite physiological signs of cyanide exposure during autopsies. The Supreme Court ordered that Vora be allowed to reopen his clinic and have supplies of sodium thiosophate released to him by the Madhya Pradesh government. The Supreme

Court also appointed an "expert committee" to prepare a scheme for medical rehabilitation of gas victims and to collect information necessary for determining appropriate compensation. Official committee members asked to be relieved of their work the following year. But two nonofficial members did submit a report, calling for the establishment of an independent National Medical and Rehabilitation Commission. Though the report itself disappeared into the bureaucratic mire, the call for an independent commission continued to circulate, anchoring critiques of existing rehabilitation schemes. The commission was to be comprised of "eminent persons," government representatives, and gas victims.

Experts and "eminent persons" have played vital roles in response to the Bhopal disaster. The legal and scientific issues have been complex and of great import to gas victims, who were and are not prepared to engage them alone. But the role of gas victims in shaping legal and scientific issues is often discounted. The role of gas victims in figuring out identities and social forms responsive to the particular demands of disaster is rarely mentioned at all. What "counts" happens in courtrooms, corporate offices, and the bland offices of bureaucrats. "Locals" aren't often invited into these spaces. Sometimes they occupy them anyway—but not without contradictions. Such is the story of the Bhopal Gas Peedit Mahila Udhyog Sangathan—the Bhopal Gas Affected Working Women's Union—the largest organization of gas victims and the only one to regularly articulate victims' demands through public protest, press coverage, and litigation.

Over time, the Women's Union has developed an ideological and strategic program that demonstrates that sustainable response to the problems in Bhopal will depend on, while needing to transform and exceed, current government initiatives. The Women's Union has resisted the settlement of the Bhopal case, demanding legal recognition of "continuing liability" for the disaster. The Women's Union has opposed all official compensation schemes, arguing that these schemes promise to produce new forms of injustice. The Women's Union has formulated and publicized alternative rehabilitation schemes, integrating medical, economic, and political concerns. And the Women's Union has insisted that "Bhopal is no isolated misery," calling for continual collaboration with others threatened by disaster. In the process, the Women's Union has articulated a potent critique of bureaucratic organizational forms even while struggling to realize an alternative through their own work.

The Women's Union first mobilized to protest the closure of a government-sponsored sewing center that provided gas-affected women with stitching orders. Many of the 600 women employed were widowed or had husbands incapacitated by the gas. Their average wage of Rs. 320 ($7) per month was often the only income for an entire family. When the center was shut down in December 1986 and job orders terminated, the survival of many families was at stake. Within four days, the women formed an organization and staged a demonstration in front of the Chief Minister's residence. Various collective actions occurred over the following months, and the center was finally reopened, giving 2,300 women jobs.

During the first few months, the most involved women formed a steering committee to coordinate activities. Then they decided to extend their work beyond the problems of the sewing center. Most had seen advocacy at work in the year following the disaster. Some had even participated in protests. But none of the women imagined themselves in frontline roles. Successful agitation over the closure of the sewing center gave them a tentative sense of authority. But a sense that they could "do it"—become advocates— does not seem to have preceded their emergence as public figures. When they speak of the Women's Union's origins, women laugh and express disbelief. But they also say that they could not have done otherwise.

VOICES OF BHOPAL, 1990
BHOPAL GROUP FOR INFORMATION AND ACTION

Interview with Mohini, age 32, resident of Mahamayee Ka Baug
Our organization, the Bhopal Gas Peedit Mahila Udhyog Sangathan, started from a sewing centre. After the gas disaster a rehabilitation centre run by an organization was started in September '85 with government help. About 600 women used to be given sewing jobs from this centre. There were 30 of us who were employed for cutting cloth at the centre and this cut cloth was given to the women for sewing at their homes. In December 1986, this centre was closed down. All of a sudden the women who were dependent on the sewing job became jobless. The 30 of us decided that something must be done to get the centre reopened. So we, along with 600 other women, marched to the Chief Minister's residence. We went on several demonstrations and had to face the police on many occasions. In April '87, 225 of us were arrested and put in jail. It was a long and hard struggle. Most of us were quite sick due to the gas. During one demonstration, a woman named Hamida Bi fell unconscious with chest pain and died 4 days later. We finally managed to get the centre reopened and now 2300 women are getting sewing jobs.

After we got the centre reopened our organization grew in number and we took up the issues of medical treatment and economic rehabilitation of the gas victims. We also campaigned against Union Carbide, organized rallies demanding punishment of the guilty officials of the company and adequate compensation for all gas victims. We opposed the unholy settlement between Union Carbide and Rajiv Gandhi's government. On five separate occasions, more than 3000 women from the Sangathan have gone to Delhi and voiced our opposition to the settlement.

We have also filed a petition in the Supreme Court challenging the validity of the settlement and now it is being heard. Earlier in August 1988, we had filed a petition seeking interim relief from the Government. On 13th March 1990, the Supreme Court ordered the Government to pay Rs. 200 [$8] per person per month to all the residents of the 36 gas-affected wards of Bhopal for three years. This amount is being disbursed but there are a lot of problems in the manner in which this is being done. We know that the struggle against Union Carbide will be a long one and we are determined to carry on with our struggle till justice is done.

Interview with Shahazadi Bahar, age 35, resident of Barkhedi
I joined the Sangathan in 1988. I was looking for a sewing job. I went to a number of places all around Bhopal. Then one of my friends asked me to become a member of the Sangathan. She asked me to come for the Sangathan meetings and talk about my problems there. So I filled a form and became a member of the Sangathan. Now I am so closely attached with the Sangathan that when I do not go for the meetings, I miss it as people miss their dear ones.

The world is very selfish. I, too, joined the Sangathan with some selfish motive. I thought I could get some sewing job through the Sangathan. But though I have not been benefited, there are others who have. Quite a few people have got monetary relief of Rs. 1000 [$40], Rs. 3000 [$120], and Rs. 750 [$30] per month. And now the provision of interim relief of Rs. 200 [$8] per month per person is a big victory for the Sangathan. This has brought in a new hope and a new determination. We are certain that we will win this battle.

The Bhopal victims are entitled to compensation. We need hospitals, medicines, jobs, clean air and water. We have to have medical treatment centers in the community itself. The bigger things are, the more they create problems. Hamidia hospital is so big but we cannot get treatment there; only those with money are treated properly. We need jobs that do not need hard physical work. I get breathless when I walk and cannot see properly. Two of my daughters are being treated for tuberculosis.

They should not have allowed Union Carbide to set up its factory. When these companies want to set up some factory, they mention some product in the agreement (with the government) and they start producing something else. Then the people in the neighborhood do not get to know what is being produced. Workers in the factory are forbidden to speak to people in the community. Such factories should not be allowed to be set up; the neighboring community must be consulted.

The officials of Union Carbide should be given the severest punishment. If someone kills just one person, he is put in jail for 20 years. And here the Carbide officials have not been put behind bars for even 20 minutes. They should be hanged. I am certain that the Sangathan will win the battle. The struggle for truth will be a success. Truth always wins; it only takes a little longer.

Interview with Dinkar Rao, age 16, resident of Kazi Camp
When the gas leaked, we were all sleeping. We started coughing and getting choked. I thought someone was burning red chilies in the neighborhood. But

my mother said it was some kind of gas. She knew, she read a lot of books. She asked everybody to stay indoors but my father did not listen. He opened the door and went out to see. Thick clouds of gas filled the room. Our parents covered us up with a quilt from all sides. So we were a little protected. But my parents took in a lot of that gas. That is why they fell so sick.

My father could not do any work after the gas disaster. He used to remain sick and in 1986 he died. My mother used to be sick also. Doctors took x-ray pictures and said her lungs had been badly damaged. Some doctors said she had got tuberculosis but we do not believe that. In 1986, she was admitted to the Jawaharlal Nehru Hospital. She used to get breathless and used to cough all the time. She could not go to office to work. My mother died in February 1988 in the hospital. Since then, I have become a full time worker in the Sangathan.

We are opposing the settlement between Union Carbide and Government of India done in February 1989. The settlement would have meant that Union Carbide officials would have been let off without any punishment. We cannot let this happen. Carbide's officials must be punished. If these officials are let off easily, they will go on killing people and making them sick. What happened in Bhopal should not happen anywhere else.

Interview with Abdul Jabbar, age 36, resident of Rajendra Nagar
I am the convenor of the Bhopal Gas Peedit Mahila Udhyog Sangathan. I am a gas victim myself; my father died because of the gas. We in the Sangathan are fighting against a killer multinational and an apathetic government. Union Carbide is trying its best to evade accountability for the genocide it has committed. It is trying to wriggle out of the situation by using its wealth and its political power. The new government at the Centre seems to have taken a strong stand against Union Carbide. But the government has yet to take effective action for medical treatment of the gas victims. We have long been asking the government to set up a Medical Commission on Bhopal. The Medical Commission would concentrate on evolving a proper medical treatment for the gas-affected people. The Commission should also look into the medical categorization that the Directorate of Claims has done. The government has to provide opportunities for people to become self-dependent.

People outside Bhopal seem to have forgotten the gas disaster. Earlier a lot of concern was expressed for the Bhopal victims but now that seems to have died down. Even today hundreds of thousands of the gas-affected people continue to suffer from gas-related illnesses and people are still dying painful deaths. Yet most people seem to believe that the Bhopal issue is over. This is indeed unfortunate.

By late 1986, activism led by outside activists was dwindling. But there was still more to be said. So the Women's Union took a place in the relay, running off what had been said before, figuring out new spaces for movement. Soon, Union women had a double-edged strategy—taking on both micro and macro articulations of the Bhopal disaster. But the Women's Union had to figure out new ways to make things come together, without claims to systemic knowledge. Mohini remembers, explaining how "at first, maybe, we thought our work was just at the sewing center. But we came to know it. The sewing center is where we see it—the gas—but it is elsewhere also. It is all over the city. It is all over the world. So our work at the sewing center is also other work. We do not understand these things. But also we do. The gas is here, and there. We can see it, everywhere, at the sewing center."

The Women's Union also had to craft new configurations of responsibility. It is not as though they only thought in terms of self-interest before. They had never imagined any roles for themselves at all beyond the confines of the family. And even at home, their roles were prescribed. They did what was expected of them. The gas tragedy demanded something else. They were called to speak, in roles without models. They had to learn new languages, which connected them to things they had never seen or even imagined. They entered public space, burdened with obligations that required them to think beyond their means.

(8.2) Breaking In

The Women's Union's Saturday meetings are held in a park opposite the Women's Hospital. On one end is children's play equipment and a large grassy area, where the annual commemoration rally is held. The other end, behind public washrooms, on a barren patch of land, is the meeting place. Women begin arriving by midmorning. Most sit in small groups. Behind them, their message is painted boldly, in English: WE WANT JUSTICE.

Many walk, unable to squander even a few rupees on transport. Most come by the pigs, oversized taxis that crush passengers together, provoking a sharp defensiveness from Union women when they are traveling together. The first maneuver is physical. Seats and standing positions are arranged for minimal contact and general surveillance, with younger women packaged within the protection of older or more aggressive friends. Among themselves, the commentary is often quietly bawdy. Translated to the offending men, it becomes a harsh righteousness, emphasized by a waving finger and an occasional slap. The tone carries well, so other passengers look on.

Some of the women are dressed in sari, others in the black cover of burqa. Many bring small children, but few are very young themselves. Husbands and other family members resist these public events with a firmness that correlates to the age of the woman requesting freedom of movement. Part of the resistance is based on concern that women have difficulty finding their own way around the city and would most likely be given wrong directions by any man they asked for assistance, as a joke. There is also

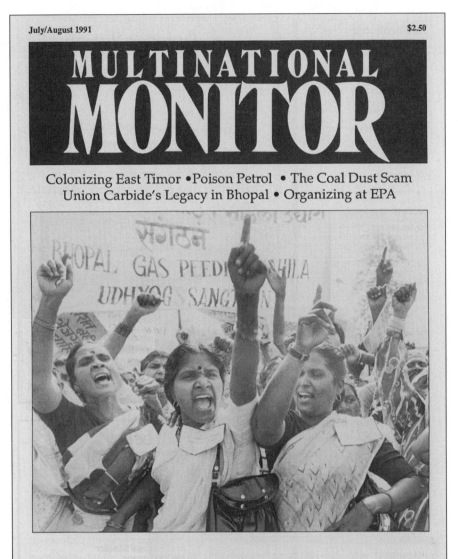

July/August 1991 $2.50

MULTINATIONAL MONITOR

Colonizing East Timor • Poison Petrol • The Coal Dust Scam
Union Carbide's Legacy in Bhopal • Organizing at EPA

THE PETROCHEMICAL AGE

FIGURE 8.1 BHOPAL GAS AFFECTED WORKING WOMEN'S UNION   Members of the
Bhopal Gas Affected Working Women's Union, on the cover of the July/August 1991 issue of
Multinational Monitor. The *Multinational Monitor* has continued to cover the disaster in Bhopal
and the record of the chemical industry.

concern about reputation. Unmarried girls could damage both their own chances and those of their younger sisters of a suitable marriage offer. Most agree, however, that it is not the women themselves who would act improperly, but the men whom they would encounter in public space.

For Muslim women, the burqa provides some protection, though their disdain for it is vehement. In the company of other women, or male relatives, the burqa is unnecessary. Nor is the burqa necessary when a woman is far from home, in the company of strangers. As one woman described, "[T]he closer you are to your own house, the more important it is to be covered. In Delhi, I may not wear burqa at all. At the park, my face is uncovered, the veil tossed back. At the kiosk, where I enter my own basti, everything, head to toe, underneath." Her resentment of the burqa mimes the metaphors of claustrophobia and blindness that permeate Bhopal: "A woman can't sweat properly. It's hot, and the skin can't get any air. Sweat pours down in great streams, down our necks, into our eyes. It covers you. The salt burns the eyes. You see even less, as you move along, stumbling over children, dogs, everything that crosses your path."

At the park, some women sit alone, watching others arrive or working on their stitching projects. Others sit in groups, joking around about husbands and other officials, telling stories about encounters at the sewing center, recounting what they have learned about other "people's" struggles and their strategies. Help is available for those who need claims forms filled out, but don't have the literacy skills. Those who can read and write find a place on the ground beneath a tree and take one form after another.

At some point, Hamida Bi's husband arrives to set up the microphone and loudspeaker. By around 11:00, Jabbar arrives on his motorcycle. Jabbar is the elected convenor of the Women's Union.

While I was in Bhopal, women usually spoke the first words. They took brief turns at the microphone to speak of the continuing injustice lived out in Bhopal, of the relationship between Bhopal and other victimized areas, and of the need for collective solidarity. The emphasis was on failures of procedural justice, evidenced by both the form and the content of official response to the needs of victims. Arguments were made "in principle," so ad hoc solutions were scorned. What the women wanted was programmatic initiative to take care of gas victims. Which meant that existing initiatives had to be swept away and replaced. What they wanted was to clean house, from top to bottom—then to set everything up differently.

Like many of India's poor, most Union women never expected to be provided for by the state in a positive sense. Nor, however, did they expect a relationship with the state by way of the negative. Insufficiency could be lived quietly. Asphyxiation could not. Demands for rehabilitation thus operate through logics of austerity rather than abundance, transfiguring conventional notions of entitlement. The emphasis was on degradation, rather than benefit. Union women demanded material resources from government programs. They do not think of these resources as "benefits," guaranteed by a government understood as provider. The government was not expected to provide, but neither is it permissible for it to degrade.

BHOPAL GAS PEEDIT MAHILA UDHYOG SANGATHAN
51, RAJENDRA NAGAR, BHOPAL 462010

Dear friends,

The genocide that began in Bhopal on 2nd December 1984 has been continuing for the last five years. The fifth anniversary of the greatest industrial disaster of the century—the Bhopal Gas disaster—will be observed on 3rd December 1989.

The year 1989 has been significant for the over three hundred thousand gas victims of Bhopal who have been maimed by Union Carbide's gases, for the over ten thousand members of the Bhopal Gas Peedit Mahila Udhyog Sangathan (BGPMUS) who are struggling for justice in Bhopal and for the individuals and groups who are supporting this struggle. This year on 14–15 February an unholy settlement was reached in the Supreme Court between the killer multinational and the Indian government claiming to represent the interest of the gas-affected people. Through this settlement an attempt was made to absolve Union Carbide of all its liabilities by accepting an amount that was one seventh of that claimed as damages. It was also attempted to drop the serious criminal charges lodged against officials of Union Carbide responsible for the disaster.

Along with other organizations seeking justice in Bhopal BGPMUS has challenged the settlement in the Supreme Court and the settlement has been in a way, stayed. Apart from challenging the settlement a few lawyers have also questioned the constitutionality of the Bhopal Act, that was passed in March 1985 and that has denied the Bhopal victims their right to sue Carbide in court. The lawyers have concluded their argument against the Act, in May but a verdict is yet to be given. Things are also moving at a slow pace on the matter of challenges to the settlement. There have been only two hearings in the Supreme Court on the matter of the settlement in the last nine months while the same court took just five minutes to put a stamp of legitimacy on the illegal, unconstitutional and immoral settlement. Attempts are being made to keep matters lingering on so that the Bhopal victims get too tired to carry on their struggle and are forced to lay down their arms and accept the settlement.

The anti-people nature of this collusive settlement is further getting revealed through the proceedings of the Supreme Court on the matter of interim relief. The BGPMUS had filed a petition in this court for provision of relief by the government in August 1988 and this is being heard since April 1989. During these hearings, in August 1989 the Directorate of Claims of the government has furnished data pertaining to the damages suffered by the gas-affected people. According to this data, out of 123,560 claimants, 51,584 (42%) have suffered no injury; 64,064 (52%) have suffered only temporary injury and only 19 persons have suffered permanent disability due to their exposure to Carbide's gases. Anyone familiar with the terrible consequences on the Bhopal gas disaster can well realize that this is yet another exercise of playing down the miseries suffered by the people of Bhopal. Several experts have joined the Bhopal victims in condemning the exercise of medical documentation on which this data is based. It is becoming very clear that since the settlement amount is so low, the extent of damages is being tailored to suit the amount of settlement received. While the Sangathan has asked for relief to be provided to 106 thousand families only 8460 families have been provided with monetary relief ranging from one thousand to three thousand rupees. It is worth mentioning that the government has spent Rs. 91 crores on the relief and rehabilitation of the gas victims in the last five years while it has spent more than three thousand crores killing people in Sri Lanka. Such small amount distributed among so few people does not provide any meaningful relief and instead causes division among people and tension within communities.

Attempts to crush the movement of gas victims go on unabated. On 9th August 1989 police rained sticks and stones on women who were demonstrating before Union Carbide's Research and Development Centre. Many heads and limbs were broken and over 150 women were injured by the cops during this demonstration that was organized by the BGPMUS as part of the "Union Carbide Quit India Week." Currently, goons have been employed by ministers in the state government to physically attack members of the Sangathan.

Despite all this, the Sangathan has resolved to carry on with the struggle for justice in Bhopal. On three occasions so far Sangathan members have traveled to Delhi and participated in protest actions against the settlement. The struggle for medical care, relief, rehabilitation, adequate compensation and punishment to the guilty officers goes on in Bhopal. We are struggling for the over three hundred thousand gas affected people of Bhopal whose bodies have been ravaged by the poisons from Union Carbide and thirty to fifty of whom are dying painful deaths every month. We are struggling to change the face of Bhopal. We are struggling so that adequate compensation is extracted from Union Carbide and its guilty officials are punished. We are struggling so that those who produce poisons for profit are taught a lesson. We are struggling so that Bhopal does not happen anywhere in the world, ever. You have played a significant role in this struggle and we hope that you will be with us in the struggle that lies ahead.

With such hopes we invite you to Bhopal on the occasion of the fifth anniversary of the gas disaster. BGPMUS would be organizing protest action from 1st to 3rd December 1989 and we would like you to participate in these actions and lend your strength to the movement.

> This fifth anniversary has a significant place in the struggle for justice in Bhopal. This is the year of the collusive settlement between the Indian government and Union Carbide. This is the year of the struggle of gas-affected women against the unholy settlement. This is the year of betrayal by the government and cruel machinations of a multinational. This is the year of courage and unity of the gas-affected people. On this fifth anniversary of the Bhopal genocide let us sing the song of life together, against the merchants of death.
>
> We remain in solidarity.
>
> Bhopal   Abdul Jabbar Khan, Convenor
> 2 November 1989   Bhopal Gas Peedit Mahila Udhyog
>   Sangathan

Jabbar's turn at the microphone was more informational at the start, crescendoing into a general critique that reiterated the emphasis on procedural justice voiced by women. The first phase focused on local-level problems and possibilities: filing claims, receiving interim relief payments, being admitted to hospitals, negotiating with doctors over how symptoms were categorized, obtaining medicines without having to pay the fees demanded by private vendors. The second phase focused on the progress of the law, including minute detail illustrating how the demands of victims were continually foreclosed via subtle bureaucratic maneuvering. The third and final phase broadened the scope of Jabbar's critique, reversing official insistence that Bhopal is "fully and finally" over. Jabbar was strikingly adversarial. His rhetoric was forceful and well punctuated. Each Saturday meeting concluded with slogans: "Victory to the victims." "Union Carbide Quit India."

(8.3) Continuing Liability

The settlement of the Bhopal case was upheld on October 3, 1991, by a ruling of the Indian Supreme Court. Shortly before, the rupee had been devalued against the dollar. This devaluation, figured in with accrued interest, more than doubled the rupee amount deposited in the Reserve Bank of India for compensation. With 15 billion rupees to work with, official distribution efforts began. Victim activists were little encouraged, working on the premise that any government effort would be toward veiling the magnitude of the disaster and correlate welfare responsibilities. Victim perspectives were not just a matter of disfaith in the political will of authorities. Nor did victims believe that the problem was just a matter of insufficient, unreliable information. The numerous occasions when government authorities had destroyed documentation and other evidence proving injury and need had not been forgotten. Neither had the "institutionalization of silence" just following the disaster, when the clampdown on information reached from the highest officials to the ward nurses at government hospitals.

Alternative approaches to rehabilitation, proposed in press releases and letters to government officials, insist on fundamental changes in the way suffering is calculated, in the way relief is distributed, and in the way care for victims is institutionalized. Most fundamental of all is the demand for "continuing liability."

PROPOSAL FOR A NATIONAL COMMISSION ON BHOPAL
BHOPAL GAS PEEDIT MAHILA UDHYOG SANGATHAN
51, RAJENDRA NAGAR, BHOPAL 462010

The purpose of the commission is to correct the problems that have impeded past rehabilitation efforts. Further, due to the long term consequences of the Bhopal disaster, it is important that there be a permanent evaluatory and administrative body that is uninterrupted by the various priorities of specific governments. . . . Without such an autonomous body to evaluate and administer aid programs, the rehabilitation effort will remain ineffective and victims will continue to suffer. Further, the pattern of rehabilitation in Bhopal will set an important precedent for dealing with future disasters in India. In appointing itself parens patriae of the gas victims under the terms of the Bhopal Act, Government of India introduced a way of dealing with mass torts that purports to be the best way to serve the interest of victims, particularly when the victims are uneducated and thus unable to successfully negotiate for themselves. Thus far, however, the role of Government of India has blatantly gone against the interests of victims and catered to the Union Carbide demand to be exonerated of all liability. . . . Government of India has further ignored the guardianship provisions of the Bhopal Act by failing to monitor the effectiveness of rehabilitation programs. Government attention has been limited to concern about press representations and has not seriously queried whether the suffering of victims has been mitigated. The National Commission on Bhopal must assume responsibility for ensuring genuine rehabilitation through an open, publicly accountable process that always privileges the interests and authority of gas victims.

Due to gas related damage to immune systems, Bhopal victims are prone to debilitating infections. Hence, provision of proper housing must be included as part of the rehabilitation package. Essential requirements include provision of clean drinking water, of sanitation systems and of pucca [hard wall] houses. Preliminary tests have shown continued contamination of soil and water in areas near the Carbide factory. Environmental clean up must be seen as part of the overall effort to provide satisfactory living conditions.

Direct distribution of cash is not a sustainable way to provide for the economic needs of gas victims; direct dole continues victims' dependence on

the government and thus continues the usurping of local control that precipitated the Bhopal disaster. A democratic and sustainable response to the economic needs of gas victims requires the generation of cooperatively managed jobs appropriate for persons suffering from gas related disabilities. The
Bhopal Commission must oversee training programs and ensure the availability of non-exploitive contracts. The Commission must also ensure that
job generation initiatives produce real results: the loans previously given to
approximately 25,000 people to help them start their own economic activity
has not enabled even a small percentage of them to become self supporting;
technical training given to approximately two thousand victims has not led
to gainful employment. The only sustained job generation scheme for gas
victims has been the provision of sewing jobs to 2,300 women, giving them
a monthly income of approximately Rs. 340 [$13].

Claims processing assistance needs to be available so that gas victims
have access to government programs, including direct distribution of money.
Because the majority of the gas-affected population is illiterate and generally
inexperienced in dealing with officials and bureaucracies, they often cannot
access available resources. Volunteers are needed to assist with forms for interim relief, for entry into educational programs, for access to jobs, etc. It is
also necessary to set up a system for continued monitoring of relief officials
so that corruption is minimized.

The Women's Union argues for equal disbursement of compensation to
all gas victims based on geographical indicators of exposure—as the only
way to avoid the creation of further stratification within the gas-affected
community, as the only way to avoid the absurdity of expecting gas victims
to prove their case in compensation courts. The Women's Union also argues
for income generation projects rather than cash relief and for a decentralized health care system that maximizes patient control over the health care
process.

The Women's Union implicitly challenges contractual concepts of compensation that assume that money can redeem harm. Women's Union proposals also challenge medico-legal logics of cause and effect that place the
burden of proof on victims as well as generalized notions of health limited
to the biological concerns of individual bodies. Acknowledging injury as
an effect on the overall health of a community compels an understanding
of "rehabilitation" much more complex than simple provision of cash payments to individuals. According to the Women's Union, caring for health
is not only a medical problem, but also a problem with all life support
systems and the exploitive logics and structures that sustain these systems.
The Women's Union's demand for continuing liability, and for a National
Commission on Bhopal to sustain it, embodies this argument.

One of the tasks that would be assigned to the commission is complete reconfiguration of the method through which the health effects of gas exposure are determined, shifting focus from diagnostic categorization to treatment possibilities. A corollary task would involve reorganization of rehabilitation efforts in ways that offset rather than exacerbated stratification, whether among victims or between patients and doctors in government hospitals. Opposition to hospital-based rehabilitation is based on a definition of health that encompasses social and economic factors often ignored in conventional programs for disease mitigation. Demands for housing provision are articulated as part of overall recovery. Demands for jobs are situated within similar reconfigurations.

Alongside demands for fundamental structural reform of rehabilitation efforts, the commission would seek to overturn the "full and final" clause of the settlement such that Union Carbide is subject to the continuing jurisdiction of the courts. The demand would be based on the argument that without legal acknowledgment of continuing liability, there is no way to ensure sufficient funding to care for victims of toxic exposure, particularly when the long-term effects of exposure remain undetermined.

The demand for continuing liability is, in part, a practical demand—for long-range rehabilitation and medical monitoring schemes, overseen by a specially appointed, permanent commission made up of gas victims, voluntary-sector social activists, health care professionals, and appropriate government officials. The demand for continuing liability is also figurative, calling for conceptions of responsibility other than those codified by formal law. The law works on the premise that "full and final" settlements are possible. And the law assumes symmetrical intersubjectivity. In other words, parties involved in disputes are considered equally capable of representing their positions—and equally capable of accessing the fairness distributed by legal judgments. A fair judgment is, by definition, one that is free from bias. To distribute justice is to actively discount difference.

The notion of continuing liability twists the logic—foregrounding the enduring asymmetry of the relationships through which gas victims will negotiate for resources, authority, and survival. Demands for continuing liability ask that acknowledgment of asymmetry be built into the law itself. Demands for continuing liability also refer to something beyond what the law can accomplish, even in its most perfect instantiation. It is a call for the law itself to be transfigured as well as for responsibility beyond the law.

(8.4) Ritual Openings

Every year the Women's Union writes a letter inviting people to commemorate the anniversary of the Bhopal disaster. Every year people come to

Bhopal and make the passage from Shahajahani Park, through the gas-affected areas to the Muslim burial grounds, to the cremation grounds, and then to the Union Carbide factory. It is a ritual that marks the passage of time without change. It is a time of grieving for individuals who have died or who are sick, but also for a loss that cannot be named. It is not only a loss of faith in the government or systems of law, though the anniversary always marks the absence of such faith. The grief is not so much a response to what should have been as it is to what will come. It is the future that has been lost, and that the anniversary provides space to mourn.

The Women's Union's anniversary letter is always densely descriptive, providing a detailed update on the ways disaster in Bhopal continues. Usually, it is printed in both Hindi and English. It is sent to other groups opposing "destructive development," in India and abroad. It is sent to individuals on a long mailing list that reads like a census of middle-class Indian progressives. And it is sent to journalists, who are expected to cover Bhopal yet again, recognizing the global significance of the disaster as well as its particularities on the ground.

THE 12TH ANNIVERSARY OF THE UNION CARBIDE DISASTER AT BHOPAL
DEC. 3RD, 1996
Bhopal, 22.11.1996

Dear Friends,

As we do every year, we are writing to you on the occasion of the anniversary of the December '84 Union Carbide disaster in Bhopal. As you are aware, 12 years have passed and the people poisoned by Union Carbide are still living in pain and dying painful deaths. Union Carbide does not shed a tear nor do government officials grieve over the slow death of an entire community. Despite odds we in the Bhopal Gas Peedit Mahila Udhyog Sangathan (BGPMUS) are continuing with our struggle against the murderous duo of Union Carbide and the Indian Government to protect our life and dignity.

As a part of our long battle, this year on December 3rd we will remember members of our families, our friends and neighbors who died untimely deaths due to Carbide's poisons and the Government's neglect. On this day we will renew our resolve to continue with our struggle for punishment to the culprits and for improvement of our condition. We request you from the depths of our hearts to come to Bhopal on this day to join us in our mourning and in our battle for justice.

Justice on the disaster has never in the last twelve years been as violated as she has been this year. And in the Supreme Court of India, too. By diluting the criminal charges against accused Carbide officials Keshub Mahindra, Vijay Gokhale, Kishore Kamdar, J. Mukund and others, the Supreme Court has paved the way for these criminals to go scot-free. In his order dated

September 13, 1996 Chief Justice A. M. Ahmadi diluted the charges under Sec. 304 (Part II) (Punishable by a minimum of 10 years imprisonment and fines) against the nine accused officials of Union Carbide India Limited (now renamed Eveready Industries India Limited) to 304 (A) (Punishable by a maximum of 2 years imprisonment or fine). Earlier this year in another case, the Supreme Court has ruled that if it takes more than two years for the trial to commence on a criminal case involving an offense punishable by less than three years of imprisonment then the case will be dismissed.

The dilution of criminal charges against Keshub Mahindra and others has now made the extradition of Warren Anderson and other foreign accused more difficult. Despite the issuance of nonbailable arrest warrants more than four years back, the Indian Government has done nothing to make these killers face trial in an Indian Court. The Officials of the Central Bureau of Investigation (CBI) do little else than express their helplessness in taking legal action against the foreign accused. It is quite clear that the Indian Government is party to the conspiracy of letting the perpetrators of the Bhopal genocide go scot-free. Why did the CBI, that takes its orders from the Prime Minister, not file an application for revision against the September 13, '96 order though it had all the factual evidence and legal arguments to demonstrate where the Judges had erred? Why is it that the Indian Government instead of arresting Mahindra congratulates him on his new joint venture with Ford, USA, speaks in praise of Mahindra's role in the globalization of plunder? In these times [when] the entire nation is losing its sovereignty before multinationals and particularly American multinationals, the Indian Government is bent on its knees before Union Carbide that has committed one of the worst crimes against humanity.

Encouraged by the attitude of the Supreme Court and the Government, the managers of Eveready Ltd. are now dismantling the Bhopal factory. The factory is supposed to be under the custody of the CBI but the Government remains mum at this blatant destruction of evidence in the criminal case. Having received no response to its complaints made to the Central Home Minister, the Chief Minister and the police, the Sangathan has got the demolition temporarily halted through legal action.

In the last twelve years the Governments of the state and at the centre have served the interests of Union Carbide while being deliberately negligent toward the deteriorating situation of the victimized people. The Government keeps on building newer hospitals in the name of providing medical relief. Soon the people in Bhopal will have more hospital beds available to them than is available in countries in Europe and USA. But hospitals in Bhopal have little apart from beds. Doctors at the Government hospitals have all opened private clinics and it is not possible to get adequate treatment without first visiting these private clinics. Equipments donated by international agencies are lying unused and hardly any investigation is carried out in the hospitals.

People who continue to suffer from exposure related illnesses hesitate to visit the hospitals because of the long queues and rude behavior of doctors and other hospital staff. In the last twelve year about Rs. 150 Crore have been spent on the medical relief of gas victims but the line of treatment remains almost the same as it was on the morning of the disaster. Neither the treatment has changed nor has there been an improvement in the condition of the people. The hospitals that require a sum of Rs. 50 lakh (US $140 thousand) to meet their monthly expenditure have now only been allotted Rs. 14 lakh (US $40 thousand) per month. For months now, distribution of free medicines to people who are too poor to buy has stopped. The Sangathan has filed a complaint to the Lokayukta on the involvement of the Minister of Gas Relief in a Rs. 1.3 Crore (US $370 thousand) drug purchase scam but the matter remains to be taken up. The Kamla Nehru Hospital was to be built at a cost of Rs. 9 Crore (US $2.5 million) but so far Rs. 20 Crore (US $5.7 million) have been spent and it is still not complete. We in the Sangathan have made repeated demands for decentralized community based health care but they have been ignored and medical relief has remained confined to the building of hospitals. And hospitals in Bhopal are no longer [a] place for medical care but opportunities for the ministers, officials and doctors to make money.

The work of monitoring the health status of the people and research on their illnesses by the Indian Council of Medical Research was discontinued in 1994. It was decided that the State Government would step in to continue with this responsibility. However, till today the Centre for Rehabilitation Studies has not even initiated any research work. There is now an alarming rise in the number of children infected with tuberculosis and people are dying of cancers of the throat and stomach. And there are no records of these deaths. ICMR studies have outlined the likelihood of manifestation of newer problems in the long term and even among the future generations. But government doctors and researchers have paid little attention to these research findings.

Ian Percival a British attorney in the Bhopal Hospital Trust is building a hospital away from the city for the gas-affected people. The hospital is being built with moneys that have been attached as part of the criminal case against Union Carbide. There would be a 30-bed ward for heart surgery in this hospital that is of little relevance to the problems of the gas-affected people. The bureaucrats who are most likely to be the beneficiaries of such facilities never tire of kow towing before Percival, an agent of Union Carbide. Quite obviously the gas-affected people cannot benefit from this hospital that is 5 kms away from the gas-affected area. It is another moneymaking opportunity for the officials and the ministers and for Union Carbide; it is the mask of humanitarianism to conceal its criminal identity.

For the last eight years we have been demanding that the Government set up a "National Medical Commission on Bhopal" with government and non-government professionals as well as representatives of survivors organizations. The Commission should have sufficient authority and resources at its disposal to carry out long term medical care, research and rehabilitation of the gas-affected people of Bhopal. The officials who came to power with each successive government in the last twelve years have been informed by the Sangathan about the necessity of a National Commission but so far there has been no serious consideration on the proposal by any of the officials let alone any action.

The present state government has broken previous records of official incompetence with respect to the economic rehabilitation of the gas victims. More than Rs. 60 crore (US $1.7 million) have been spent from the public exchequer and yet less than 100 persons have found gainful employment through official initiatives. The Chief Minister has repeatedly assured us that the sewing centres that had been closed by the BJP government in July '92 will be reopened. These have been, as we have learnt to our dismay, hollow assurances. The 2300 gas-affected women who have lost their jobs and livelihood are now tired of reminding the Chief Minister of his promises. The 152 work sheds built in the Govindpuur Special Industrial Area at a cost of Rs. 8 Crore (US $2.3 million) that were supposed to employ 10-thousand gas victims have not employed a single person. Half the work sheds have been converted into barracks for the Rapid Action Force and the rest have been distributed among influential personas. Today in Bhopal there are at least 50,000 people who have been so incapacitated due to Carbide's poisons that they can no longer pursue their usual jobs. Gas-affected women suffering from breathlessness and persistent cough cannot roll bedis or cut betel nuts to earn a living and are facing starvation. For many families it is a painful decision that has to be made each day—whether to purchase medicine or food.

The gas-affected people's hope of getting medical treatment and improving their condition with the money they receive as compensation has been in vain. Over 95% of the claimants received only Rs. 25 thousand (US $700) as compensation. Out of this about Rs. 10 thousand (US $280) are being deducted against the interim monetary relief paid by the government. The rest of the money received in compensation gets expended in repayment of debts. Nearly half the cases of death claims have been rejected and compensation has been denied wrongfully. The ignorance of the judges about the effects of the gas exposure on the human body and the inability of the claimants to pay bribes has been the principal reason behind the wrongful rejection of death claims. The Government has stopped monitoring or recording of exposure related deaths since December 1992.

Friends, the continuing death, disease and misery caused by Union Carbide is an issue that concerns not only the six hundred thousand people in Bhopal. It is an issue that concerns people all over the country and indeed all over the world. If the officials of killer Carbide go scot free, it will encourage hazardous corporations to continue to make profit by imposing risks on the life and health of the people. We urge you to join our battle for life against those who threaten the survival of our planet. We appeal to you to support our demands and help us keep our resolve.

Our demands:

1) The Indian Government must take immediate steps to extradite Warren Anderson and the authorized representatives of Union Carbide Corporation, USA and Union Carbide Eastern, Hong Kong.
2) The Central Bureau of Investigation (CBI) must file a revision petition against the Sept. 13, 1996 order of the Supreme Court diluting the charges against Keshub Mahindra, Vijay Gokhale, Kishore Kamdar and other accused officials.

3) The Indian Government must set up a National Medical Commission on Bhopal for long term health care, monitoring and economic rehabilitation of the survivors of the disaster.

4) The toxic contamination of the soil and ground water in and around the Union Carbide Factory must be investigated by a public body composed of government and non-government scientists. Union Carbide must pay for the rehabilitation of the degraded environment and for provision of safe drinking water to the neighboring community.

5) The Government's expenditure on the relief and rehabilitation of the gas affected must be reviewed and the Five Year Action Plan of the state government should not be sanctioned without the approval of the survivors' organizations.

6) Immediate steps must be taken to review cases of compensation in which gas victims have been denied appropriate compensation.

7) A memorial for those killed by Carbide's poisons must be constructed and December 3 must be declared a National Day of Mourning.

8) Action must be taken against the management of Eveready Industries Limited for destruction of evidence through demolition inside the Bhopal factory.

We request you to come and join in the activities on the occasion of the 12th Anniversary in Bhopal. Please reach Yadgar E. Shahajahani Park, 1.5 kms from the Railway Station, at 10 am on December 2, 1996.
We look forward to your participation and support.
In Solidarity,
Abdul Jabbar Khan
Convenor

Journalists in India have covered the Bhopal disaster every year, often fleshing out their stories with the detail provided by the Women's Union's annual letter. Some come to Bhopal itself to interview ministers, doctors, and Jabbar Khan—the convenor of the Women's Union. Most take photographs of the graffiti on the walls of Bhopal demanding justice—and that Warren Anderson be hung.

Warren Anderson was the CEO of Union Carbide in 1984. Every year at the anniversary rally, a huge effigy of Anderson is burned. In effigy, Anderson looks rather Indian—with black hair, a curled mustache, and Asiatic eyes. The effigy stands for something that is Indian, American, and more. It is a way to embody a dysfunctional global system—and to indict the concepts and institutions that legitimate it.

News stories on the anniversary of Bhopal often read like litanies listing the many ways rehabilitation programs in Bhopal have failed. The anniversary rallies commemorate these failures—alongside the failures that occurred before the gas leak, leading up to it. Time is folded together. The origin of suffering in Bhopal is difficult to locate.

The Women's Union returns—again and again—to specific moments that could have set up the future otherwise. The moment when Union

Carbide received government approval to produce pesticides in an area adjacent to residential colonies. The moment when gas victims began to be counted and categorized, and the moment when official categorization schemes were sanctioned by law. The moment when the pursuit of liability was shut down through "full and final" settlement of the Bhopal case.

The members of the Women's Union insist on remembering these moments to contest the origins of things. In part, their objective is to expose the role of the government of India in the production of the disaster. In part, their objective is to expose the magnitude of the disaster and how it continues. The Union women want the historical record set right. But what they want cannot be accomplished merely by filling in what has been forgotten, once and for all. What the Women's Union wants is for the Bhopal disaster to be remembered differently—in a way that is not conclusive, in a way that admits continuing liability, in a way that acknowledges how the future folds into what has come before.

In other words, the Women's Union wants historical recognition not only of the facts of the disaster, but also of the structure of disaster. Official recognition of all that gas victims lack is important. Even more important is official recognition of the structural contradictions within which gas victims are positioned.

Gas victims were not allowed to opt out of the settlement of the Bhopal case because they were deemed juridically incompetent. The settlement itself nonetheless threw the burden of law back to gas victims—who have had to seek compensation before the claims courts set up in Bhopal. The asymmetry of the social relationships within these courts in extraordinary, and only hints at the asymmetry of the global system in which the courts themselves are situated. Faded, watermarked papers indicating a categorization or a long hospital stay are the way to truth and money. Many gas victims cannot read them themselves.[1]

The Women's Union does not only argue for more comprehensive calculation. It also wants recognition of the brutal asymmetry in which gas victims live and struggle to survive. It is historical perspective that they want—and that the settlement has effaced.

(8.5) Law of the Grassroots

When the members of the Women's Union decided to broaden their commitment to a range of issues shared by gas victims, they began to feel the need for male leadership. They felt a man would be taken more seriously, would have more experience, and would have greater freedom of mobility. Mobility became increasingly important with the prioritization of legal struggle and the subsequent need to travel to Delhi to present demands.

The man chosen to lead the Women's Union was Jabbar Khan, a gas victim who had participated in the organized protests of 1985 and in various relief projects in the colonies near where he lived. Jabbar has a charismatic style that quickly won the support of many women. Under his leadership, membership grew, and issues became both better defined and more widely disseminated. Weekly public meetings remain a significant source of information for the gas-affected community. Updates are given on the state of the legal proceedings, on health care options, and on possibilities for income generation. The meetings also provide practical help for dealing with the relief bureaucracy, particularly the problem of claims processing for those who can't read or write.

The insufficiency of official rehabilitation schemes has been harshly apparent since 1984. But the possibility of building alternatives has depended on resources and sanction from the government of India. Work for "no more Bhopals" has also required engagement with the law. So the Women's Union has had to maintain a dual, often contradictory, agenda: building rehabilitation at the grassroots, while engaged with the law, which has focused attention on Delhi and fostered a style of organizational leadership that reinstitutionalizes expert authority. The Women's Union's role in the legal process has been crucial, even if unsuccessful when judged according to the final legal judgments. But prioritizing legal initiative transforms the institutional structure of the grassroots. ← explain

The Women's Union's turn toward the law in 1987 was provoked by rumors of a settlement of the Bhopal case. Judge Deo of the Bhopal District Court is credited with initiating the discussion, reiterating American Judge Keenan's claim that settlement would be in the best interest of victims. Suggestions that rapid closure on the case was the only means to justice— in the form of relief and compensation—acquired the status of the obvious. An actual trial, it was assumed, would take many years, during which gas victims would be left to fend for themselves. By 1987, details on a possible settlement began to circulate. Union Carbide would pay the Union of India somewhere between $500 and $650 million as "full and final" compensation, in exchange for the dropping of all charges—including criminal charges. Oppositional response was vehement. The rumored settlement was denounced in the Indian Parliament. Ralph Nader wrote a letter to Rajiv Gandhi. Protest rallies were held in both Bhopal and Delhi. Members of the Women's Union traveled to Delhi for the first time, staging a day-long *dharna* at the New Delhi boat club—in an area reserved for such protests. The protest worked. Judge Deo set a new trial date and passed an order directing Union Carbide to pay $270 million in interim relief.[2]

Victims celebrated. News of the order quickly spread throughout the city. Colored powder was thrown, and people wept with a renewed hope.

The International Center for Justice in Bhopal, based in New York, issued statements suggesting that Deo's decision could provide a model for world-wide reform, delinking interim relief to victims of industrial disasters from the process of establishing liability—so to ensure that immediate relief is available without compromising final legal claims.

Judge Deo's order has been widely discussed as a possible precedent for handling mass injury cases in ways responsive to the urgency of disaster, while avoiding overly hasty processes of law.[3] Deo's order is also infamous because of the way it harbored the future by way of the negative. Being an interim order, it could not be decreed. So Union Carbide refused to pay and challenged the order before the Madhya Pradesh High Court in Jabalpur—on grounds that Judge Deo was not authorized to pass such an order under the provisions of the Indian Civil Procedure Code. Carbide's lawyers said the order introduced a "perverse" and unenforceable concept of law, which would only delay justice for victims. Union Carbide also demanded that Judge Deo recuse himself from the case, since his order was evidence of blatant prejudice. Carbide's application for recusal was widely interpreted as a move toward claims that Indian courts had not provided due process, which—according to provisions in the dismissal of the Bhopal case from U.S. courts—would exempt Union Carbide from an Indian judgment. The government of India took contempt of court action against Union Carbide's lead lawyer, Fali Nariman. The contempt actions would be cleared along with all other charges in February 1989.

By September 1988, the Indian Supreme Court had admitted two cross-appeals against the order for interim compensation, filed by the Union of India and Union Carbide. The Zahreeli Gas Kand Sangharsh Morcha—a coalition of progressive activists formed shortly after the gas leak—was granted the status of intervenor in the proceedings, setting an important precedent for the role of voluntary organizations in the legal proceedings. But no other parties—including the Women's Union—were allowed to participate. On the first day the cross-appeals were heard, Chief Justice Pathak again spoke of settlement. Apparently, there were talks between lawyers for Union Carbide and the government of India. There were few leaks, and no attempt to inform victims of what was being negotiated on their behalf. The announcement, on February 14, 1989, of a full and final settlement came as a shock. But the content and structure of the case were familiar. The amount of the settlement was $30 million less than the rumored settlement of 1987. And all further proceedings against Union Carbide were quashed.

Union Carbide described the court's decision as "fair and reasonable." Members of the Women's Union helped "bash" Carbide's Delhi offices.

Telephones were thrown through windows. Desks were overturned. News accounts told of the "militant mothers of Bhopal."

The Women's Union does not position itself within Gandhian traditions of nonviolence. But recollection of the property destruction following the settlement of the Bhopal case is reflective. Many women have told me that they thought it was the right thing to do, though it may not be right ever again. They were outraged. And they had to respond. Bashing windows was all they could do. And it helped them understand the limits of the language of law.

The material effects of the Women's Union's interventions in the Bhopal case are evident at every turn. Most acknowledge the Women's Union's role in securing interim relief payments for victims, even if six years after the gas leak. Most also acknowledge the role of the Women's Union in the review of the settlement of the legal case. Though the settlement was upheld, criminal proceedings against Union Carbide were reinitiated. Equally important, protesting the settlement provided a venue for demonstrating that the compensation scheme on which it was based is inadequate. The Women's Union has not overturned the system, but it has forced it to acknowledge its own contradictions. But the price of focus on the law has been high. Prioritizing legal initiative transforms the structure of grassroots organizations, sliding authority more and more toward a central figure able to represent local problems in the standardizing categories that the law demands.

Leaders of grassroots legal initiatives become links between an organization's members and official institutions, or metropolitan support groups. A phone call is received or a package delivered by mail, indicating a new development in the case that requires response. Often there is little time for consulting other members of the organization before reply. Even if consultation is possible, there is little room for dissent. Presentations to the courts require a unified position—a reduction of different perspectives to the one line that carries the force of collectivity. Presentations to the courts also require a reductive approach to the problems under review. The diverse ways injustice materializes in small encounters with hospitals and ration shops must form a single, if variegated, image. The particularities of experience become the subject of anecdotes, if included at all. The arguments made in legal petitions can occasionally be supplemented by stories, which both confirm and exceed the argument's logic. But the main argument must collapse everything into a unifying image.

Jabbar—as convenor of the Women's Union—was responsible for collapsing diverse and unwieldy realities into a single position that bore the Women's Union's name. Success at this task required high skill and insight that reached far beyond the law. The task could not be done in isolation. Jabbar relied heavily on consultation with activists from outside Bhopal, based in Delhi, Bombay, or even New York. Logics had to be imported from outside, turning Bhopal into theoretical reasoning on the proper role of law in society. Basic questions about the role of law in society became integral to the Bhopal story: Should law ensure compensation for those injured when the intentionality of those responsible remains in question? Should compensation be conceived as an act of restoration, implicitly sanctioning any unevenness in access to resources and opportunity that prevailed prior to the injury under review? Must the law punish the guilty, or is it sufficient to set a precedent strong enough to deter future risk? How should the law define acceptable risk when its decisions will be read as license for or against certain ways of economic development?

Incorporation of these broad questions into Bhopal has been crucially important, as has been Jabbar's leadership. And Jabbar has kept the women of Bhopal behind him, both literally and figuratively. The contradictions are obvious, but difficult to judge. An image of Jabbar from my trip to Bhopal in late 1996 is suggestive. Jabbar remained constantly on-call—advising people on their dealings with the claims courts, interceding when people were harassed or denied access to medical care, demanding that jobs be created and that the criminal case proceed. Money was very short. Rumors that interim relief payments would soon stop made Women's Union members hesitant to contribute monthly dues of Rs. 5. But Jabbar had gotten a cellular phone. People could now find him, anywhere, anytime, to ask for his guidance or support. He carried with him what many people consider a sign of the technological systems he is known to critique.

(8.6) Endemics

1990 was a year when the Women's Union had considerable support throughout Bhopal. The Women's Union was credited with securing interim relief and was seen as the link between Bhopal and Delhi that made change a possibility. The effects of policies that would divide and rule Bhopal were not yet on the surface.

In 1990, the settlement of the Bhopal case was a concrete reference point for demanding change. It focused the Women's Union and provided a way of speaking about Bhopal that connected it to the outside world. Critiques of the medical categorization scheme on which the settlement was based kept challenges to the settlement tied to the local level. Critiques of the way science had been used to legitimate injustice tied Bhopal to the broad challenges to dominant development models that were sweeping across India.

Critiques of the way the settlement was finalized without the consent of the plaintiffs tied Bhopal to even broader discourses, about rights and the excessive license of the state.

The anniversary commemoration in 1990 seemed to bring all these issues together, along with many people. A contingent of activists protesting the Narmada Dam came to Bhopal—as did members of the Chattisgarh Mukti Morcha, a union of mine workers from eastern Madhya Pradesh led by the charismatic Niogi. People came from Delhi, from Bombay, and even from the United States. The Women's Union prepared food for everyone. A tent was set up in the park, as were huge pots for boiling potatoes and frying puris. The equipment was rented from someone who usually provided for weddings. The Women's Union acquired significant debt, but it created a space for believing that a grassroots environmental movement, strengthened by regional alliance, could take shape.

At some point during the night, the women stopped for prayers. Muslims and Hindus joined hands to form a circle. One woman would take the lead until she was interrupted, and the voice would shift to another register, perhaps even to another religion. Hamida Bi became entranced. Her prayers were cadenced from the beginning, but she stared straight ahead. Slowly, the rhythm picked up, her body swayed faster, and her eyes seemed somehow to turn both inward and upward. The other women kept hold of her hands and watched her carefully. Later they told me that this was the way Hamida Bi was when she gave her thoughts to the disaster. I only knew the story Hamida Bi had told me when I asked what had happened with her family. She told me of her daughter's baby daughter, who survived the night of December 3, but continued to have terrible difficulty breathing. Hamida Bi told her daughter to put Vicks salve on the baby's chest and cover her with a blanket. When they lifted the blanket, the baby was dead.

By 1991, all sense of possibility in Bhopal was shattered. The settlement was upheld in October. During the rainy season, Muslim slums had been cleared and read as an index of communalist bias that threatened to shape the future. While interim relief continued, its distribution had become mired in corruption and bureaucratic inefficiency. Niogi—leader of the Chattisgarh Mukti Morcha—had been killed. Goondas known to have been hired by the industrialists of Chattisgarh had come into his house in the dead of night and shot him in the face. Rajiv Gandhi, too, had been assassinated. Curfew was imposed throughout Bhopal, reminding of the riots in 1984 following the assassination of Indira Gandhi. Bhopal had not rioted in 1984. By 1991, the seeds had been planted.

The anniversary procession in 1991 began with an argument between some of the women in the Union. Some wanted to chant slogans as they marched through the streets. Others insisted that the code of silence fol- lowed every year before should be continued. Jabbar stepped in to impose the silence. Then the procession passed shops that had not respected the bandh—closing their doors to show support and respect. A few of the women threw stones. Jabbar again stepped in. The woman he reprimanded most harshly was a Hindu. People said he had better be careful. The possibility of retaliation was mentioned. The possibility of communalism within the Women's Union itself bubbled to the surface for the first time. The reality of conflicts between Jabbar and middle-class activists in Bhopal could no longer be ignored.

Collaboration between the Women's Union and middle-class activists is a crucial component of efforts to keep ghettoization of "Bhopal" at bay. Until recently, there were not even telephones to help keep the conversation alive. Acute shortages of the tools of communication are most visible, how- ever, within Bhopal itself, in the everyday traffic between the homes of gas victims, the claims courts, the hospitals, and the racketeering middlemen waiting on every corner.

Most members of the Women's Union can neither read nor write and thus have trouble helping even their own families wade through the "paper proof" of victimization. Jabbar Khan handles Hindi-language publications with great skill, but needs help with translations into English. Middle-class activists help offset the resource deficit, sharing language skills to offset the uneven authority that different languages carry.

But many middle-class activists have argued that the Women's Union is undemocratic beyond repair, precluding collaboration. Others, including me, felt that refusal to collaborate would be conformist, an insistence that there is only one way, one style. To abstain whenever organizational dy- namics were disputed seemed to avoid the political challenge of working together, across nationalities, classes, religions, and even ideological dif- ferences. It seemed naive to hope to start from "community" and to deny responsibility for constructing it.

Meanwhile legitimate critiques of the Women's Union's organizational dynamic proliferated, throwing into high relief how work for and within the grassroots is driven by asymmetry—at every turn, on every level, in the inequalities among languages as well as among peoples, among the disempowered as well as when authority encounters its margins.

I admired most of the Women's Union's tactics, particularly its creative engagement with public protest. Many of the women gloried in the stories of sitting on the lawn of the Supreme Court, littering the landscape with their bodies, interrupting the proceedings of officialdom with their slogans. Or interrupting Babulal Gaur in his divali celebration. It was the Festival of Lights, so they arrived with candles and lined the path to his door. He was the Minister of Gas Relief. By his authority, victims were allowed, or denied, admittance to hospitals, to jobs, and to all future possibility. The women tell the story of their call on Gaur with an ardor for detail. They say he smiled and gave the traditional greeting, his face glowing with the purifying orange of turmeric—like a sweet mango as it begins to rot.

The Women's Union's strategy of street protest worked both within and against traditional forms. So did the Women's Union's organizational structure, but the retrofits made more noise. An internal hierarchy sustained the Women's Union's organization, headed by Jabbar Khan, who was elected to his position by Union women. Jabbar advanced much of the Women's Union's work, both locally and in Delhi, taking advantage of a cultural authority and right of mobility that women members did not have. His style was charismatic, and therein effective. It was also paternalistic, and sometimes autocratic.

Many outsiders accused Jabbar of acting the part of a "film star politician." At times, it was difficult to deny the similarities. And there were occasions when Jabbar definitely crossed the line.

There is no question that Jabbar purposefully invoked traditional structures of authority and purposefully tapped the power of established institutions. And he did minimize the need for interruption within his own organization. But Jabbar did know that politics must be located in many places at once, requiring continual negotiation of disparate fields of reference and means of legitimacy. He knew that the whole truth of Bhopal could not be told, making us dependent on unreliable modes of description and imperfect categorization schemes. He also knew that languages, like people, are unequal and that differentials of power must be strategically engaged, rather than denied. And he knew that things are always lost in translation, reminding us that law and justice will never coincide.

Jabbar found more repose than most middle-class activists I knew. We were disturbed by the contradictions, almost to the point of paralysis. Jabbar skated over the interfaces, moving with pragmatic logic across different conceptual orderings of the problem at hand. Jabbar could have provided a model, had he not been dismissed for being too contingent, too tied to established structures of history and power, too much a product of the disaster he worked to ameliorate.

(8.7) Interruptions

her place

From the start of my time in Bhopal, I was highly involved in and committed to the work of the Women's Union. Most days were structured by the Women's Union's sense of relevance and strategy, materialized through written response to official statements regarding the health and legal status of gas victims, through public meetings, and through continual negotiation over the Women's Union's organizational structure.

Saturdays were spent at outdoor meetings. Information was circulated. Commitment was fired and expressed. Social relationships were codified and transfigured. Between times were spent writing, organizing street demonstrations, and working to secure jobs, medical care, and cash relief for victims.

Occasionally, we left Bhopal to participate in demonstrations in Delhi or to visit other victims of "destructive development." I also went off on my own, trying to keep some grasp on a research project redefined to encompass the breadth of grassroots environmentalism in India. Even then the Women's Union directed my outlook. Visits with other women's organizations left the most forceful imprint. So much work remained to be done before women would have the mobility, confidence, and authority necessary for extensive participation in Indian politics. Seemingly small moves nonetheless evoked grand possibilities. Women's involvement in politics, no matter how unauthorized, interrupted business as usual.

Fieldworking within this level of involvement had a definitive methodological effect. It also provoked sharp conflicts of conscience, though not of the expected sort. A sense of epistemological betrayal provoked by the competing demands of research and politics was not the problem. I felt no obligation toward neutrality and no compulsion to routinize my perspectives according to dictates from elsewhere. Unlike those involved in similar research strategies during the 1960s, I had not been taught that objectivity is the criterion of research validity or that partisanship compromises comparative insight. Instead, I learned to fear truth claims abstracted from specific sites of articulation and to disdain all effort that privileged procedure over substance.[4]

Involvement with the Women's Union both refined and restrained my angle of vision. Perspectives were always compromised by political imperative, of often dramatic urgency. This meant that any hope of comprehensive understanding was emphatically foreclosed. Many of the restraints were practical. Because I was seen marching through the streets with Union women, I was not welcome at government hospitals, so I knew of their inadequacies only through the effects on the bodies and speech of victims. Because I continually worked to formulate coherent Women's Union perspectives, I persistently downplayed differences within the Women's Union itself. Because I worried so much about the structure of relations between the Women's Union and the Bhopal Group for Information and Action

(BGIA), I distanced myself from issues surrounding the Women's Union's internal organizational dynamic and leadership.

My motivation for aligning with the Women's Union was simple, even though derived circuitously, rather than from any experience with successful "community-based" moves for social change. To the contrary, my own experience with "community" had often been stifling, demanding loyalty to entrenched social forms and discouraging efforts to interpolate one's identity within new fields of reference.[5] Though my own experience was coded by ethnicity rather than blank victimization, it still left me wary of any demands for solidarity or any promise that collectivity is a good in and of itself. What I could believe in, however, was the need for interruption—the introduction of the extraneous into our ways of thinking and acting in order to break the flow of stale or unquestioned truth claims. Feminist teaching about the work of the margins was influential, as was haunting recollection of the ways good people and good ideas could turn malevolent. The name of Heidegger was particularly resonant, warning against philosophies at the service of politics and against nostalgia for Spirit or People. The call was for engagement, but tentatively, recognizing politics as a balancing act at necessary odds with the clean, straight surface of certainty.[6]

BHOPAL GAS PEEDIT MAHILA UDHYOG SANGATHAN
51, RAJENDRA NAGAR, BHOPAL 462010
PRESS RELEASE
13.4.91

To commemorate Earth Day 1991 (April 22), Bhopal gas victims ask for a renewed focus on the boycott against Eveready batteries. Union Carbide produces Eveready batteries; consumer rejection of Eveready batteries through a boycott is a means to protest Union Carbide's abuse of people and environment.

"Bhopal" has become a symbol of the problems of industrial society. Revitalization of the Eveready Boycott will both demonstrate solidarity with the struggle for justice in Bhopal and be a statement of public concern about the way global society is "developing." Industrialization means that safety of people and environment is determined by corporate quest for profit. Bhopal is only one example of the suffering caused by corporate greed and of the importance of regaining local control. Revitalization of the Eveready Boycott will be a statement of non-cooperation with corporate enterprise that continues colonialist patterns of outside control and local exploitation.

Revitalization of the Eveready Boycott on Earth Day is of particular importance due to recent corporate attempts to take over the environmental movement. Through expensive advertising campaigns, corporations have tried to convince the public that they are concerned about the safety of people and environment. Yet, the record of corporate abuse is endless.

Witness the destruction in Bhopal, at Exxon's Alaskan oil spill, in Italy's Seveso. Bhopal gas victims believe this hypocrisy must be exposed and the environmental movement preserved as a forum for debate on the critical issues of sustainable and just development. Earth Day focus on the Eveready Boycott will highlight this concern for the integrity of environmentalism and remind of the urgent need for noncooperation with transnational corporate exploiters.

Boycotts are a way for the public to punish a corporation through market mechanisms. As a response to the Bhopal disaster, this public punishment is especially important since the courts have been unwilling to hold Union Carbide liable. Despite overwhelming evidence that the gas leak was a direct outcome of negligent management practices, gas victims have been denied a trial. Disclosure of the cause of the Bhopal disaster through a trial would glaringly display the way corporations put profits before people. Further, a trial would force the judiciary to take a stand on issues of "development" through dependence on multinational investment.

Court action against Carbide would disrupt the pernicious habit of multinational corporate investment in the Third World to avoid the costs of safety. It would be a statement to the world that India refuses to sell the lives of her people in exchange for the glamour of technological progress and participation in global capitalism. Such a statement could massively contribute to the prevention of industrial disasters all over the world. However, such a statement would also shatter the blind faith in progress through technological advance that members of the establishment use to justify the increasing exploitation of the masses to serve the rich.

Courts most often are not an instrument of justice but a means of national development goals that consistently ignore the needs of most citizens. In this regard, the Bhopal case is not unique. Though it involves the worst industrial accident in history, the issues are being dealt with with the same disregard for human suffering that has guided legal decisions throughout the capitalist era. Legal decisions are most often political decisions according to the influence of powerful corporations. It is thus imperative that we understand the courts as a support system for corporate activity and thus see that autonomous, extra-legal action is a necessary requisite of justice. Boycotts are one form of autonomous, extra-legal action.

Boycotts punish corporations by directly reappropriating profits made with each sale of a consumer good. Further, every time an individual refuses to buy a certain item, the explanation for not buying educates those around about important social issues. Hence, individual refusals to buy slowly damage the public image of a corporation and become catalysts for dramatic shrinkage of the population eager to buy. The public education that occurs through a boycott is perhaps even more important than the financial setback suffered by the corporation. Long-term changes in patterns of corporate behavior will only be possible if the mass public becomes more vigilant in condemning corporate abuse. . . .

Perhaps the clearest lesson of Bhopal has been the need for extra-legal struggle. It is clear to many observers that the relief received by gas victims has not been the result of an efficiently operating legal system but due to

sustained protest by gas victims. Victims themselves often argue that unless
they had organized and marched to protest deplorable conditions, compensation and health care would be even less available than it is today. Further,
attention to victims' dissent against the February '89 settlement has only
come through vigilant agitation. Clearly, continued and extended public
pressure is essential to the on-going struggle. Mass participation in the Eveready Boycott is a way we can directly contribute to this goal.

Most of my daily work with the Women's Union was clerical. I wrote, in
English, in response to both the Indian Supreme Court and the local Relief
Ministry. There was no question that the writing needed to be done. I had
the resources required for doing it, so I did. It didn't matter much whether
I agreed with either the logic or the strategy of the words I wrote. I was
responding to marginalization, not truth per se. The truth did matter, however, and the task of constructing and legitimating it was long and arduous.

Many days I spent all my time at my computer, away from the tumult of
Bhopal's streets. Like an armchair anthropologist, I waited until material
was brought to me for translation and interpretation. Jabbar would come,
or a few of the women, and describe what required response. The chair
before my computer was one of few in the house, so I often sat above
Jabbar and the women, feeling like queen and servant, fraud and devotee.
Staying focused could be difficult. Jabbar could run at least ten arguments
a minute, laced with facts, figures, and an occasional exclamation of disbelief. If women were there with Jabbar, they would offer rich examples
that confirmed and fed the logic we were trying to build. The women also
had logics of their own, though often they couldn't cut through the dense
web of words Jabbar provided as our ground.

Most often, I simply wrote what I was told to write and pushed for justification only when I could not make narrative logic hold. Interpretive
moves were necessary to determine which arguments would be most efficient and which facts would best sustain them. My task was to produce an
organizing principle that held at bay randomness incongruent with the language of politics. The disaster had to become communicable, so I struggled
hard to "say it straight."

Translations in many directions were always necessary. I could follow the
Hindi, somewhat. Jabbar could translate from Hindi to English and back
again, somewhat. Sometimes other members of BGIA would be around
to help. Even then the translations were never perfect. They were kludge
jobs, working with available resources to forge workable, even if imperfect,
schemes.

The basic facts were soon lodged in memory, though always in need of elaboration specific to the particular turn of politics that provoked our response. What counted as basic facts, however, remained contested. One example: in the immediate aftermath of the gas leak, many working to support gas victims insisted that the sum of affected people was 250,000; insistence on this number as a limit was a response to the distribution of free rations to over 600,000 people, which caused surpluses and black markets, while turning relief initiatives into overt efforts to build electoral constituencies. An even more fundamental critique of the 600,000 figure was based on the seemingly arbitrary modes of categorization exercised throughout the relief effort. While 600,000 received free rations, only those earning less than Rs. 500 ($36) per month were awarded Rs. 1500 ($107) as cash relief. Victims earning over Rs. 500 per month were left to fend for themselves, no matter how sick they were or how much debt they incurred when they fled the city during Operation Faith.[7]

The government never disclosed how these categories were established. Critiques of the 600,000 figuration of the gas-affected community were, then, valid—in 1985.[8] By 1990, the 600,000 figure was a routine reference, validated by the political schemes through which victimization in Bhopal had come to be configured.

When I first began inscribing the 600,000 figure into most everything I wrote for the Women's Union, I was not aware of its transformation over time—within the articulations of those working with and for gas victims. I assumed that this number had remained relatively stable, at least since the official designation of thirty-six gas-affected wards in late 1985. But, by many accounts, these wards had a combined population of only 300,000.[9] Where did the figure of 600,000 come from, and was it "accurate"? Given deficits of both time and research resources in Bhopal, I had little opportunity to pursue these questions.

The figure of 600,000 worked well as a stable point of reference in arguments against further extensions of the area considered gas affected, run alongside persistent criticism of the way many legitimate claimants had been excluded from registration and resources—because computer printouts had misspelled their names or had listed their addresses or ages incorrectly; or because they were minors, yet could not be listed on the claims of their parents; or because their residence addresses had changed, often due to forced relocation, during which ration cards and other proof-telling documentation were destroyed along with their houses.

Another number requiring stabilization referred to the quantity of people who had been medically examined at any given point. We relied on official figures, as prepared for the courts, because they were sufficient to sustain critiques of the out-of-court settlement of the case. The settlement amount of $470 million was legitimated with medical categorization data said to be representative of the magnitude of injury and disability requiring compensation. We tried to crack this legitimacy by consistently pointing out that only a fraction of gas-affected people had been examined when the settlement was announced in 1989, and even when it was upheld in 1991. We participated in the authorization of these official numbers for specific strategic ends. In other turns of strategy and narrative logic, we overtly undermined these same numbers, insisting that the protocols relied on for official evaluations of health status were grossly unscientific. Though this bifocalism did not involve logical contradiction, it was tricky to sustain— like a juggling act with no fewer than five balls in the air, each requiring its own spin and balance. The performance was often clumsy, but the show did go on—every day, for different audiences, on any stage available.

Quantification of the problems in Bhopal was necessary. The Women's Union had to mark its position within the contest, so I helped its members hold their line. At the outset, I couldn't remember the numbers. Before coming to Bhopal, my suspicion of statistics and other quantification devices had licensed an almost complete loss of memory. Numbers didn't offer me any information, and scale could be determined by other means. "Bhopal" revoked this license.

With time, I would learn the numbers on Bhopal, from many different sources; the effect was inundation, not stabilization. Unlike its promise, quantification produced neither consensus on "the nature of the problem" nor ways of rendering the complex into practical programs. Instead of providing stability, quantification brought an unending onslaught of "relevant indicators," often marked by the violent extremes of diverse ideologies, all stormily contested.

Inequality was evident at every turn—among languages and cultures as much as among people. It now reminds me of what people say about the translations between Arabic and English. Arabic has changed the most, while English has operated something like a gold standard—providing a stable, authoritative frame of reference even when creoled in Boston or Bombay. The grammar of law and science has similar brawn. Like the English language, it resists inflection by anything foreign, forcing meaning to accord with established syntax. All that exceeds established categories and rules soon becomes subsumed within them. In Bhopal, descriptive metaphors can be pulled from the gas and its dispersal into bureaucracy. The gas asphyxiated, forcing people from their homes and the semantics of everyday life into the corridors of hospitals, claims courts, and banks; local idioms could not render their diagnoses into meaningful statements. Now syntaxes of law and science shape the everyday, claiming to have rules, categories, and definitions for all that crosses their path. The Women's Union insists that there are things that don't fit, that the semantics they know can't be contained within the space science and law have offered.

The Women's Union has never asked me to write for a return to a past when all was well, for restoration of a time when syntax and semantics lived in harmonious accord. The Women's Union does not have a past that can be drawn on for inspiration, entitlement, or images of rehabilitation. The origins they cite are in the gas, in a rude interruption of everyday life already marked by poverty and estrangement. The gas was an obtrusive beginning, begetting a "manufactured community" that has no language of its own. Jabbar Khan reminds us often of the challenge: "Before the gas, Bhopal was not a place of struggle; now we must make it one."

My participation in the truth claims of the Women's Union was not without anxiety. We wrote in a language of realism, inflected by evangelical fervor. We posited certainty about our facts and insisted that the enemy could be identified, and must be punished. We denounced those who wavered in their certitude, emulating the unquestionable expertise of those we challenged. Our tone was often shrill and our style strident. Rank desperation countenanced these moves. Many of the women had been widowed by the gas. Others, while themselves ill, supported large households of people requiring constant care and expensive medicine. They desperately needed a different kind of response from the authorities, so their interruptions needed to be loud and, at times, rude.

My anxiety about these truth claims was not due to doubt about their veracity. While I could not "believe" in the certitudes I wrote, I nonetheless considered them highly legitimate—and necessary. It was a matter of focus. For multiple reasons, I could not know whether the content of our claims was entirely without error. I nonetheless could be sure of the legitimacy of their form and intended effect. Even if I didn't know that we were entirely right, I did know that official descriptions of the disaster were systematically wrong. The need for certitude was thus displaced, attaching itself, by way of the negative, to the center rather than the margins.

NINE

Anarchism and Its Discontents

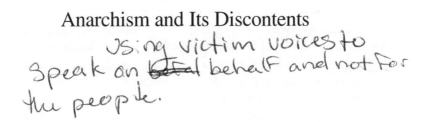
Using victim voices to speak on their behalf and not for the people.

(9.1) Speaking Of . . .

The challenges faced by the voluntary sector in Bhopal have been enormous and incongruent. Government initiative has been crucial, but has operated according to logics that could be blamed for causing the disaster. Victims have urgently needed health care, but not at the expense of emphasis on economic rehabilitation—which is beyond the scope of the emergency room model that has oriented the work of government hospitals. Victims have needed help proving that they fit within the categorization schemes through which compensation has been distributed, even while those categorization schemes have demanded systematic critique. Victims have needed a voice in both legal and medical arenas, yet have lacked necessary language skills—and so have had to rely on "middlemen," and most middlemen in Bhopal have deserved censure as "self-interested, bribe-taking goondas." The importance of sharing expertise has persisted, but roles for middle-class progressives have hardly been straightforward.

Grassroots rhetoric emphasizing the importance of community control has had particular salience in Bhopal. Any possibility of realizing community control, however, would have required intensely collaborative effort. Administration and governance of rehabilitation could not simply have been shifted to committees of victims. Victims themselves have a great deal of relevant expertise, but not of the sort necessary to coordinate and implement a rehabilitation scheme for 600,000 people exposed to toxics.

Continuing disaster in Bhopal entangled actors once separated by geography, race, and class. Distributions of wealth, risk, and authority have remained forcefully uneven—locale has mattered. But the politics of change has globalized, forcing all local initiative into conversations with power for which there is no indigenous idiom.

The Bhopal Group for Information and Action (BGIA) originated within these double binds and has continued to try to work within them—translating disaster into workable, even if imperfect, expressions of justice. The paradoxes that shaped BGIA during the time I was in Bhopal were not abstract. Daily encounters with gas victims and the official rehabilitation apparatus horrifically materialized ideological critiques of the state, of bureaucratic rationality, of "the clinical gaze," and of all established ways of organizing change.

Working at the local level, BGIA could not sanitize its critiques. Every word and every action were compromised by a context heavy with corruption, frustration, and the fatigue of continuing disaster. Our role as activists could not be imagined in terms of ideological purity, or even integrity. Work on the ground in Bhopal was rife with contradiction—and at constant odds with idealized conceptions of how progressive social relations should be embodied.

Outside volunteers had undertaken many forms of work in Bhopal from the outset—directing efforts to provide medical relief and job training, organizing victims, producing documentation that countered government plans and data.[1] At first, BGIA concentrated on the last of these tasks, hoping to avoid involvement in irresolvable questions over how its members should position themselves in relation to gas victims. Like many other middle-class progressives in India, founding members of BGIA were acutely aware of the difficulty of setting terms of engagement with the poor that did not mime those of colonialism. The challenge was to collaborate, without coercion, leveraging the resources of middle-class status to speak of, but not for, victimized sectors of society.

This challenge was difficult enough to articulate, much less to realize in social practice.

BGIA chose a strategy of deferral. The strategic and ideological challenges of community organizing were tabled, as were the challenges of actually running clinics or job provision schemes. Instead, BGIA focused on writing. Using the voices of particular victims to counter grand generalizations regarding health, economic security, and other rehabilitation responsibilities, BGIA consistently interrupted official descriptions of the disaster. The goal was to create forums within which victims could speak, their local realities could be described, and their position in the global order could be located. The challenge was to avoid representing victims in the paternalistic, patronizing ways of many journalists and elected politicians.

BGIA's strategy of deferral did not hold. In part, this was a matter of time. By the time I came to work with BGIA in early 1990, there were no other middle-class, English-speaking activist groups working on Bhopal at the local level. Rigid delineation of what BGIA would or would not take on was therefore impossible. Rehabilitation had proven disastrous, so we worked to establish a clinic, structured to provide medical relief and monitoring, as well as jobs and local control. The litigation continued in ways that structurally silenced victims, so we helped victims prepare materials addressed to the courts. The work of documentation remained central, but was in no way exempt from the effects of ill-shaped social roles.

The conundrums of activism in Bhopal were daunting and raised perennial questions for all involved: How can middle-class progressives work for justice in collaboration with those denied it? Can middle-class progressives acknowledge the gross inequalities that they both represent and work within, while relying on purposefully egalitarian styles of engagement? What styles of leadership are appropriate? How can expertise be deployed without reproducing status hierarchies? How can dissenting opinion be respected without paralyzing collaborative work? To whom, or what, is the middle-class progressive responsible?

(9.2) Powers of Practice

Throughout the years I was in Bhopal, I worked and shared a household with other members of BGIA. Theoretically, the group was structured by anarchist tenets denouncing bureaucracy, hierarchy, and all traces of authoritarianism. Practically, it was a reckless and clumsy attempt to enact a social form for collective effort that upheld our criticisms of the established Left.[2] Anarchism provided both guidance and excuse, throwing into high relief the importance of finding new ways to organize social change.

Our goal was to displace established forms of society and culture through everyday maneuvers that forged "lived alternatives." Theoretical denouncements of the state, and of established means of opposing it, did motivate us. More important, however, was our everyday context. At every turn, authoritative discourses and institutions operated like the notorious *lathi,* the long cane wielded by Indian police. The reports and programs instituted within the courts and Relief Ministry didn't harm with one definitive statement. Most often their arguments were convoluted, their instructions were vague, and their realizations were corrupt. The effect was not that of a bullet, but that of scattershot or a rain of blows from all directions, which stops only to begin again. This monstrous process debilitated victims beyond reason and crazed all those who witnessed their plight.

Structural contradictions took their toll on our morale, as well as our efficiency. Microlevel issues of health care, vocational rehabilitation, and housing had to be addressed alongside macrolevel issues of international law, technology transfer, and trade liberalization. Navigating the interface was treacherous—and obligatory. We had to work to rehabilitate victims. We had to work to establish the legal and organizational structures necessary to prevent similar disasters in the future. Health and economic rehabilitation required government resources. Preventative measures required a continual critique of both the conception and the practice of government in Indian society. The temporal rhythm of these efforts differed as much as their orientation. Response to government initiative was infused with a sense of crisis, inflected by graphic images of the material effects our work could have on the lives of gas victims. Work for preventing the Bhopals of the future was less directed and often seemed rather sterile. But it helped maintain fragile links to hopefulness, which we relied on heavily. Sathyu, our in-house sloganeer, voiced the refrain: "Bhopal can never be rehabilitated, only educated."

We wanted to challenge the modern institutional order and its exclusionary practices. We didn't trust knowledge, so we deferred to the social. In principle, we rejected theory and threw our lot into experience. We knew that every word spoken and every message received were imprinted with concepts we hoped to implode. We couldn't transcend what we critiqued, so we tried to displace it. Anarchism was our means, a way to a responsible politics rooted in Bhopal itself—within disaster, rather than ideal. The goal was to "Remember Bhopal"—in practice.

Our focus on the means of politics could have been traced to Gandhianism and the insistence that if one behaved like the rulers, one could not but become like the rulers. Instead, we traced our roots to recent developments in the women's movement and to the argument that one did not have to work within and under the party to be revolutionary. Autonomy, from the electoral Left as much as from the state, was the goal. The organizational means was to be nonstructured, collectivist, and oriented toward "internal critique" as a first step toward a society rid of hierarchy and its abuses. Publications from the women's movement insisted that we take nothing for granted, examining "the whole question" and all the questions that went with it. The challenge was to rebuild from the ground up, finding new ways to organize our identities, our work, and our hopes for the future. Systematic discontent provided the starting point; anarchism, rather ironically, provided the structure.

Everyday life was complicated, as well as all too simple. Boredom, underscored by a sense of urgency. Intense work, occasionally interrupted by hard-edged play. An acute sense of isolation, interrupted by visits from other activists and short trips to Delhi. We argued constantly, to entertain ourselves, to shift blame around, and as a way to strategize.

Anthropology was among those things due suspicion—and perhaps censure. My housemates knew the stories of fieldworkers sent out to count and categorize, easing the work of the colonial state. They were relatively sure that anthropology could not be practiced otherwise.

I shared these critiques of anthropology, but identified myself within its discipline nonetheless. My arguments that anthropology could be done differently were partly accepted and partly ignored. I came to Bhopal with a critique of science drawn from feminist critiques of the exclusionary force of honorifics. Humanism. Progress. Freedom. I feared them all. Others in my household concurred, speaking often of what all could be justified through reference to tradition, nationalism, and even socialism. Stories of justified terror and legitimized exclusion threaded our politics together, cutting through factional disputation over primary determinacy and revolutionary end.

Our reliance on storytelling was crucial. Theoretical language was scorned, derided as an exercise of whitewash that grossly simplified both explanation and solution. With time, I learned to talk through these stories, privileging the details and forgetting abstract principle.

Stories about Communist parties were told frequently. Stories that drew out images of meetings conducted from a long, cloth-covered table at which sit rows of elderly men—none of whom questioned themselves, none of whom could be questioned. And stories like that about a young Kashtakari woman from a tribal area near Bombay, gang-raped by men of the CPI-M. Her crime: association with an independent organization that implicitly challenged the centralized authority of the party.[3]

I was told that there was a time when many activists insisted that these men "weren't really communists." By the late 1980s, hope of salvage was cut apart, scissored by stories that cut heroes to size and shredded all faith in their tactics. Demands for party allegiance, Gandhian moralism, family honor. I was told they were cuts of the same cloth. Our task was to refute them all.

There was one holdout among us. A Gandhian. He focused all of his energy on natural farming projects, believing that change was possible only through disengagement. His goal was to get out of the system, both economically and spiritually. When he had malaria, he refused medicine, and we argued over whether he had the right to refuse collective judgment. He refused revelry and was accused of obsession with purity and control.

It was a strongly partisan world. Intense commitment was expected and assumed. Partisanship was nonetheless questioned. Allegiance was suspect—something in need of acquittal. We came together from different social and political experiences, sharing neither ideology nor tactical orientations. This wasn't supposed to be a problem. If sufficiently critical, aware of the history of failure in which we worked, our differences would no longer matter. People spoke of being "beyond ideology," in a space where revolutions had no model, obliging brave new moves of society and culture. The future had to differ from the past, but it had to be worked out through daily moves that created new people, new modes of alliance, and—only as an after-effect—new ways of thinking.

The contradictions were harsh. We worked to form an anarchist collective. We worked against surveillance, bureaucratic rationality, and all forms of determination. Our method duplicated much that we critiqued. Every moment, every move, was subject to evaluation and judgment. Even the most mundane activities. Washing clothes. Peeling vegetables. Sweeping the floor. Our reasoning quickly became ridiculous. Was vegetarianism Brahmanic or a show of solidarity with the poor? Was aversion to filth "bourgeois"? Asceticism was scorned, but practiced, in selected domains. Vanguard behavior was expected, but all expertise associated with control and all leadership equated with arrogance and abuse.

We struggled hard to hold the group together. Emotions ran high, in familiar configurations. All of us came from rigidly traditional family structures—most, of high caste. Memories of rebellion were cast in vivid detail. Ripping off the Brahmanic thread. Dropping surnames that identified one's caste origin. Refusing arranged marriages. Refusing the credos of our scientized education. The desire for something different was extraordinary.

Analysis of how we were doing was unrelenting. Since assessment was a matter of interpretation, it was a vehicle for vehement dispute. But discord was the aesthetic. Discord was considered productive—and evidence that we were not co-opted by the singular logic of established political agendas. Impasse was provoked by any show of allegiance, not disagreement.

There was no demand for conformity of belief. One invited denouncement, however, if questions slid too far into personal anxiety. Skepticism was bad faith. Doubt, the beginning of the end. Despair, inexcusable.

There was no blueprint, or any tolerance for neutrality. Ideology provided neither coherence nor any means to mark our differences. In public, we were all composure, girded for evaluation, a unified front. Within our house, only chaos, fraught with disfaith on more levels than could be acknowledged aloud. Books were strewn everywhere. Steinbeck. Doris Lessing. Bakunin. Feyerabend's *Against Method. William Blake, Visionary*

Anarchist. Much of this reading came from anarchist outlets in England and the United States. It was supposed to guide our efforts toward "Indian alternatives."

(9.3) It Isn't Famine

The stories BGIA told to carve out a space for its work originated in many places and languages. The effect was chaotic. There was, however, a persistent theme, not so much within the stories, but in the way they were told. The stories were told as interruptions. They were a way for us to question established truths, through detail that embodied something different. They carved out a space in which it made sense to engage institutionalized power, if only to reveal its contradictions. BGIA's own contractions were often revealed as well.

At its best, BGIA's narrative strategy was a resource for the Bhopal Gas Affected Working Women's Union. We could explain while critiquing logics of law, while offering rhetoric that helped secure the Women's Union's authority within discourses of human rights. We could help the Women's Union draw on the charter of rights built into the Indian Constitution, while telling stories about the systematic violation of basic rights in India since independence. The story about the suppression of the Telgana revolt during the late 1940s was a favorite. Peasants in nearly four thousand villages in what is now Andhra Pradesh armed themselves, seized land, and demanded that the independence promise of democratic rule be realized. We also told stories about the "snuffing out" of armed struggle in West Bengal during the 1960s, when middle-class activists working to dislodge the control of their own social class were branded as traitors to the nation. Stories about failures of "development" in the 1980s were closer at hand, but still unfamiliar to many Union women. They were urbanites. Struggles for rights to water and access to ancestral lands were foreign to their experience. Connecting these stories to images of an emerging grassroots environmental movement of which the Women's Union was a part was the next move.

The retelling of stories about people's struggle filtered conversation within the BGIA house. The stories drew both from Marxist histories and from "newer" histories cast to illustrate the illegitimacy of the state and its development goals. Bhopal was a key reference in these newer histories, and they, in turn, provided important narrative support in our efforts to articulate how disaster in Bhopal continued to unfold. But these newer histories also confounded the logic of our work. They emphasized the need to rebuild civil society so that the state would cease to operate as primary guarantor of rights and welfare. The state was supposed to leave citizens alone, respecting the autonomy of local culture and economy. In Bhopal, such autonomy was not possible. Local culture and economy were a product of the disaster, manufactured to specifications laid out by the state in plans for development and then rehabilitation. Design flaws were obvious, but what was most needed was for the state to assume *more* responsibility. "The state" was another double bind.

After all, a state-sanctioned production facility had literally asphyxiated those it was promised to serve. To argue that the role of the state in Indian society had been excessive and abusive was quite legitimate. To argue that the goal of politics was autonomy from the state was a different matter. The state and its institutional practices could be said to have "caused" the disaster. And the official rehabilitation regime in Bhopal was, in fact, a "mere continuance" of these institutional practices. But this repetition could not license bald opposition. Delinkage from the operations and institutions of the state was neither possible nor desirable in Bhopal, whatever the legitimacy of claims that colonialism was alive and well in every bureaucratic initiative.

In Bhopal, translating between ideology and practice was a relentless process. Members of BGIA concurred with the most sweeping critiques of the state in Indian society—and with demands for more state involvement in the welfare of Bhopal. Duplicity? Perhaps, but without deceit. Every show of police force at Women's Union demonstrations reminded us of the contradictions. So did the most routinized institutional practices.

A report produced by BGIA in 1990 mapped out the double bind embodied by the state with particular detail—through a survey of outpatients at two government hospitals. The goal was to assess the risks that accompany drug-centered therapies. The findings were disheartening. The overall line of treatment differed little from that provided in the immediate aftermath of the disaster, emphasizing temporary relief and the efficient processing of patients. Many of those interviewed complained that the doctors wrote out their prescriptions "without even taking their pulse" and without listening to all of their symptoms. Patients also complained that doctors came to their offices late, behaved unsympathetically, and then encouraged them to have certain investigations done at private clinics, often run by the government doctors themselves. The prescriptions that patients were given often could not be filled at the hospital dispensing counters due to "unavailability." Meanwhile rumors circulated about large hauls of medicines being stolen from hospital stores, with the likely connivance of senior officials. Thus, despite promises of free medical treatment, the average family of patients surveyed was spending almost a third of its total income on medical treatment. Of those surveyed, 372 responded to the questions related to occupation and income. The average monthly income was Rs. 750 (about $30 at the time). The average monthly expenditure on medicines and treatment: Rs. 230 (about $9).

But the insufficiency of government medical care was not the only problem. There was lack, but also glut. The 522 patients surveyed were given 1,562 prescriptions. Of those medicines prescribed, 67 percent were considered *unnecessary* by the survey team because they were based on "irrational" medicine—medicine in which "no logic is followed and treatment does not have a scientific basis." The report identified four indicators of irrational treatment: (a) drugs indicated for a particular symptom are not prescribed, (b) drugs prescribed are not indicated for the symptom presented and can lead to aggravation of the symptom, (c) drugs prescribed are in quantities far in excess of the required dosage, and (d) hazardous drugs are prescribed. BGIA categorized treatment of gas victims as "irrational" in 457 out of the 522 cases surveyed. Out of the 522 cases, 277 involved prescription of drugs banned in several countries and considered bannable by the Indian government.[4]

This BGIA report, like many others, found injustice in both shortage and saturation. The state did both too much and too little. Narrating this paradox into the stories that simply pitted the state against civil society didn't quite work.

(9.4) Rebuilding as Subversion

In the immediate aftermath of the disaster, the government of Madhya Pradesh promised that it would organize relief on a "war footing." The promise can be traced in an article published by the *Times of India* in March 1994. The article told of competitive bidding between two public-sector computer and electronic firms for a 40 million rupee (somewhat over $1 million) contract for a fingerprint technology system to be used in Bhopal to minimize chances of impersonation by fake claimants. The bids were being considered by a committee including former members of the Indian Supreme Court, the Union Law Secretary, the Secretary of the Chemical and Petrochemical Ministry, and the Chief Secretary of the Madhya Pradesh government (Alvares 1994, 176).

The priorities evident here are maddening and cruelly absurd. Hearing of such priorities sometimes pushed the BGIA household into cynicism and paralysis. Other times it galvanized defiance—defiance shaped by the absurdities of its context. The idea that cucumbers could be sold as a political statement against Union Carbide is but one example. Other moves of defiance are even more difficult to interpret. Were proposals for a nongovernmental health monitoring program absurd? What about proposals for clinics staffed by victims, in their own communities? And what about plans for non-drug-centered therapy—as a response to injuries allopathy could not seem to fix, but also as a statement of defiance against the logics and institutions of Western medicine, particularly multinational pharmaceuticals? Was it absurd to entangle these different levels of critique, or imperative?

BHOPAL GROUP FOR INFORMATION AND ACTION REBUILDING AS SUBVERSION—INDEPENDENT INITIATIVES TOWARDS REHABILITATION OF THE GAS AFFECTED PEOPLE OF BHOPAL

Rehabilitation is a neglected area for revolutionary thought and practice. Mass disasters (destruction in Iraq, Chernobyl, Armenian earthquake or Bhopal) have given rise to situations that call for innovative political intervention in the aftermath of such disasters. Yet there is scanty evidence of such intervention, if at all. We believe that through intervention in such situations simultaneous with one's fulfillment of the essential task of rehabilitation, it is possible to act and think in ways that can encourage mutual cooperation and expose the illegitimacy of the State. The following agenda for long-term voluntary action on Bhopal has been drawn up with such ideas in mind.

1) Growing Without Pesticides

If there was any need for it, the industrial genocide in Bhopal demonstrated that pesticides are hazardous. The aim of "Growing Without Pesticides" will be to demonstrate further that they are unnecessary. The idea is to purchase agricultural land near Bhopal with the gas victims (members of the Bhopal Gas Peedit Mahila Udhyog Sangathan) contributing to the common pool from the compensation/settlement amount they receive from Union Carbide. And then participate in pesticide free (natural or organic) farming to grow vegetables. Cucumbers can then be sold as political statements against Union Carbide. *why cucumbers?*

2) Production and Distribution of Nutritional Supplements

We know of gas affected people who buy so called "health additives" (produced mostly by multinational companies) in the hope of putting some life into their toxin-ravaged bodies. Voluntary and even government-sponsored studies have highlighted the need for nutritional supplements for the gas victims. The aim of this project will be to fulfill this need (to an extent possible) and encourage cooperative efforts in the community. The idea is to promote the formation of a cooperative (either in a community or in the organization of the victims) that would produce low cost nutritional supplements that incorporate traditional recipes. A number of groups have demonstrated the possibilities of production of such supplements from locally available materials through low technology methods. Efforts need to be made to ensure that such a project is self-sustaining in terms of resources.

3) Participatory Health Care

The gas-affected people of Bhopal will need health care at least for the next twenty years. Given the large-scale requirements of such health care, resource constraints make it difficult to conceive of adequate non-governmental response to such a situation. It should, however, be possible to act in a few critical areas.

a) Long Term Monitoring of Health Status

Long term monitoring of the health status of a representative population of gas victims will go a long way in documenting the continuing damage caused by Union Carbide's gases. It would also help in identifying the nature and extent of health care required by the gas-affected people as the years go by. Both medical and non-medical persons can be involved in such an exercise and it should be possible to train volunteers among the gas-affected people so that monitoring is largely handled by them. Monitoring of health status will entail the use of some equipment like spirometers and these have to be arranged in sufficient numbers through contributions from solidarity organizations or individuals.

b) Screening of Medicines

There is substantial evidence of indiscriminate prescriptions of hazardous and/or unnecessary medicines by private and government doctors in Bhopal. It should be possible to involve victim activists in screening of such medicines prescribed to individual gas victims and provide necessary counseling by advocating avoidance or suggesting substitutes.

c) Promotion of Non-drug Centered Therapy

The medical response to the health problems of the Bhopal victims has predominantly been drug centered. Dispensation of symptomatic drugs (providing only temporary relief, if at all) has resulted in the consumption of hazardous and/or unnecessary drugs by the gas-affected people, posing, as we have found in a study conducted by us last year, serious risks. The relevance of breathing exercises and yoga in the treatment of gas victims has been demonstrated on a number of occasions but governmental response on these has largely been gestural. It is possible, we believe, for a group of volunteers (at least one of whom is trained) to involve a small number of gas victims in the training and popularization of non-drug centered therapy.

d) Drug Based Health Activity

Given the nature of injuries suffered by the gas victims, the kind of medical problems that ail them and the state of the dominant system of medical treatment, it is essential that voluntary intervention include health activity based on prescription and dispensation (for a price) of medicines. The involvement of activist gas victims is an essential prerequisite for such an activity. Six months after the disaster through the JANA SWASTHYA KENDRA (People's Health Center), a group of voluntary doctors, non-doctors and gas victims had initiated such activity. Though it could not be sustained for long, it demonstrated the possibility and the relevance of such efforts. A small group of volunteers (professional medical persons and non-professionals) can initiate an activity in which rational drugs bought in bulk quantities are prescribed and sold (on a non profit basis) to gas victims who come for such help. Activist gas victims, who will be involved from the beginning will, hopefully, "take over" the activity at some point of time. Through such health activity, it would also be possible to monitor the health status of the gas victims provided proper records are kept. Efforts on the lines discussed above have been initiated in Bhopal and decidedly it will require active and long-term support both in terms of physical participation and contribution towards the necessary resources.

4) Vocational Rehabilitation

As in other such situations, an overwhelmingly large section of the population affected by the disaster consists of people who had been, prior to the disaster, earning a living through hard physical labor. A substantial majority of such people have been, due to gas exposure, incapacitated to continue with such work and most carry on with their traditional jobs like pushing hand carts, carrying loads, rolling bidis, construction labour only because their economic condition forces them to. The government's efforts towards providing jobs in accord with the health condition of the gas affected have been scandalously inadequate and singularly lacking in innovative ideas. It will need a few volunteers with some experience in such matters to initiate income-generating activities (garment stitching, health food production, running fair price co-ops have been suggested) among the gas victims. Widespread support, especially in the marketing of products, would be required for such an activity and it can be initiated only after such support has been assessed.

The Call

Given the spectrum of misery and the reasons for hope presented by the Bhopal situation, the list of possible independent initiatives can grow. However, to facilitate participation of groups/individuals (based in Bhopal and other parts of the country) in putting the ideas into practice, we suggest the formation of a network to begin with, which can be called "Friends of Bhopal's Valiant Victims." We hope that the need for rehabilitation activity in Bhopal will be appreciated by members of such a network and that such activity will be carried out with a spirit of subversion.

(9.5) Writing Alternatives

In Bhopal, there is a constant need for words; satisfying one demand provides only short reprieve before the call to language comes again.[5] In BGIA, we wrote our arguments and descriptions, using every trick of language we could muster, recognizing that "what counts" depends on the specific ways allegiance crosscuts any audience. Letters to the editors of Indian newspapers, to the Prime Minister, and to the World Court. Pamphlets for journalists and students. Affidavits intended to overturn American denials of jurisdiction. Books for children. Position papers, and poems.

Evaluation of the efficacy of our representations was a daily effort. The effort was not that of the avante garde—an attempt to speak a completely other language unco-opted by prior definition. New things had to be said, but through a negotiation with older forms. The affidavit is a genre that had to be worked within, as was the genre of the newspaper editorial or that of a position paper. Writing an argument to persuade court deliberations required tactics that would have been ineffective or inappropriate in a pamphlet addressed to students. Fictional forms may have seemed the best way to provoke understanding of how disaster was lived, but they could not have been sent to a judge.

Figuring out the right way to write was always a practical challenge. It also forced us to try to understand the societies, cultures, and political economies in which we worked. In asking how to cast our arguments to students, for example, we had to consider what it meant to be a student in India in the late 1990s. While preparing legal documents, we had to keep in mind the crosscutting, globalizing forces within which our words would be read.

As someone who came to work with BGIA from the world of advertising, Vani Subramaniam was particularly adept at putting language in the world. Her years working for advertising firms in Delhi—a career catalyzed by the increasing presence of both foreign and domestic corporations in India—made her part of the growing middle class. In the late 1980s, she began contributing her skills to organizations like Saheli, a women's collective in Delhi, and BGIA. Her lack of "any kind of very Left background," as she puts it, was a double bind, "both good and bad." She says it sometimes made it harder for her to figure things out. But it also made her acutely aware that the work of progressives could not be cleanly differentiated from what it challenged—whether corporations or conventional middle-class sensibilities. Nor should it be, if middle-class progressives want to reach beyond the "already converted."

Interview with Vani Subramaniam/Delhi, December 1996

V. SUBRAMANIAM: Whether it's a leaflet for Saheli or a postcard for Bhopal, what one is trying to do is linked to advertising. If the skills are only developed within advertising, there are limitations. You have to be so simplistic, so personal; you have to crawl into people's lives with an end product. It's irritating. But, somehow, I think it helps you focus on the basics. It simplifies the logic and tells you that if you really want people to listen to you, you damn well speak in the language that people would want to read. Which is basic at one level, but I just feel that it has taught me that. It's also given me attention to detail, the kind of detail that would bypass anybody who was not in the business. Those things I'm glad for. . . .

[I started by] talking about the need to portray some of the issues differently. I had been collecting things over the years related to many issues, whether it's Amnesty International or Exxon or spastic children, or any of a million things like that. All worthy causes, but something that people respond to relatively easily because there's an element of pathos, and it's quite clear there's a disadvantage, and there's you who has the advantage, and you can do something—whether in real terms or in cash terms. And it's quite simple to do that work in terms of advertising. What becomes difficult is to portray issues of primarily corporate or State failure—in a way that people will accept them. Because so much of their lives, either professionally or personally, is dependent on them, or identified with them. Which is what I was doing in the rest of my life, in the [advertising] agency. So I realized that on the one hand I was contributing to the build-up of those corporations, yet that didn't give them the right to do what they wanted on their own terms.

. . . So that's how I started this alternative communication work, if one can call it that. The postcards were the first thing, and the first year was the one with the statue, and the one with the child. Then we decided that the next year it needed to be more focused on medical issues. By the first year we had decided that it all had to be something outside of the general grammar of Bhopal work.

K. FORTUN: What does it mean to get out of the predictable grammar of Bhopal? What were you trying to avoid?

V. SUBRAMANIAM: In a sense, it came naturally to me. Because I'm not an activist who comes from any kind of very Left background. So I don't come with that baggage, which is both good and bad. There are lots of people who come from a heavier Marxist background, say, who are a lot sharper about the implications of things, which sometimes takes me a while to figure out. I used to call it the difference between getting into it from a "political background," and from a more "humanitarian background." A disgusting term, but I just use it to make that difference.

So for one, I just don't have the jargon or thought or whatever. And, in a way, this is terribly important. If you're reaching for middle class understanding—forget about support, yeah?—just understanding or sympathy— then you have to speak the language that people speak. It's ridiculous to keep talking in your jargon, because then you'll just keep talking within sangathans [movements]. I felt that that was important, to make that transition. While it's true and while it's dramatic, when you start every write-up with "On the night of . . ." I feel it becomes very difficult for people to absorb it, after a while. Because that's what they read the morning after, that's what they've been reading for ten years after. It's not a derogatory comment at all. I just feel that in some forms, you need to do it in a different way. Which is what I was trying.

And I was trying it with other things, too. So, for example, even in the work on the Narmada dam, that's how our discussions started. I was saying, "look, there's been a lot of noise against this big dam, but I think that what we probably need to do is talk more about what kinds of solutions are possible." Because the middle class is, by and large, quite clear that development has to happen and somebody has to pay the price—and as long as it's not them they don't care, you know? I just felt that had to be done, and that's how we actually started.

K. FORTUN: Did you get any opposition on grounds that it was too middle class, or too slick?

V. SUBRAMANIAM: From some people, at first. Definitely. Especially that it was too slick. Because I had been exposed to it a lot earlier in agencies. I had seen the most boring subjects made into the most exciting things. And I feel that there is always a way to do that. And if you've got the training, you can damn well put it to more exciting use.

Theek hai [okay?]. Professionally, I still want to keep my earnings independent from whatever else I'm interested in. So I make money from many of the same corporations that the organizations I work with are working against. For example, one of the clients I had done some work for—we later had a campaign against. He knows exactly that I'm the same person. So in that sense it's a bit frightening, because you have burned some boats behind you. You can't keep the two worlds separate, and I refuse to do so.

. . . Times change: when I first started doing alternative work and people asked me what I did, and I would say advertising, it was the worst thing you could say to anybody. You had to prove your intention, almost. And so I was very conscious in those days that I come from this commercial world, and I needed to be careful what I said and wrote.

Suro [Surajit Sarkar, Subramaniam's husband] was reading to me from this Stefan Zweig biography of Balzac, and talking about his past and the stuff he had written earlier when he was trying to make two ends meet, all the garbage he used to write. And Zweig says that, you know, "despite all his maturity later, it always shows through." I didn't sleep for nights! And it's a fact: you can't just erase a part of you. That is one of the problems that happens when you're trying to straddle two different worlds—bring them together, yet keep them distinct.

But that's not the case now. I don't get that reaction of horror, because I think the entire disillusionment with the NGO sector, and funding, and corruption in politics has become so apparent. The same people who would earlier say, "are you sure you want to continue this advertising work?" now say "why don't you just continue this advertising work?" And I don't take it badly. I feel that it is a change in the times. It's a pretty sad indicator, actually—that you can't have faith. But you can't. That's a fact. You can't say that one world is clean and one world is not, and what is unfortunate is that this world of funding agencies and groups is so dirty—it's so problematic. So if I can continue to do this, to be financially independent and yet have enough time and energy for this other work, that will be one of the biggest things in my life. . . . (Fortun and Fortun 2000)

BHOPAL. THE SUFFERING CONTINUES. SO DOES THE STRUGGLE.

FIGURE 9.1 POSTCARD OF STATUE AND MARCHERS A postcard to commemorate the anniversary of the Bhopal disaster. The statue at the bottom stands outside the gate of the Union Carbide plant in Bhopal. The postcard was designed by Vani Subramaniam and produced by the Bhopal Group for Information and Action.

The postcards Subramaniam produced for BGIA circulate widely. One card commemorates the struggle for justice in Bhopal. The card itself contrasts two images. At the bottom there is a photographic image of a statue that stands outside the Carbide plant, of a woman running with her children. The statue is cast on a white background. The figure is at the center

'CHILDREN OF BHOPAL'. A PAINTING BY KISHORE UMREKAR.

FIGURE 9.2 POSTCARD OF CHILD   A postcard to commemorate the anniversary of the Bhopal disaster. The postcard was designed by Vani Subramaniam and produced by the Bhopal Group for Information and Action.

FIGURE 9.3 POSTCARD MEMORIAL   A postcard to commemorate the anniversary of the Bhopal disaster that builds on the symbolism of other memorials. The postcard was designed by Vani Subramaniam and produced by the Bhopal Group for Information and Action.

and facing straightforward. Somehow, a terrible stasis is invoked. Above, cut out images of a protesting crowd are silhouetted against a red sky, moving forward with a loose gracefulness.

Another postcard foregrounds a painting by Kishore Umrekar, a Bhopal artist. Umrekar himself is severely handicapped. He has been in street theater performances in Bhopal in which he served as the figure of India. In the painting on the postcard, the colors of the Indian flag billow behind a child, but they look more like saris hung to dry against a backdrop of smokestacks. The meaning of this variegated image is unclear, but the child gazes directly at the viewer nevertheless. A third "re-counts" the number of dead in Bhopal, drawing on the force of monuments like the Vietnam War Memorial, rendering quantifications into personal names.

The call to make new connections between things is, of course, sometimes literal. One of Subramaniam's later post cards asks us to "connect the dots" that would hold former Union Carbide CEO Warren Anderson accountable. The message requires little interpretation: PUT THE BUTCHER OF BHOPAL BEHIND BARS.

FIGURE 9.4 POSTCARD TO CONNECT THE DOTS   A postcard to commemorate the anniversary of the disaster. The back of the card is printed with a message to the Prime Minister of India, encouraging him to request extradition of Warren Anderson (CEO of Union Carbide at the time of the disaster) from the United States so that he could be criminally prosecuted in India. The postcard was designed by Vani Subramaniam and produced by the Bhopal Group for Information and Action.

(9.6) At Odds

The urgent tone of life in Bhopal had many effects. The imminence of the tragedy weighted every word and move with moral import. Even the smallest gesture seemed duty bound and gravely consequential. The ways in which people related to each other seemed to have particularly fateful significance. So we stretched our imaginations and habits into shapes purposefully out of sync with the world around us. But the urgency of the situation also called for efficiency and effectiveness in our engagements with institutionalized power, and this sometimes led to assertions of expertise and authority that duplicated the very abuses of authority that we sought to challenge. Often there was simply no time to create the feedback loops through which expertise could have been reciprocally produced and transfigured. Our politics, too, became operations of control.

With time, I would learn that it was not only in Bhopal that the social roles of environmentalism remain so at odds with the ideals that motivate them. Most often deficits of moral conscience or political will are not the problem. Throughout India as well as the United States, I have met many, many environmental activists deserving the greatest admiration. The best among them struggle constantly with questions about how they should position themselves in the complex social dynamic of environmental politics.

My own way of understanding and engaging the Bhopal disaster has, of course, changed over the years. While I was in Bhopal, for example, writing about the contradictions of being a middle-class progressive hardly seemed a priority. To the contrary, it would have seemed much too personal. And decisions about what should be written about, and what should be left off the record, would have been almost impossible to make. Such decisions never became easy. But I have come to think that writing about advocacy in practice—about ethics in the real world—is a crucial part of the political project of responding to Bhopal.

In Bhopal, we continually ask how roles for victims should be configured. And we knew that building new roles for victims would have to bounce against the roles that had already been carved out for them. Doing something different depended on nuanced understanding of what already was. History mattered. Building new roles for middle-class progressives will no doubt require a similar approach. To *Remember Bhopal,* we will need to query ourselves as much as the system in which the disaster continues to operate.

(9.7) Toward a Politics of Disaster

When the sorrow of my heart
came up and filled my eyes

I had no choice
but to listen to a friend's advice:
I washed my eyes
with blood.
Now every face and form
has turned to blood.
Blood red the golden sun
blood red the silver moon
blood red the morning's laughter
blood red the night's weeping;
every tree a column of blood
every flower a bloodied eye
every glance a sting of blood
every image flowing blood.
So long as this river of blood
keeps flowing
its colour is the red desire
of martyrdom, pain, anger, and passion;
but it is stopped
it twists and turns
with hatred, night and death,
the colour of mourning
for every colour.
Do not let this happen, friends:
bring me a flood of tears instead,
the pool
in which to wash this blood
this blood in my eyes
my blood-soaked eyes.

Faiz Ahmad Faiz, "When the Sorrow of My Heart"

The BGIA household was always chaotic and beset by controversy. But it also created a space for friendship and expressions of care forcefully out of joint with the world around it. The poetry of Faiz Ahmad Faiz was a shared passion. We often listened to readings on scratchy cassettes. Someone would translate for me, while insisting that the English ruined it. The cadence pulled you in. The voice was both comforting and driving.

In Faiz's poetry, the figure of the beloved redoubles. The beloved may be a woman, a people, or a political cause. History redoubles as well, becoming a river of blood that flows across every face and form. One begins to see the world through blood-soaked eyes. Every tree. Every flower. Every image. The past inhabits them and is the way to different futures.

THE PASTS AFFECT ON THE FUTURE LIVES ON.

TEN

Communities Concerned about Corporations

*Victim Silence
within organizations.*

(10.1) Constitutional Constraints

The Bhopal disaster has provoked a great deal of commentary in various academic disciplines and within the popular press. It is often presented as a matter of "risk management," situated within a configuration of multiple "stakeholders." The discourse of stakeholders, now relied on by all sides of the political spectrum in engaging with environmental issues, emerged in the attempt to recognize the complex interaction of different parties in late-industrial societies. The "stakeholder model" recognizes differences in objective social position that must be adjudicated. Thus, there appears to be a recognition of conflict that could serve as an important addendum to environmentalist discourse grounded on universalistic claims of a natural, proper relation between humans and their context.

But there are problems with the stakeholder model, most evident in its performative effect. Minority players in environmental conflict continue to feel marginalized; capital continues to determine what is both feasible and desirable; the "manageability" of environmental risk is rarely questioned. The stakeholder model does provide a place at the table for citizens and others whose allegiances are "beyond the fence." But it is not good manners to insist that there are, indeed, divergent views of reality and tastes for change. Some oppositional voices keep talking anyway.

Communities Concerned about Corporations (CCC), a coalition of labor, community, and shareholder activists, is an example. CCC was formed in 1992 in response to political-economic trends that exacerbated the possibility that the Bhopal disaster could be replayed in the United States.

Members represented local groups across the country. During the years when there was funding, CCC came together once a year to forge a "new social contract" with corporations. My first meeting was CCC's second meeting, held in Port Alto, Texas, in September 1992. I had just returned from Bhopal. The stories told and concerns addressed threw the structural parallels between India and the United States into high relief. Reckless, closed-door management of chemical production did, indeed, seem a global phenomenon. The Texas backdrop was particularly sobering. The Gulf Coast region is home to the world's largest petrochemical complex. During the preceding few years, there had been major "mishaps" at several plants in the area. Widows were at the CCC meeting, as were representatives from the Oil and Chemicals Workers Union Local 4-449, based in Texas City.

CCC's third meeting, in June 1993, was in Dickinson, Texas. Forty-one people participated in the two-and-a-half-day meeting: Sixteen workers from seven different unions. Five representatives of victims' organizations. Fourteen community activists. Three shareholder activists. Three full-time labor/environment organizers—who, like myself, didn't "fit" into any of the recognized constituencies. Their role was crucial, but it was difficult to articulate. Differentiation of CCC from mainstream, "professional" environmental organizations was important, so ties to "struggle on the ground" were emphasized. Shareholder activists also fit uneasily with the "grassroots model." But they embodied other important themes: the need to mirror the complexity of environmental problems with complex social alliances; the need to work from both outside and inside (which shareholders were able to do via their access to the annual meetings of corporations); the need to rebuild corporations, rather than destroy them, because of the jobs they provide; and, centrally, the need to conceive environmental problems in terms of ethics, rather than straightforward self-interest. Practically, this translated into efforts to coordinate and learn from struggles in different locales.

In 1993, target projects included the formulation of protocols for community-labor plant audits and the organization of "circuit rider meetings"—involving visits by CCC members from different constituencies to particular communities, where they would try to help organize and educate similarly diverse coalitions at the local level. Most ambitious was the plan to draft a proposal for the "Corporation of the Future," specifying the organizational form necessary to sustain the new social contract CCC was seeking.

The story of the emergence and development of CCC is a story about frustration with the environmental regulations enacted in the 1980s. A key provision of these regulations called for citizen involvement in risk management decisions, to be enacted through participation in Local Emergency Planning Committees (LEPCs). Many members of CCC participated in LEPCs. Some even participated in the Chemical Manufacturers Association's Responsible Care program, which in many locales replaced LEPCs with Citizen Advisory Panels (CAPs).

Nan Hardin says that all she learned was that Union Carbide wasn't going to tell her what the solvent TF-1 actually was. All she had the right to know was that a polychlorinated biphenyl (PCB) removal plant was going to be located near her home in Evansville, Indiana—near enough to the Ohio River to raise grave concerns about the area's drinking water supply. Diane Wilson says all she had the right to know was that she would not be invited to become a member of her LEPC, despite the particular knowledge and concern she has as a professional shrimper in the Gulf of Mexico. Wilson repeatedly requested admission. She was told, "Oh, honey, that's an industry thing. You can't do that." Wilson argues "that even the most ignorant person can see that this is an inside deal." Most members of CCC concur. Some tell stories of LEPC meetings that turned into turf wars between different agencies or emergency crews—"requiring" industry to step in as the coordinating force. Others tell of meetings where the drone of technicalities silenced all questions and where the greatest concern seemed to be about offending plant managers. *god forbicle.*

Workers say that all they've had the right to know has been about plant closures for relocation overseas, while remaining plants replaced union labor with contract labor. Both put surrounding communities at risk. When—as in Bhopal—plans are being made to relocate a facility, maintenance is often neglected. When—as in Bhopal—trained workers are replaced by undertrained, inexperienced workers, routine procedures are prone to fatal mistakes. The latter was the case at an ARCO plant in Channelview, Texas, where the husbands of two founding members of CCC worked. Greg Davis and Roy Shipp were killed on July 5, 1990. Nine other contract workers were also killed, as were five ARCO employees.

The story is like Bhopal in too many ways. CCC member Chris Bedford explained what happened: Tank 68720 processed wastewater from the propylene oxide/styrene monomer (PO–SM) unit prior to disposal of the wastewater by deep-well injection. After an accident involving a much smaller waste surge tank in 1980, tank 68720 was redesigned. One new feature was a continuous nitrogen purge that mixed nitrogen with the oxygen that built up in the tank through the decomposition of hydrocarbons. Another new feature was vent compressor C-6801, installed to compress the hydrocarbons into liquid form so that they could be pumped back into process units for reuse. Hydrocarbon vapor in the presence of oxygen . explosive—so it was crucial that both these features functioned properly. On the evening of July 5, 1990, vent compressor C-6801 was not working. The nitrogen purge had been turned off on July 4. The specific reason the nitrogen purge was turned off is not known. But "it is known that operators in the unit had received no formal training in the operation of Tank 68720. Workers learned by word of mouth, a situation that was complicated by the plant's very high turnover rate. Operators did not know of the critical role of the nitrogen. The oxygen sensors designed to warn them of danger were not working" (Bedford 1996, 2).

According to ARCO's own records, vent compressor C-6801 was out of operation for maintenance 41 percent of the time between January 1 and July 5, 1990. Since April, it had operated only erratically, reputedly causing unpermitted venting of vapors to the atmosphere (which, if detected, could have cost ARCO up to $10,000 in Environmental Protection Agency [EPA] fines). ARCO chose not to shut down the PO–SM unit. Instead, workers worked overtime under the direction of "step up" supervisors—hourly personnel who function as temporary managers when regular managers are not available. Most of ARCO's experienced managers were off for the holiday. But Greg Davis and Roy Shipp had been working since 7:00 A.M.—past the legally mandated quitting time at 11:00 P.M. Shortly after 11:00 P.M., Davis, Shipp, and other workers clustered around tank 68720, trying once again to get it fixed so that they could go home. The temperature inside and outside the tank had continued to rise throughout the day. And the tank started rumbling. Then it blew up. For a time, the entire plant was under threat.

Bedford interpreted the lesson:

> In a time when industry is asking—demanding—that we trust them to be self-regulating, in the name of the elimination of so-called wasteful regulation, the ARCO Chemical Corporation explosion at Channelview, TX offers a sobering dose of reality. An industry that rewards managers almost exclusively for production and profits inevitably treats safety and environmental concerns as poor stepchildren. Without outside regulation by communities and inside regulation by hourly workers with union protection, explosions like the one at Channelview will continue to occur. (Bedford 1996, 3)

In Bedford's analysis, the right to know is crucial, so legislation is important. But enforcement can't be left to the government. CCC's task was to build an institutional structure that "gives real meaning" to the right to know.

The problems with the promise of right-to-know legislation have been conceptual, as well as practical. Most generally, knowledge has not equaled

they had sufficient time and expertise to wade through
ɔn provided under these laws, risk bearers have not had
uthority necessary for effective intervention in corporate
Instead of "empowerment," LEPC meetings and worker
ve generated a profound sense of structural impediments
unat not only keep citizens from achieving greater safety, but also make
them pawns of the system. CCC's response was not a critique of democracy
and law per se, but an insistence on the need to retrofit what the U.S. Constitution had become.

CCC grounded its work in constitutional claims to the sovereign rights
of the U.S. people, as opposed to those of corporations. A key reference for
the group was a pamphlet titled "Taking Care of Business," which documents how corporations came to be recognized by law as persons and how
the corporate charter has evolved from an instrument of public accountability into a license for reckless pursuit of profit. Authors Richard Grossman and Frank Adams explain that "when we look at the history of our
states, we learn that citizens intentionally defined corporations through
charters. In exchange, for the charter, a corporation was obliged to obey all
laws to serve the common good and cause no harm" (Grossman and Adams
1993, 1). Elsewhere Grossman elaborates by insisting that "a sovereign
people worthy of the name defines corporations, instructs their directors on
appropriate behaviors, and prohibits perpetual special privilege." Grossman cites poet Gary Snyder to emphasize that citizens do, indeed, have
room to maneuver: "To be truly free, one must take on the basic conditions
as they are: painful, impermanent, often imperfect—and then be grateful
for the impermanence and the freedom it grants us. For in a fixed universe
there would be no freedom. And with that freedom we improve the camp-
site, teach children, oust tyrants" (Grossman 1996, 2).

FOUNDING DECLARATION
COMMUNITIES CONCERNED ABOUT CORPORATIONS

The people of the United States of America charter corporations, control
their right to operate and retain sovereignty over them. Our labor and our
lives create the wealth which makes possible corporations and the profits
they derive.

While we support the right of people to own private property and we acknowledge that shareholders have property rights in corporations, we recognize that many corporations received tax breaks and direct legislative subsidies from the American people and that corporate activities can negatively
affect workers, shareholders, communities and the living environment.

Therefore, we demand that corporations must produce and operate in an ethical manner that serves and protects our common interest. We uphold that: *and products?*

- Corporations shall invest in environmentally sustainable production in the communities where they already operate, creating jobs for workers and paying taxes to citizens whose skill, labor and community infrastructure have made possible corporate operations and profit making in the past.
- Corporations shall respect the autonomy of the workers employed by them and recognize workers' right to organize and bargain collectively as essential to the safe and ethical operations of corporate facilities; invest in workers' skills to ensure long term economic corporate and community prosperity; acknowledge the obligation of all workers to act ethically to protect the health of themselves, their communities, and the living environment and to ensure the economic survival of their employers and their jobs.
- Managers and Directors of Corporations shall be accountable to the shareholders who invest in them and to the communities who subsidize and nurture them for all policies and operations affecting the achievement of ethical production—including environmental responsibility and social justice concerns.
- Corporations shall not harm the living environment that sustains all economic activity.
- Corporations shall be accountable for the damage they cause, restoring health to workers, community members and the living environment as best as possible and fully compensating those injured by corporate operations and policies.

We, the undersigned, in order to preserve and heal our people, our lands and our corporations hereby constitute ourselves as Communities Concerned about Corporations. We do so to redefine and transform the corporation to make it operate ethically and safely and in the long term interest of our communities, our workers and families, our stockholders and our common environment; to establish the above as principles for the economic redevelopment of our nation; to preserve and extend our democracy; and to enhance the future and freedom of all.

Approved, September 9, 1992
Port Alto, Texas

CCC's conception of the need and possibility for change underwrote a straightforward statement of purpose for the commemoration of the tenth anniversary of the Bhopal disaster:

We demand our democratic rights. We demand a new charter for the Union Carbide Corporation. We want the company to be a part of our communities' future but only on condition that its managers give priority to hiring local workers, and respect their rights, protect and restore the environment, and share the risk inherent in hazardous industry. Above all else, we demand that workers and neighbors play a major role in deciding the future of Carbide facilities in our communities. This is the democratic way. This is the American way.[1]

The strategy of charter revocation linked CCC to a fundamental irony of the Fourteenth Amendment to the U.S. Constitution: formulated to protect emancipated slaves, the Fourteenth Amendment turned into protection of the corporation as a person in ways that often override the rights of human persons, particularly in minority and low-income neighborhoods.

CCC worked to resolve the contradictions of the Fourteenth Amendment. Other contradictions it worked squarely within. Members of CCC knew that the U.S. Constitution has delivered unevenly, particularly to minorities. Awareness of the failed promise to protect labor unions was also particularly acute. The Bhopal disaster extended the disfaith. Nightly concern that a cloud of toxic gas could come into your home and blanket your children does not do much for old-fashioned patriotism—especially when the EPA stands by to watch.

Members of CCC were harshly aware of disjuncture between themselves and the U.S. government. Many tell of the nightmare of this recognition. They became concerned about toxins in the air or water near their homes or about the possibility of catastrophe like in Bhopal. They contacted a local government official. They learned that he plays golf with guys who work for the company. So they contacted state officials and learned that those guys are awarded for attracting business to the state. Some officials are quite eloquent, insisting that the chemical production in the state is contributing solutions to the worldwide problem of hunger and to the special prosperity of the United States. "We've got a choice," one state official said, "learn to live with chemicals or live like squirrels." He proceeded to remind his audience that "everything is made of chemicals anyway. Our bodies. Even potatoes—which are also poisonous if you eat too many of them."

But the harshest recognition came through dealings with the federal EPA. "It broke my heart," one woman told me. "I was so sure that as soon as they knew, the problems would be fixed." It didn't happen that way. There is no Love Canal in the story of resistance to the hazards of chemical production. Presidents have not flown to town to save the children. Hazardous wastes are simply easier to denounce than production itself. Communities can "just say no" to waste. Production is more complicated. Jobs are at stake. The local baseball field is crowned by a plaque thanking Union Carbide for its contributions. Most everyone in town is entangled in what they critique.

Tensions between "jobs and the environment" often produce choices that many considered "barbaric." "There are some choices that are uncivilized, period," one worker told me. "Being given the choice of joblessness or hazardous working conditions. That's hardly the sign of an advanced society. Die. Die. Take your choice. And they use those lines about us wanting to go back and live like monkeys. We're even behind the monkeys. Monkeys don't sell their children's health for a job." Residents living near plants, but not employed in them have expressed a different, but related logic. "We are asked to drink PCBs and 'manage' when our children begin vomiting after playing outdoors for too long. To complain is to risk a plant closing that could cost hundreds of jobs. We are asked to choose between our own families and our neighbors. That's hardly American."

CCC's refusal of possessive individualism resonates with the "we" of its constitutionalist rhetoric. Members of CCC recognized themselves in the collective pronoun. And they insisted that it meant something different than what it had become. When they challenged corporate appropriation of the constitutional "we," it is as though they were insisting that they had been misrepresented. They were the "we," but what they heard in mainstream usage was not what they meant. And they didn't appreciate having words put in their mouths. "The way the law works. That's not what I mean when I think of myself as American. It's shameful." [2]

(10.2) Language Learning

Imagine a public meeting somewhere in Louisiana. The room is crowded; people lining the walls hold babies and clutch the hands of children. The women at the back look particularly tired; many of the men still wear oil-stained workshirts and jeans that have been washed a million times. There are a few teenagers.

The look of people at the front of the room is different. The men in white, no-iron dress shirts are from town hall. The button-downs with tiny stripes are on men from the company. The young, well-scrubbed one is from the EPA. A few wear tie-dye and sandals.

The meeting is called to order. Most everyone knows the rules by now. After precisely two hours, everything would be over. Commentary from the audience would have been heard in two-minute slices, cut off by the mayor, standing in as moderator. The issue, too, is well understood: they want to site a hazardous waste dump nearby, promising that it will bring in much-needed revenue and won't even be noticeable from the road. The material will be shipped in from somewhere in the Northeast, where there isn't any more space for disposal. Louisiana is expected to be hospitable.

Not everyone is convinced. People move aside in the back, making way for the man with the aquarium, already filled with waters from local wells. Another man follows with a bucket. Despite protest from the mayor, the aquarium is set up on the table on the stage—at the end, but still in plain view from every angle. The man with the bucket announces that he is carrying fish, which he intends to put into the aquarium. By the end of the meeting, the fish will be dead.

The well-scrubbed one from the EPA bounces to the edge of his chair. The older woman from the Audubon Society visibly flinches, takes off her glasses, and prepares to speak. The lawyer from the Sierra Club begins, without the calm that seemed so cultivated just moments before. "You can't do this," he says. "Think about it. Why are we here? To kill those fish is to kill everything we believe in, everything we fight for. A few fish. It seems so insignificant, so trivial in the larger scope of things. But to kill them would be to continue our society's addiction to short-term goals, failing to see that these few fish are precisely what we are after—what motivates us, guides us, symbolizes our values. . . ."

It begins to sound like a lecture. A low murmur comes from the back, becoming a steady chant. "Kill the fish. Kill the fish. Kill the fish" (Edelstein 1988, 167).

This story was recounted by Lois Gibbs to illustrate the difference between traditional environmentalists and toxic environmentalists. It was April 1984; her audience was the New Jersey Grass Roots Environmental Organization, a coalition of groups working locally throughout the state.[3] They didn't need knowledge of Bhopal to be concerned. Already, they knew about disaster. An important case had been decided in New Jersey in 1983, finding that Jackson Township had created a "dangerous condition" in operating a local landfill in a "palpably unreasonable" manner. Residents of Legler, New Jersey, were awarded damages for emotional distress, medical surveillance, and diminished quality of life and were reimbursed for the costs of a new water system.[4] And 1984 was the year the California Waste Management Board commissioned the infamous Cerrell report to advise it on how to overcome political opposition to the siting of waste management facilities. Cerrell Associates found that

> the state is less likely to meet resistance in a community of low-income, blue collar workers with a high school education or less. . . . All socioeconomic groupings tend to resent the nearby siting of major facilities, but the middle and upper socioeconomic strata possess better resources to effectuate their opposition. Middle and higher socioeconomic strata neighborhoods should not fall within the one-mile or five-mile radius of the proposed site. (Cerrell Associates 1984, 43)[5]

What New Jersey activists didn't know in 1984, and perhaps sought from Gibbs, was what to call their concern. Like others in the toxics movement of the early 1980s, those in Gibbs's New Jersey audience were struggling for words. Their central concern was human health. Their goal, as Gibbs put it, was to "plug the toilet." Lois Gibbs ended her lecture in New Jersey by noting that the traditional environmentalist would not have done what needed to be done—kill the fish (Edelstein 1988, 167).

Environment and Development in the USA:
A Grassroots Report for UNCED

Compiled by the Highlander Center, New Market, Tennessee 1992
Jean True, active member of the Bridge Alliance in Henderson, Kentucky,
and chairperson of Kentuckians for the Commonwealth's Toxics Committee,
1987

Before the Unison PCB removal plant came to Henderson, I had never really been involved in community issues. I guess that like a lot of people I
assumed that government was taking care of itself and my elected officials
might know how I felt by mental telepathy.

I realized how wrong I was the first time I went to a public meeting about
Union Carbide's Unison plant. That was in 1985, during Carbide's leaks in
Institute, West Virginia, and not long after so many people had died because
of the Union Carbide plant accident in India. Many people were concerned
about the company and came for the same reason I did.

What happened at our zoning board hearings really frightened me. There
were so many citizens there, concerned and angry, but most of our elected
officials weren't there, and those who were asked few questions. It was as if
someone had already decided to hand our community over to Union
Carbide.

There were too many inconsistencies and unanswered questions, like: If
western Kentucky is in such danger from earthquakes, why do we keep putting toxics plants in a high-risk zone? Why were the risks to the river and
groundwater being ignored? Why wouldn't anyone tell us what the major
chemical was that would be handled at the plant, a secret solvent called
"TF-1"? Two and a half years later, we are still fighting our governments to
find out what it is and why it is being emitted into the air in our community.

I learned the hard lesson that everyone learns when they get into these
battles: the system is not working and only people can make it work.

In Henderson, we thought if we were very polite, if we found all the right
documents and studies, if we made the right requests, then government
would respond. It just didn't work.

One of our biggest mistakes was focusing too much attention on the U.S.
Environmental Protection Agency in Washington and Atlanta. Time and
time again, we watched that agency sidestep, bend or totally ignore laws,
regulations and their own documents. In the end, Jack Ravan, EPA's region
administrator, handed Union Carbide full operating permits. Shortly afterwards, Ravan left EPA to work for Rollins Corporation, a company which
runs one of the incinerators receiving PCBs from the Henderson plant.

We started the Bridge Alliance because we saw the need to get people
from both sides of the Ohio River working together. People from both Indiana and Kentucky stood to be affected by the Unison plant because of the
possible emissions, transportation problems and our common drinking water
source, the Ohio River. The Bridge Alliance wanted to educate the public
and work on political action and legislation. . . .

I think a lot about Denise Giardina's book *Storming Heaven,* which describes the brutalizing of the people and resources of eastern Kentucky and West Virginia. In the book, an organizer talks about the "fullness of time" when people finally took risks to stand together and form the mining unions. I think the fullness of time has come again for Kentucky. The KFTC (Kentuckians for the Commonwealth) Toxics Network is part of an exciting groundswell of people reaching out to each other, speaking out, taking risks to make a system work again. People have to become a part of government again.

One of the things I've learned from KFTC is the importance of developing an organization with a respect for individual members and their part in a democratic process. One of the reasons for our national waste crisis is that a small group of people has cut the majority of Americans off from vital information and a real voice in their own communities. Sometimes grassroots groups unwittingly mirror this problem by turning away from others in their communities that they consider too "uneducated," too "unsophisticated" or too "radical." They wind up being a tiny group fighting a big battle against huge odds.

Every person who lives and breathes with us in a community has a right to information, a voice in the process, a share in the responsibility and dignity. I think this is what KFTC is all about.

From watching KFTC's successes, I believe it is a mistake to look at Washington or the EPA for many solutions. Kentucky's toxic waste battle will be fought and won right here in the state of Kentucky and in our own hometowns. We failed to do that in Henderson. Hopefully, we won't make the same mistake twice.

I have two sons, Jamie and John, who have had to live with me through the Unison fight, the phone calls, meetings and other work. Sometimes, the work takes me away from them; sometimes, they have been with me carrying signs on a picket line.

My husband, Jim, has spoken out on these issues, too, and gives a lot of support and respect to what I do. He considers this part of my "job" even though his paycheck supports our family.

I think we both feel that our kids will learn from what we do; they must see that we have to make some sacrifices to change our world. They have to see that we're not afraid to speak out, and that public officials are not gods but human beings with a moral obligation to be held accountable for their actions as our representatives. Our kids are really what this is all about. We owe them a future.

The toxics movement that emerged in the United States in the early 1980s was both a supplement and a challenge to mainstream environmentalism. It emerged in response to awareness that communities of people already marginalized by race and economic status have been particularly burdened by the health hazards of industrial production. Their concerns were not addressed within conventional conservationist strategies, which many argued "conserve the status quo," preserving wildlife to preserve upper-class access to leisure without addressing the uneven distribution of the risks and reward of industrial production. Since the late 1970s, with community activism in Love Canal as a landmark case, communities affected by toxic waste were increasingly involved in efforts to define and control the risks to which they are exposed. In the process, the discourse of environmentalism became laden with the concepts and rhetorics of civil liberties—prompting collective action to pursue citizens' right to know, speak, and decide for themselves.

By the late 1980s, strategies for collective opposition to the siting of toxic waste facilities were well developed. Strategies for opposing toxic threats associated with petrochemical production facilities were just beginning to be formulated. CCC recognized that it lacked a language. And it was committed to finding its terms on the ground, within the experience of toxic exposure. So it decided to launch a labor/community organizing drive at and around the ARCO plant in Channelview, Texas, that had exploded in 1990. Local members of the Oil, Chemical, and Atomic Workers Union (OCAW) were part of the campaign. But CCC itself assumed leadership. The most obvious statement was that unionization was important not only to protect wages and worker health, but also to protect the health of surrounding communities. And CCC was intent on demonstrating that all of its constituencies—including shareholders—believed this.

The decision to organize at ARCO was made at CCC's 1992 meeting in Port Alto. In a follow-up memo to CCC members, Chris Bedford explained that "the decision was made to undertake a specific labor/community organizing campaign to develop the language for and to work out the issues implicit in the CCC coalition. The thought was that we had to get 'engaged' in a real struggle for responsibility with power in order to proceed with our work." To accomplish this, Bedford, then convenor of CCC, moved to Houston, along with Wendy Radcliffe, an environmental activist from Charleston, West Virginia. One of their first moves was to meet with local religious activists. The meeting was a fiasco. Religious leaders insisted that union organizing was not necessary, unworkable, and "something we don't do in Texas." Some were also particularly concerned that "the church not be used as a front." One of the local leaders who met with Bedford and Radcliff called the national leadership of his church, pulling into the fray people in charge of the church's investments—people who had previously shown support for CCC. Another religious leader in Houston was thought to have passed documentation of CCC's plans to ARCO management. Not long after, CCC lost the support of Texans United, a regional division of the National Toxics Campaign. Leaders of Texans United thought that overt involvement with a union drive would compromise their ability to negotiate with companies on behalf of residents.[6]

Bedford interpreted these fiascoes as evidence that "a silence existed around the issues that led to the formation of CCC." Part of the problem was that "far too much of the language used in the labor/community/shareholder/victim interchange was created by a legal and regulatory process designed to separate different constituencies. Defining a problem in a narrow, specific way is limiting and dividing." So efforts to "refine the themes and language" of CCC's work intensified, with a considerable sense of urgency, since the coalition was threatening to crack apart.

CCC had had a working relationship with the Interfaith Center for Corporate Responsibility (ICCR) since its first meeting in June 1992. Already, ICCR had helped arrange meetings between community representatives and executives at DuPont, ARCO, and Union Carbide. And CCC had supported grants written by ICCR to begin involving local church communities in their shareholder resolution work. But their "languages of engagement" were not only different, but also at odds. Thus, the "Great Petrochemical Corporate Accountability Tour" was organized to introduce national ICCR representatives to issues on the ground in Texas. Five national and local religious representatives participated.

The tour began at 9:00 A.M. on March 19, 1993, at OCAW Local 4-449's union hall, in Texas City. OCAW members were ready to talk. They had just won most of what they bargained for in a three-year contract covering 40,000 oil and chemical workers. The contract included significant health and safety provisions. But OCAW also knew that its bargaining success was partly due to cutbacks that left the industry unprepared for a strike. Unlike in previous years, industry had had to acknowledge that there was not adequate skill and management to run the plants without union presence.

OCAW members told tour participants story after story about fights with management to win common-sense health and safety precautions. But they also were able to recount a commitment recently made by AMOCO to work with the union on a list of 300 items needing to be addressed to make the plant safe for workers and the surrounding community. The group then went on a plant tour at AMOCO, which also provided lunch. AMOCO management made it clear that shareholder involvement in the tour was their motivation to participate.

After lunch, the tour stopped at the Phillips plant that had exploded in 1989, then at the ARCO plant. ARCO had formally refused permission to tour the plant, so a memorial service was held at the plant gate. That evening tour participants had dinner with five of the families who lost husbands or sons in the ARCO explosion. Afterward they watched part of an HBO special called "Death on the Job," which has extraordinary footage of the catastrophes at Phillips and ARCO and of the "almost Bhopal" in Texas City in 1987.

Saturday was again supposed to be filled with meetings. On Saturday morning, participants were supposed to meet with mid-level managers of the petrochemical industry in a "neutral" forum on the University of Houston–Clearlake campus. According to Bedford, "[O]ver 50 invitations were sent and personal contact was made with a dozen executives." But no one showed up. In the afternoon, tour participants were supposed to meet with an African-American community at Barrett Station, near the worst Superfund site in the area. This meeting was canceled after community members were told that their participation could jeopardize a pending settlement with ARCO and EXXON to secure a safe water supply for their community. Sunday didn't go much better. After a full day of meetings, it was agreed to leave all reference to ICCR off all literature related to the ARCO campaign.

Later reflection on the tour by members of CCC recognized both successes and failures. The effort to establish links between ICCR's national activities and local community efforts didn't work. But links between ICCR and OCAW were strengthened, turning into plans to pull together worker-owned proxies (through employee stock ownership plans) to support ICCR work on environment and justice issues. And workers learned that they could tell their stories effectively to outsiders—contributing terms to the language CCC was trying to develop—which had begun to be referred to as "ethical production."

CCC's struggle to develop a language for political engagement with toxic production remained difficult. CCC was involved in union organizing. It supported a hunger strike by Diane Wilson to protest toxic dumping in Gulf Coast fishing waters. It issued calls for revocation of Union Carbide's corporate charter. It circulated documentation explaining innovative legal strategies. None of these initiatives was free of controversy. But a sense of possibility was sustained. The commitment to develop new political idioms from "the bottom up" remained strong.

(10.3) Bottoms Up

In an introductory letter, Lois Gibbs explains that "CCHW [Citizens' Clearinghouse for Hazardous Waste] is about family, survival and justice," working on the principle that real change comes from the bottom up, from people at the grassroots who stand up and fight for themselves. The pam-

phlet that the letter prefaces goes on to explain that CCHW members define
"clean" as "put back the way it was . . . restoring a site to its condition
before it was contaminated." Commentary on the need for hazardous waste
reduction emphasizes that "wastes can be handled in ways that not only
don't pollute, but generate profits."

The pamphlet's glossary goes on to offer more definitions:

> ACCEPTABLE RISK: What somebody else says is OK for you. CERRELL
> COMMUNITY: Named after the 1984 Cerrell Report, a community that's
> rural and low-income, where people hold to traditional values and try to
> mind their own business. A perfect place for a new waste site! COST-
> BENEFIT ANALYSIS: A mathematical process designed to make your
> community bear the risk while somebody else makes all the money. HYS-
> TERICAL HOUSEWIFE: A woman leader who challenges government or
> industry polluters. For some reason, there's no male equivalent of this term.
> NIMBY: "Not In My Backyard," industry's term for democracy. The true
> spelling is NIMBI, which means "Now I Must Be Involved!" NO STATIS-
> TICAL SIGNIFICANCE: There aren't enough dead people around to im-
> press regulators enough to take action. PUBLIC HEARING: An event where
> the people speak, but the public officials don't listen. Also known as an exer-
> cise in futility. RISK ASSESSMENT: Science's best guess at the death toll.
> Like a captured spy, torture it enough and it will say anything. RADICAL:
> What you're called when you ask uncomfortable questions. SANITARY
> LANDFILL: A contradiction in terms (oxymoron).

CCHW's definitions demonstrate savvy awareness of the force of words,
and deep cynicism about government. CCC also emphasized the need for
new political idioms, built from the experience of people in toxic communities. But CCC's relationship to organized labor twisted the formulae. Lois
Gibbs articulated the goal in restorationist terms: clean is "put back the
way it used to be." Corporations, hopefully mandated by government, must
do the work. For CCC, "clean" was a process, rather than a point of return.
Clean (enough) is what could happen *within* the production of toxins if the
social structure of production was rearranged. Unions were necessary, but
so was union collaboration with people outside the plant, including environmentalists. The long lockout at a BASF plant in Geismer, Louisiana,
that began on June 15, 1984, provided a model.

BASF is the second largest chemical company in the world, and the largest German chemical company. During World War I, BASF was part of a
group of companies operating under the name of I. G. Farbin. Its profits
increased with the sale of poison gas as a weapon of war. During World
War II, I. G. Farbin enjoyed huge government subsidies to help maintain
German autonomy. A new plant, built during the war, was called I. G.
Auschwitz. Over 300,000 inmates from the concentration camp nearby
worked at the plant. Nearby death camps also had links to I. G. Farbin.

Zycon-B, a Farbin product, was the gas used in the death chambers. Twelve I. G. Farbin executives were prosecuted as war criminals. They were charged with persecution on political, religious, and racial grounds; extermination; imprisonment; and other inhumane acts. All I. G. Farbin executives disclaimed knowledge of the crimes they were accused of.

After the war, I. G. Farbin was reorganized, with BASF at its center. Its campaign to destroy unions began in the 1970s. In the late 1970s, 2,000 of the 20,000 BASF workers in the United States were members of OCAW. By 1988, the number was down to 200.

The BASF plant in Geismer, Louisiana, is the largest of eighty BASF plants in the United States. It is part of a twelve-plant complex in Ascension Parish and part of a string of ninety-three plants along the Mississippi River between Baton Rouge and New Orleans. Its hourly consumption and production of chemicals is extraordinary: 12 tons of phosgene, 41 tons of chlorine, 30 tons of ethylene oxide, 35 tons of isocyanates, 15 tons of hazardous wastes.[7] In 1984, 370 OCAW members worked at the Geismer plant, which had been OCAW organized since it was built in the late 1950s. Local 4-620 had been in contract negotiations for many months when the lockout was announced. A lockout is a company strike. Workers are willing to work, but are not allowed to do so. The Geismer lockout would become the longest lockout in labor history, extending into 1988, when less than half of the workers were allowed to return to work. The Geismer lockout was the eighth lockout OCAW had experienced in Louisiana in a decade. It also was the most disastrous. The effect on workers and their families drew the attention of local religious groups and the support of many environmental groups never before associated with labor campaigns.

OCAW's environmental initiatives in Geismer began a year after the lockout began, when workers gathered to mark a map with places inside the plant where they remembered toxics having been dumped. Previously, they had not known of the long-term repercussions. By 1986, OCAW was working with the Sierra Club, the Louisiana Environmental Action Network, and local community groups to produce a report on air pollution near the plant. In early January 1987, a second Sierra Club–OCAW report was released, and OCAW began helping organize residents in Ascension Parish. In June 1987, OCAW marched with Greenpeace in a protest at the state capital, in Baton Rouge. A new alliance was in the making. Workers learned that "the environmentalists aren't just a bunch of kooks." Greenpeace learned that workers harbor invaluable information, and the possibility of safe plant operation. BASF initiated an aggressive public relations campaign, including billboards welcoming tourists to Ascension Parish. OCAW had its own billboards, welcoming visitors to Bhopal on the Bayou.

CCC embodied the legacy of the Geismer lockout. The significance was

FIGURE 10.1 BHOPAL ON THE BAYOU   A billboard on a highway leading to Geismer, Louisiana, in the mid-1980s, during the longest lock-out in labor history. Photograph provided by the Oil, Chemical and Atomic Workers Union and reprinted with OCAWU permission.

particularly visible in the year preceding the tenth anniversary of the Bhopal disaster, during which CCC conducted an intensive research campaign in communities across the country where Union Carbide has plants. The project culminated with a commemoration in Charleston, West Virginia, a few miles from Institute, where the sister plant of the Bhopal plant is located. An extraordinary amount of data had been collected and compiled in a "research compendium" almost 300 pages long (Draffan 1994). Extraordinary social relationships had also been built.

The key event of the commemoration was in an auditorium. Mike Lombardi was the first speaker. He told of his everyday fear sitting in the control room of the AMOCO plant where he worked, knowing that bolts had been tightened by untrained contract workers. And he told the story of a fellow worker who had been burned to death after being spewed with gasoline from a flange improperly installed by a contract worker. Diane Wilson spoke next, about the explosion at Union Carbide's Seadrift plant, just a few weeks after it was declared the safest plant in Texas. The last speaker was Rehana Begum, a member of the Bhopal Gas Affected Working Women's Union.

The stories told during the commemoration brought connections between them to the surface. They also called for a reconfiguration of environmentalism. Bhopal would be at the center. December 3 would displace Earth Day, as Corporate Clean-Up Day.

FIGURE 10.2 AN INVITATION TO FIGHT BACK   A poster designed and circulated by Communities Concerned About Corporations, for a major event in Charleston, West Virginia, commemorating the tenth anniversary of the disaster.

Remember Bhopal

Get involved with the cleanup!

Polluting Corporations must:

✔ Eliminate toxic discharges through zero discharge technologies and substitution of non-toxic processes and products.

✔ Restore communities victimized by environmental injustice.

✔ Guarantee workers' rights to organize unions, refuse unsafe work and protect the environment.

✔ Compensate victims of toxic production.

✔ Disclose all information affecting public health and safety.

✔ Stop replacing skilled and/or unionized workers with unskilled contract workers.

Corporate Cleanup Day December 3

Fight corporate abuse in your community and workplace

Contact Communities Concerned about Corporations, Nan Hardin and Harriett Kimmel, Corporate Cleanup Day co-chairs
210 West Jennings St, Newburgh IN 47630; phone (812) 853-5336.

FIGURE 10.3 REMEMBER BHOPAL/CORPORATE CLEANUP DAY   A poster designed and circulated by Communities Concerned About Corporations, for a major event in Charleston, West Virginia, commemorating the tenth anniversary of the disaster.

(10.4) Intimate Connections

Dichlorodiphenyl-trichloroethane (DDT) was introduced in the United States in 1942, becoming one of the "secret weapons" contributing to Allied victory in World War II because of its capacity to prevent insect-borne disease. When introduced for civilian use in agriculture, DDT was lauded as a crucial component of the "war on insects." Along with other insecticides, it came to signify potential plenty, in both the West and the developing world (Wargo 1996, 73).

Then, in 1962, Rachel Carson published *Silent Spring,* using DDT to explain bioaccumulation of toxins in food chains and human fat. Carson also revealed that DDT was present in human breast milk and could be linked to reproductive failure of fish and wildlife. The threat of chemical contamination was compared to the atomic threat. Chemicals—as culture—became quite complicated.

The chemical industry, scientists, and politicians responded to Carson's arguments with powerful arguments of their own. Monsanto Chemical Company, which began producing DDT in 1944, responded to Carson with a parody titled "The Desolate Spring" (Monsanto Company 1962). Robert White-Stevens, writing for the *New York Times,* was more direct:

> The major claims of Miss Carson are gross distortions of the actual facts, completely unsupported by scientific, experimental evidence, and general practical experience in the field. Her suggestion that pesticides are in fact biocides destroying all life is obviously absurd in the light of the fact that without selective biologicals these compounds would be completely useless. The real threat then to the survival of man is not chemical but biological, in the shape of hordes of insects that can denude our forests, sweep over our crop lands, ravage our food supply and leave in their wake a train of destitution and hunger, conveying to an undernourished population the major diseases (and) scourges of mankind. If man were to faithfully follow the teaching of Miss Carson, we would return to the dark ages, and the insects and diseases and vermin would once again inherit the earth. (White-Stevens 1964, 1, quoted in Wargo 1996, 80–81)

Secretary Freeman of the Department of Agriculture made the connection to economics:

> Perhaps one way to indicate just how important these benefits are is to point out what would happen if we did not have them. Production of commercial quantities of many of our common vegetables would be drastically reduced, including corn, tomatoes and lima beans. . . . Commercial production of apples would be impossible. Peaches and cherries would almost disappear from our markets. Any number of diseases that infect grapes, cranberries, and raspberries would drive these fruits off the market. Commercial production of strawberries, peaches and citrus would be impractical. . . . The production of eggs, chickens and other poultry in the South and Southwest

would no longer be possible. Economic production of beef in the South would be virtually impossible. . . . Our total supplies of meat and milk could be drastically reduced. . . . The percentage of stored food condemned for spoilage or other damage would spiral upward with consequent reduction of supply and higher prices to the consumer. . . . Only families in the higher economic brackets would be able to enjoy many of the most nutritious foods now available to all. . . . The lowering of national dietary standards in itself would be a severe restriction to public health. Economic repercussions—affecting not only the farmer but also the processor, retailer and consumer—would be equally serious. (Senate Subcommittee on Reorganization and International Organizations of the Committee on Government Operations 1963, 94–95, quoted in Wargo 1996, 81)

Professor Frederick Stare, chairman of the Department of Nutrition at Harvard's School of Public Health, took on the task of reassuring the U.S. populace. An excerpt from his newspaper column was reproduced by Mississippi Congressman Jamie Whitten in his own book, titled *That We May Live:*

The current hysteria about agricultural chemicals has seeped in under the doorsills of American homes.

A woman recently said to me, "I feel like Lucretia Borgia every time I put dinner on the table. Am I poisoning my family?"

That concerned woman, interested primarily in the health and well being of her family, deserves to have an end put to her confusion about agricultural chemicals, particularly pesticides. Her bafflement stems not from stupidity, but from the claims and counter-claims of self-appointed experts who usually don't know what they're talking about.

They are usually extrapolating from some findings on birds, bees or fish, or the unfortunate result of some child inhaling or swallowing large quantities of some pesticide. Such findings just don't extend to the use of agricultural chemicals in growing, protecting or preserving of foods.

Let's set aside all arguments about how or why the current controversy started and concentrate instead on letting facts speak for themselves.

One irrefutable fact the critics of pesticides have been unable to answer is this true statement: there is not one medically documented instance of ill health in man, not to mention death, that can be attributed to the proper use of pesticides, or even of their improper use as far as ill health from residues on foods. . . .

In spite of this lack of evidence, many people now have the impression that pesticides contaminate our food supply and are harmful, probably lethal. This gap between fact and fancy must be closed or we will do ourselves great harm by allowing disease and famine to rule the earth.

Are pesticides poisons? Of course, that's why they work. They are poison to the insects, worms, rats, weeds and other pests against which they are directed. Because of strictly enforced regulations and tolerance levels, however, the hazard to man from pesticide residues on foods is almost nonexistent. They are dangerous if you handle them carelessly or leave them around where small fry may "play house" with them.

> You can have full confidence in your foods. They are not full of poisons, *[really]*
> as some food faddists would have you believe. They are nutritious and the
> quality is much better than it was a generation ago.
>   Eat and enjoy them. . . . (Smith 199, 390–91, quoted in Whitten 1966,
> 103)

The arguments used to counter Rachel Carson were indeed forceful. Critics said that nature (the biological) had to be and could be controlled by culture (chemicals), that facts could speak for themselves, and that science was unquestionable. They acknowledged that pesticides are indeed poison, explaining patiently that "that's why they work." But they also insisted that expertise and regulatory enforcement turned what was poisonous into prosperity. Pesticides were positioned in stark opposition to disease and famine. Reliance on experts was the means to avoid devastation and chaos.

Carson's critics could not, however, stop the way Carson's message had "seeped under the doorsills of American homes," transforming the most domestic activities into threats. The fear and skepticism that would spawn toxic environmentalism in the 1980s had begun to circulate.

One of Carson's most lingering metaphors is of a grand-scale human experiment conducted with no knowledge of potential dangers. Experts only make things worse because each sees "his own problem and is unaware of or intolerant of the larger frame in which it fits." The public, Carson argued, was "asked to assume the risk that the insect controllers calculate," blindly following the blind (Carson 1963, 13). Members of CCC and others in the U.S. toxics movement during the 1980s understood what she was talking about. They also recognized that the problem Carson had identified was cultural and political, as well as scientific and technological. At issue is what "counts." *—?*

Carson argued that "under the philosophy that now seems to guide our destinies, nothing must get in the way of the man with the spray gun. The incidental victims of his crusade against insects count as nothing: if robins, *[Yes.]* pheasants, raccoons, cats, or even livestock happen to inhabit the same bit of earth as the target insects and to be hit by the rain of insect-killing poisons no one must protest" (Carson 1962, 83). By the 1980s, the man with the spray gun had lost his heroic stature. The cultural logic nonetheless remained in place. The side effects of chemicals—including the Bhopal disaster—continued to be discounted. The man with the spray gun was replaced by the man from public relations. *← reference?*

(10.5) Just Cause

One way CCC distributed its message was through *Just Cause,* a lively, informative newsletter with a long feature story, brief updates on work by members of CCC, and announcements of upcoming events of interest. The editors, Nan Hardin and Karen Lynne, separated the articles into three categories. Some of the articles described the complex of problems confronted by local groups; some of the articles focused on innovative solutions; some of the articles revitalized histories of labor and community activism that could provide inspiration and example. Hardin and Lynne's sense of timing was important. They did not see the articles in *Just Cause* as statements of previously codified positions. Instead, *Just Cause* was a place where environmental activists could learn what they thought *through* writing.

CCC faced the challenge of crafting new identities for its members in many ways. CCC also faced the challenge of explaining the importance of its work to outsiders. Some outsiders overtly opposed CCC. Others simply didn't understand the problem or how it implicated them.

It is from labor organizers within CCC that I heard the most explicit description of how one organizes the unorganized. The hardest catch is a young worker who doesn't come from a union family and thus doesn't tend toward unionization as a matter of tradition. Often the young worker seems unconcerned. Most often he is male. He has a job, now. His body seems invincible. He may even brag about the fumes he can inhale without being affected. He is rigidly proud of his independence. Then he is told a story about another man not unlike himself. At first, he shows vague interest. He doesn't seem to notice the parallels. The story may be about a man who worked for the company for twenty-five years and then was laid off following a poor performance report by a moody manager. Or it may be about a man who got sick too young, died, and left his wife and children in poverty. Or it may be about workers at a plant that was closed down for relocation overseas, on only a few days' notice. Or it may be about "give-backs"— lower pay, longer hours, fewer benefits—sacrifices said to be necessary to keep a plant open.

Sometimes it is necessary to be overt: These men were not treated fairly. And it didn't just happen. These were strong, competent men who had shown great loyalty to the company. The next move by the organizer is even more direct: "This could be you, buddy. Don't kid yourself."

Some workers never respond. Others take their time. It is as though the stories have to percolate a while and perhaps be supplemented by an image from a film or by other stories, overheard in a bar or at a doctor's office. Some, finally, begin to ask questions. They begin to realize that maintaining their independence may be more complicated than they had thought. For the most part, most had never even thought about it. The world may or may not have seemed hostile; they were simply in it, not wanting much, just to be left alone, thinking this was possible if they just minded their own business. It's not like they wanted to conquer the world. They just wanted the television and microwave to work.

Union organizers told me that it was not as though workers in the 1980s were naïve about the American dream. Workers' talk was peppered with black-humored remarks about the rich getting richer and everyone else working at McDonald's. But this didn't make unionization seem any more relevant.

Union organizers did not set out to tell these workers what to say or think. Instead, they talked to them about their experience and the experiences of others in similar positions—drawing out relationships and repetitions, throwing the world's divisions into high relief. The fantasy of just getting by, just minding one's own business was cut apart.

What the organizer gave the worker was not so much the truth as a structure of significance—a way of connecting seemingly disconnected data points. In the process, the organizer intervened in a fantasy of independence—independence imagined as anonymity. The worker thought he could merge with everyday life without purposeful organization. The organizer reminded him that everyday life is already organized. What the union promises is not salvation, but articulation. Union alliance with coalitions like CCC adds even more links.

(10.6) Structural Problems

Reaction against citizen concern that a disaster like Bhopal can "happen here," in the United States, has been vehement from the start. But now it is contained in the pages of the business press, not in the official statements and glossy brochures distributed by chemical companies. The business press helps maintain the stage on which the great drama of contemporary environmentalism is performed. It is opposed and assisted by so-called radical environmentalists. The business press insists that all is lost unless productivity (of profits, in particular) is given absolute priority. Deep ecologists insist that all is lost unless economic development is stopped, or even reversed. Together the business press and radical environmentalism operate like the proscenium arch that separates a stage from an auditorium. You can't be in the play unless you cross into a space structured by harsh opposition.

An article published by *Fortune* magazine in September 1994 illustrates one of the boundaries (Dowd 1994). The main billing: "Environmentalists on the Run." The subtitle: "Business leaders, local officials and angry citizens are demanding an end to rules based on silly science and bad economics. This time, they just might win." The article tells the story of Alar, the growth regulator used on apples that the EPA deemed a carcinogen in 1989, after Meryl Streep and *60 Minutes* set off a panic that had parents pouring apple juice down the drains. United States apple growers lost some $100 million. Five years later the American Medical Association, the National Cancer Institute, and the World Health Organization concurred that Alar poses no real threat to human health. The risk is now said to be "less than that incurred by eating a well-done hamburger or a peanut butter sandwich" and is compared to the much greater risk incurred by playing high school football (Dowd 1994, 91).

Alongside other stories about the silly science of environmental regulation, hard figures state the economic facts: "America's cleanup bill is considerably larger than those of its principal competitors. U.S. environmental spending amounts to 2.2% of GDP, vs. 1.6% to 1.8% in Germany and 1% to 1.5% in Japan" (Dowd 1994, 91). These facts ground the article's lead prediction: that with the exception of legislation to reform the Superfund law, no major environmental legislation would get through Congress in 1994. Instead, three reforms, dubbed "the unholy trinity" by opponents, would dominate the agenda: (1) all new rules must be preceded by cost-benefit analysis, (2) no rules if the government is not willing to pick up the costs, and (3) compensation when rules severely limit the use and value of private property (Dowd 1994, 91).

Fortune's rhetoric depends on an opposition between environmentalism and economic growth. So-called radical environmentalists have often shared the structure of argumentation, even while inverting the terms. Like the business press, "radical" environmentalism polarizes environmentalism and economic growth. There is little attempt to work within the opposition. One side of the binary is privileged, the other discounted. The choices are harsh: jobs *or* the environment, private property *or* the common good, science *or* concern. Groups such as Earth First! demonstrated the strategy, arguing that we must fundamentally limit established patterns of both production and consumption, whatever the costs. The biosphere itself is the threatening, aggressive force that should concern us.

Three particularly stunning articulations illustrate the logic. As famine ravaged Ethiopia in the mid-1980s, Earth First! founder Dave Foreman argued that "the best thing would be to let nature seek its own balance." In other words, we should let the Ethiopians starve, recognizing that there is a natural logic to the order of things, not unlike that of Adam Smith's market. In the same interview, Foreman expressed his version of patriotism: "Letting the USA be an overflow valve for problems in Latin America is not solving a thing. It's just putting more pressure on the resources we have in the USA. . . . Send the Mexicans home with rifles" (Dowie 1995, 210). Then, in May 1987, the *Earth First!* journal published an article by "Miss

Ann Thropy." The argument: "I take it as axiomatic that the only real hope for the continuation of diverse ecosystems on this planet is an enormous decline in population. . . . If the AIDS epidemic didn't exist, radical environmentalists would have to invent one" (Dowie 1995, 210).[8]

It is unclear whether Earth First! has given the Bhopal disaster the same high marks for productivity that it gave the Ethiopian famine and AIDS. Commentary on these statements has nonetheless circulated widely among environmental activists, in both India and the United States. Many recognize that the logic of Earth First! opposes, but sustains that of magazines like *Fortune,* and the business press generally. Some environmentalists, even if not themselves habitants of the radical periphery, consider this radical opposition productive. Anonymous voices from mainstream environmental organizations often recognize Foreman and his cronies as "real resources," who create a space within which groups such as the Wilderness Society can sound "eminently reasonable" (Dowie 1995, 210). Strategic differentiation may also explain the disjuncture between the business press and business itself. By the early 1990s, corporations could not longer be heard as antienvironmentalist. They may have seen the amendments to the Clean Air Act as a threat, but they could not simply oppose them. Corporations continue to remind audiences that we cannot expect a "risk-free world" and that everything is dangerous if mishandled—"even water, if you stick your head under it." But they rarely baldly oppose environmental safety and economic growth, at least in public performances. If Earth First! creates a space in which the Wilderness Society sounds eminently reasonable, the business press creates similar space for business itself. As a vice-president at Dow explains, "One way to lose a reputation for combativeness is to stop being combative, and that's what we've done. Another thing we're doing is to spend more time finding out what people are thinking" (Hoffman 1997, 98).

What people are thinking is heard within the discursive structure created through the opposition of the business press and groups like Earth First!. Concern about risk is force fit into logics that categorize all said into either support or condemnation of economic growth. Groups like CCC have little room to maneuver. Concern about air pollution is heard as suggesting that we should "go back and live like squirrels." Concern about workplace hazards is heard as lacking concern about competitiveness and being ignorant of "how the world works." Robert Kennedy, a Union Carbide CEO, has explained the problem with a story about the pencil. "Pencils are dangerous. They have sharp points. A child could stick one in its eye, or in its ear. I doubt that we could get the pencil approved today" (Pickering and Johnson 1991).

(10.7) The Many Faces of the New Expert

Symbolic analysts.[9] Knowledge workers.[10] Cultural producers. Reference to people in these roles recurs in descriptions of a world order built around new communication and information processing possibilities. Social roles have emerged that resist categorization in terms of class. Some in these roles make a great deal of money. Some make very little. Sources of income vary. Some are salaried. Often one person will have contracts with businesses, government agencies, and nonprofit organizations. Some are politically conservative. Some consider themselves radicals. Many do some pro bono work. Some consider all their work "voluntary," since they support themselves through projects funded through grants. Their means of production include telephones, computer networks, airplanes, and other technologies that have accelerated connections among people, places, and once disparate "data sets." The social and rhetorical skills necessary for interacting with people in diverse settings are also crucial. The product of their work is new links—among people, ideas, phrases, and events. These are the new experts. Environmental problems are surrounded by them. Coalitions like CCC are catalyzed by them.

The new experts that have catalyzed toxic environmentalism complicate understanding of how and where advocacy happens. How, for example, should one position Ward Morehouse, who played an important role in founding CCC, as a trajectory of work on the Bhopal disaster since 1984? Morehouse founded and has maintained the International Council for Justice in Bhopal, a coalition of groups that has issued periodic statements regarding the Bhopal case, responding to specific issues faced by gas victims, but also to the need to remember Bhopal so that it doesn't recur elsewhere. Morehouse often quotes Milan Kundera, speaking of Bhopal as a struggle against power—as a struggle of memory against forgetting. For Morehouse, the struggle against forgetting has taken an array of forms, with many different purposes—suggesting how diverse the enterprise of advocacy can be. Morehouse has supported victims' roles and interests in the Bhopal litigation. He has supported activists working in Bhopal and has organized their travel to the United States and the United Kingdom so that they could tell the Bhopal story themselves. He has organized public hearings on industrial hazards and human rights, bringing together testimony of victims from Bhopal, Minimata, the Texas Gulf Coast, and elsewhere.

Morehouse is a full-time activist. His work is supported from the proceeds of Apex Press, which distributes both its own publications and others, on topics ranging from appropriate technology to workplace safety to economic justice in "the new economy." Morehouse's own books on

Bhopal have been published through Apex, as has an account of the disaster by T. R. Chouhan, a former worker in Union Carbide's Bhopal plant. Chouhan's book was printed in India, through coordination with The Other India Press in Mapusa, Goa. Morehouse's effort to build Apex into a viable "alternative press" has served many purposes. It has supported his own work—much of which involves raising money for grassroots initiatives. It has provided an outlet for distributing the story of Bhopal in many forms. It has provided a place for people to work where daily work relations are understood as one more way to "Remember Bhopal."

Experts like Ward Morehouse have created new configurations of self, nation, and world, updating what it means to be a progressive advocate. They broker in symbols and try to sell new definitions of health, security, and fairness. They help define what counts as relevant, and who and what should be seen in relation. Ward Morehouse provides an extraordinary example, as do many of the middle-class activists I worked with in India. So does Dave Henson.

In the late 1980s, Henson, too, was a full-time activist. He helped organize and secure funding for meetings between people separated by geography, experience, and ideological tradition. He visited people in their home communities to go with them to meetings with the EPA or a local corporation. He understands the different ways advocacy can be organized.

The organization of advocacy within grassroots communities and between community members and outside experts is rarely straightforward. Some outside experts mime the role of the traditional schoolmaster. Others are passive aggressive, obliging people to hold hands and speak of unity, while sharp differences of interest and perspective go undressed. CCC once had a convenor of this sort. Members ousted him. They didn't like his style. But they also resented his control over what the coalition could come up with. They didn't like it when some guy from Washington, D.C., acted as though he already knew what would come from a meeting before it ever took place.

Dave Henson had a different approach—and an overt commitment to help people understand why the organization of expertise matters. He works within the traditions of the Highlander Center, a rural Tennessee-based organization involved in adult education since the 1920s. A Highlander approach is one that assumes that people know more than they realize. The task of the organizer is to draw out this knowledge, triangulating what is learned from one person with what is learned from others and from history. The collection of oral histories in oppressed communities is an important strategy. "Lay epidemiology" is another. Sessions in which participants map out their place in the economic web are also important.

INTERVIEW WITH DAVE HENSON, Occidental, California, March 1995[11]

K. FORTUN: If you were to meet a stranger, how would you describe
yourself?

D. HENSON: I always have trouble with this. In a way, I think about the
audience before I answer. If it's a family member, or a family friend, it's almost an arbitrary cross-section of where I grew up, a suburb of California.
I say I work with the environmental movement and the social justice movement. I think these two concepts are graspable immediately. People often
ask me if I work for Greenpeace—that's always interesting, because that's
what they see as exciting and radical, if they're supporters or not. It's sort of
a cowboy, American ideal.

I say that I have, because I have worked with Greenpeace, then try to spin
it out a little more, saying that I work with communities fighting against specific problems in their area—toxic pollution, a corporation wants to put an
incinerator in their backyard. We try to develop community-labor alliances
because workers and community members are equally affected.

I answer for general folks real generally. I claim participation in a movement; I try to reduce it to environmentalism and social justice. I like to walk
away thinking that I've sparked a sense that these two things are related—
that the environment is about people, too.

K. FORTUN: Do you call yourself an activist?

D. HENSON: Yes, activist, and organizer. I try to put out both those words.

K. FORTUN: In your head, is there a difference?

D. HENSON: I think so. An organizer, I think is more intentional, it's more
of an identity. This has been my lexicon. . . . Someone who perceives of
themselves as an organizer is intentionally involved with strategic planning,
has to have a certain detachment in order to be effective—detached from the
issue at hand, seeing the issue in historical context, or the development of
issues around that issue, nationally and internationally. Basically, be aware
of context. An activist can be that too, but also can be someone out protesting something right up front—the plant behind their house that just blew up.
They got involved, they're six months into it, and now they're an activist.

Although I see that we could switch these words around. But I'm interested in the distinction between people—not just those who are victimized
being activists because they're called out on specific issues, but—almost—
stages of development in self-identity as an activist, into an organizer.

K. FORTUN: Your definition of an organizer—how does it relate to your
notion of an intellectual? Detachment, responsible for an empirical reading
of the world . . .

D. HENSON: I think that the difference is the grounding in practice—at the
start—rather than theory. Sometimes they can both meet in the center. Personally, I clubbed my way through school. Never got a college degree. Went
to law school, didn't finish. I like school a lot but as a student, I was always
more of an activist. So I thought I should stop trying to pursue formal education, admitting that what I am best at is organizing. I want to dive into an
issue, or a set of issues, helping people figure out what they're doing. Starting from there but still liking to read and discuss with people working at

an academic level—to help me understand what I'm doing. . . . These kinds
of conversations we are having today.

K. FORTUN: Do you think progressive academics are an important part of
the configuration of Leftism in this country, or do you find yourself more
antagonistic? Academics are more part of the problem, than any help . . .

D. HENSON: I try to experience this, and probably everything, with two
heads. One is as organizer-sociologist—the detached analyzer of what's going on in context. The other coming from my personal experience. I value
the former much more, because I think anecdotal experience is just that: anecdotal. The people I've met, the processes I've been through, engaging academics and communities, it's given me a lot of experiences. But nonetheless,
it's still anecdotal. I can step back and learn from a lot more people as this
organizer-sociologist, and make a better conclusion.

Many people I work with are disdainful of academics, because theory
and practice are so separated. All it takes is one or two bad experiences for
people to classify that way. It's also a stereotype, so it's easy to slip into.
Then I know people, like myself, who have reflective discussions about
activism, evaluation, strategic thinking separate from the issues you're pas-
sionately involved with at the moment. I think that there is a quick understanding that there is a role for academia.

For me, that role is a sense that there are theoretical—no, not theoretical,
well-articulated languages to describe what I'm doing and . . . I know what
I'm doing, kind of. . . . But it's only when I stop and think about it, and talk
about it do I actually get fiber on which to identify myself. I think that's
critical. When I talk to students, on college campuses, I try to paint a picture
of this fiber: What are you as an activist? Because so many people come and
go. An organizer for three years, and then leave it. How could that be?
Doesn't it become part of your fabric, or identity? And, within that fabric,
there are different processes by which we are naming what we are doing. I
rely on other people to take the time to do it—this naming.

Yet, I have so many experiences with people who slip over the line. . . .
Highlander is a great example. . . . People come to Highlander wanting to
write a Ph.D. on popular education or community history. They study what
people are doing, get a Ph.D., and the people feel that someone has ripped
them off—took up their time without them getting anything out of it. They
never saw that person again. Someone who now is a doctor at some university. But, you see something different with the organizer-sociologist head—
you can see that person constructing a curriculum for 25 years of teaching
that's based, in part, on experiences they had at Highlander. Actually, it's really important, for the 700 some odd individuals who come out of that program. They are going to be important people in society.

K. FORTUN: While I was working in Bhopal, writing for the movement
there, my job was to help produce descriptions of what was going on. And I
think my academic training helped me do this. On the other hand, I've very
sympathetic to the critiques of academics that you mentioned. But it bothers
me when I encounter anti-intellectualism in the movement here—the sense
that scholarly work is not relevant to the Left project. Maybe we need to
think more about how scholarly work can circulate in the way you find useful, how academics can spread their skills 'round, just like other resources.

D. HENSON: A couple of things. I just bought someone's master's thesis, which is supposed to be about different models of grassroots leadership. This guy Tom Shaver, don't know him, a friend of a friend told him to call me because I might be interested. In large part, he based his work on Kentuckians for the Commonwealth, one of the premier grassroots groups who have documented their training processes. They've put out great manuals that I'm using in my training in Central Europe right now. I read a brief description of the thesis, and it appears that he has gone around interviewing a series of people, picking out what's helpful. Looking at their manuals, consolidating, contextualizing. Giving things names. Giving historical relevance. Spitting it back in 150 pages. That's something that I think I can learn something from. But I'll read it before I recommend it to anyone else. He's made it available, did a mailing to activists. I like that, when someone puts back.

K. FORTUN: Helping people develop a strategic sense of expertise must be an important part of your educational work. Expertise, necessary but not necessary. Strategic, but part of the problem.

D. HENSON: Absolutely. We do a lot of work on how to use, contain and work with experts—how to make sure experts don't take over your strategic process. Medical doctors can come in with a propensity to urge you to do a health study that targets urine samples so you can get liver studies, because that's their specialization. We do it with charts, visually, putting medical advice in its place, as part of a strategic plan, in relation to other things, like grassroots organizing, building alliances with farmers, etc. The advisor is a part, not the whole.

One needs knowledge, but with a begrudging attitude. When you've got good experts, you're really proud of them. They're your ace in the hole, your big cannon to validate your experience. Larry and Sheila Wilson were here yesterday, from Yellow Creek. They are the staff directors of the environmental health program I've worked with at Highlander. They just won a lawsuit, after fifteen years in court—an eleven million dollar settlement with this tannery that has destroyed their community. In their own family, one of their sons, Wade, who is fifteen, is severely altered—his personality and constitution, by the toxins. The number of people who have died from this is unbelievable. Their daughter, who is sixteen, has just had her second miscarriage. This one had no spine; the brain came out of the back of the head. Severe abnormalities that mimic the abnormalities in the fish and the animals, the offspring of the sheep and the cows. A personal horror. Just five weeks ago, she had this abortion—five months into term. They're very poor, an Appalachian holler community. Yet they rose up, with a team of lawyers.

To listen to Larry tell about the case, and what happened in court—it was only a fifteen-day trial. It's fascinating. He's constantly badmouthing the attorney, but when the attorney nailed them on something he was "my attorney, up there fighting for us." But he was very aware when the attorneys would step over the line—and do something without consulting the community. His central pride was that he sat with the attorneys in the courtroom. When they were questioning somebody, they would always stop and ask, "what else do you got?" The community sat behind and would pass notes.

K. FORTUN: So, do you think of law as an important tool for grassroots work?

D. HENSON: I think it has always been one of the best tools. Really, there are three tools: People power—boycotts, strikes, organizing people to exert direct power. Legislative—we change the law; we can elect people to change the law. And the third is legal remedy. But I think we've over-relied on suing to get what we want. In California, the old growth issue is a good example. The main strategy for protecting the last five percent of ancient forests in the state, and in the country, has been to file suit. Blocking cutting through suits that have very little to do with the actual issues at hand. The spotted owl, the snake. The way they misfiled this or that. Nickel and diming. A successful strategy for the last ten years, but one that, ultimately, is losing. Because, of course, the corporations tighten up their ship and learn how to circumvent.

In the meantime, we don't have a movement. People in the community value those trees; they would mandate that Pacific Lumber not be able to log the headway. But just a few days ago they announced that the largest privately owned patch of ancient forest is going to be logged. We've failed because we've relied on legal mechanisms. People who understood the issue turned it over to law firms. Fund raising went toward the litigation, rather than building community organization that could transcend the issue, building a Left movement that has a consciousness and an identity on a variety of issues. They did the exact opposite. They centralized the process in an arcane way, reducing the movement to the staff at a few organizations, reducing relations with the public to a fund raising appeal. Rather than grassroots organizing. Issue by issue you find this happening.

When you're involved in the issue, it does seem the most prudent thing to do: "Look, they are going to cut next Thursday. What are we going to do? Let's file a suit. Get a six-month waiver. Then file again." That's the problem of not having a movement, a party, ultimately, which could have people systemically working on this, finding alternatives that include both political and economic strategy.

K. FORTUN: This argument is made in the environmental studies literature I teach with: Once a movement takes up litigation, participation falters. But a remaining question: can the law become a reference point for movement building? An example I think of is Institute, West Virginia, where the sister plant to Union Carbide's Bhopal plant is located. A low income, minority community. They are at sea. Outsiders tell them to go knock on doors, but they don't know what to tell people they want them to do. A lawsuit could be supported. Otherwise, how do you name a goal, when the problem is so complex, bifurcated, and amorphous? It does seem a little much to ask a working mother to take on the chemical industry, writ large. Asking them to take on a specific lawsuit, a specific issue seems more doable.

D. HENSON: A lawsuit held people together in Yellow Creek for fifteen years. Meanwhile, they also ran for school board elections, and other things, took over an incredible Appalachian old-boy power structure: incredibly corrupt: I'll gravel your driveway Thursday if you'll vote for me. That's just the way it worked. They flipped it on its head. They ran people to take over the

school board, the water board—that has all the money, because it's all federal, because there is no local tax base. They focused on the lawsuit. They
met regularly for fifteen years, to talk about the lawsuit. A lot of people
stuck together for a long time to actually build a community power base. It
was entirely grassroots; it never got polluted; no one helped them. No one
took over the case. They kept it as a community organization, using direct
Highlander ideology: we're going to keep control of this thing. An almost
provincial disdain for academics and other outsiders.

K. FORTUN: Explain the different ways this could be understood, through
the different approaches of "activists" and "organizers."

D. HENSON: As an activist, all we do is file suit. Nothing else is possible.
My other head tells me that we have to develop infrastructure to our movement, so that we have political structures that can ride out the urgency.
What's the goal? To build power. To build community based power. Democratic power that people identify with. Not on an issue-to-issue basis but in
an ideological way, a cultural way, a spiritual way. Actually feeling like the
future is in transformation, beyond the immediate interest in closing the incinerator. Something indicative of a whole set of problems. This takes political education. This takes organization. Often outside organization. Methodologically, we are learning. Popular education. Academics often call it action
research. Participatory methods. Transparency. Accountability. Words like
that.

K. FORTUN: I haven't yet understood where notions of community alliance
come from: are they brought in from outside, a pragmatic outcome of local
work, what? In Texas, for example, there seems to be a very regionally specific sense of civic responsibility, but not through alliance. It's not a collective culture. There are many codes of decency but you wouldn't go to a
meeting to enact them. There is almost a sense that if you need other people
to help you pull off morality, something is wrong with you. So it's not necessarily nefarious . . .

D. HENSON: I do think it stems out of the fetish for individuality. It's easily
explainable, how it happens. But once seeded, the idea of alliance goes real
deep. It's like we know as a species that we are communal, as well as individual. We have to reinvent it constantly. Bring back the communal. Not exclusively. But finding the bounds between our American private space and
our communal sense. But you're right. Maybe it's also a class thing. In
poorer, or more rural areas, people are much more on their own—rugged
individualists. You take care of your own deal. You're neighborly. You kick
in when you can. Everybody prides themselves on that: "we're a community
that gathers together when there is trouble." Someone dies and everyone
brings a ham. But the idea that you would have political alliance. . . . Politics
is a heresy. That you would step out and speak out. People have a sense that
they have a right to their rights, but organizing has a bad name. But it only
takes a little seed, to get past the first hurdle, and then the obvious thing is to
get together to figure it out. People first start asking: whom do we call next?
Rather than: it's us.

So the next problem is the way liberal democracy deals with complaint.
Sure, come on in to the EPA regional office or the corporate public relations

office. We'll take care of your concerns. Have a cup of coffee. How's the
family? Compromise. It's a beautiful system. Keeps the lid on everything.
The machine is a marvel.

But I think the uptake is cyclical. At times, the morphine works, making
us forget the next ridge. Highlander's educational programs work directly
against this, teaching people to exercise community power, which is called
"community." It comes and goes as a notion. But there no longer is an Eastern excuse; political involvement can't be called pink, anymore. So you can
actually say that you are a community organization, or alliance, talk about
the means of production in various ways—and its history. That's very exciting. It's a big difference. People don't want to meet with the EPA director,
they want to exercise power. Mechanisms of compromise still occur, but
with less persistence.

K. FORTUN: What is it that's cyclical? People's disenchantment with the
EPA's cup of coffee?

D. HENSON: What's cyclical is people remembering that we are a communal species; that alliances give off results. The more power you have in your
meeting, before you call in the expert, or the government, the better results
you'll get. It's like that movie "The Organizer." The guy comes into town
on a train with a suitcase, an Italian community. He just brings in the seed.
Ultimately, the film might have been anti-communist. . . . In any case,
people are upset because there is an outrageous situation with the local mill.
The organizer moves in, shacks up at someone's house, starts seeding the
idea that if we all get together. . . . I think what I'm saying, what's cyclical,
is that people remember. It's a deep remembrance. That this stuff works. We
don't have to identify with the Wobblies to remember. We're primates. So
it's pretty much down there. And people have experience with it. Whether
in the PTA or Little League. You have to cross over the big hurdle of embar-
rassment—I don't know a better word—for getting involved politically.
People here are quite civically active. But when it gets political, when you
start talking about corporations, it's embarrassing.

K. FORTUN: If you were talking to people in Texas City, attempting to be
a catalyst, to organize, how would you convince them that political participation is important? What do you say when they ask "why?"

D. HENSON: It isn't that it's better to participate, than not. The function of
participation in healthy communities is the question. It's not good or bad,
but it's the kind of participation and the level of participation that's in question. Participation for participation's sake is a strange notion.

K. FORTUN: Are you saying that participation is always, already there . . .
that the choice is one of quality, not quantity?

D. HENSON: Yeah. Be deliberate about it; be intentional about it. Participation isn't always democratic, and democracy itself can mean different things.
Accountability is the key point.

K. FORTUN: You often use the term "accountability." Can you define it
for me?

D. HENSON: I mean that in representative democracy, where at one point or
another each of us is called to act on behalf of a group—testifying, going to
a meeting, taking some action—in economic terms, doing economic activity

on behalf of the whole—accountability is a valuation. The plant that's running upstream is accountable to a community in terms of its effects on a community, and that it's not accountable only to its stockholders: that's what I'm saying is a better democracy. The people who should be involved in a decision about any particular action are those that are affected by the decision. And the activist creating that effect should be accountable to those affected. That's the principle to start from.

So you set it up by saying: Wait a minute. You live in this town. That plant is polluting the river that you live on. What they do, how they do it, what they produce, how they produce, who benefits from it, who suffers from its side effects: the people who make those decisions should not be shareholders in Chicago. It should be the people who are directly affected. And let's sketch out who those people are: they're the consumers, they're the people who log the wood that builds the fire to build the material, they're the people who live downstream, they're the people who work in the factory. Thus, economy ought to be held accountable to the people in which the economy circulates.

That's the big democracy question: the people ought to be the decision makers. There's no longer any more room on the planet to exploit and pillage. Colonists could go into some wilderness with no people and just cut the trees because the trees would be fine. We all have to manage it together now. We're all in this together. Nobody's separate from anybody else. And that's the long term. Then there's the short term: all right, the tannery gets their permit next Thursday. What are we going to do? Are we going to sit here and talk about the big picture or are we going to actually do something? We always forego the former, the long term, in order to engage now. And what happens is that after five years of struggle on that incinerator, we win. Or we lose. And either way, we disappear. Unless Greenpeace sets up an office, or Citizens for a Better Environment, or the Sierra Club, or somebody with a structure to carry on something. But it's likely, again, to be staff-based and expert-heavy and this and that. So I think the other discussion is actually the stuff with which we reform the terms of our community, the terms of our participation in community.

Instead of seeing consumers as the primary unit, or parents, which is the other primary form of engagement—through school and sports and things. And it's not church, either. But it's equally cosmologically-framed. I don't have a plan or a structure that works, but that's what we need to be doing at every little place we are.

Real grassroots people get radicalized, just like in the STP program at Highlander: get people around a table for three days to talk about common experiences, give it time, and the light bulbs go on. I think it's in us to understand that we all have a common experience. Solidarity is a natural thing. We are all in this together. Actually, I can take some of your load. I can suffer over here for your benefit. Because it's in our long-term interest. It's like we're genetically coded to get that, because, otherwise, we'd all be taking each other's food, and we'd all starve. I think that's the basis, in nature.

In this country, the grassroots haven't built a network with power. It's not like India, where the grassroots actually confronts state power. The American party structure won't allow it. It will placate things way before then. It will be made all right enough that it won't be worth your time to spend all of it organizing. So academics, outside organizers, urban people—they have roles as catalysts. We need to bring in the idea of alliance. Get people to remember. Remind people about solidarity. We have a grand tradition of solidarity in this country. You can always bring something out of people's local experience. Bring in Highlander's idea of getting communities to do oral histories. Ask the old people. You'll find amazing stories about how people organized. Create a culture. Remember it. A culture of resistance, and of organizing. A culture of questioning fundamental economic order. Which used to happen—in the thirties, the twenties. It was on the national plate for working people. It got truncated entirely after World War II. Not by design, but in a kind of organistic way: Capitalism figured it out that if we call all this communism, link it to Stalin, ride that wave. . . .

K. FORTUN: This is a good point to define more terms. What's the difference between "the political" and "the civic"?

D. HENSON: I think it's been defined for us by the post World War II anti-communist, strategic outlay of American economy. It's like everything else. Was it a decision by some white guys in a back room in Washington? I tend to fall down on saying it's the organism of capitalism. It's an organism. It responds. Not with a brain. You lop off the top and it doesn't matter. It naturally found its way, like a watercourse, down to thousands of different little acts by individuals. Putting the idea of political activism in the realm of communism. Any critique of the economy: the corporation shouldn't have carte blanche to do what ever it wants, any kind of union organizing. Anything that complains about the system is in the realm of communism. Other than that, you are welcome and encouraged to get involved. Not just the Little League and the Garden Club, but the local school board and even a development council: taking part in deciding whether a mall should be built on main street. But as soon as you start talking about property rights, about lack of citizen involvement in what's happening economically, and environmentally—the organism disallows it.

K. FORTUN: So, the "civic" has disallowed economic assessment?

D. HENSON: We have a great political democracy in this country. It works. We don't have any form of economic democracy. This argument has been stewing in our movement for a long time. By "political" I mean civic democracy, participation in the details of decisions about the governance of our lives. Even things that are extremely consequential—what our children will be taught, what our individual rights are to speak, act, travel. On the criminal code, the civil code. It's open to discussion and participation. We can write referendums and do amazing things: cut taxes in half, pass environmental laws. But when you approach anything that has to do with the levers of the economy, its absolutely off-limits. There aren't civic structures built to do that.

K. FORTUN: Now that the Red threat is gone, is there more space to maneuver?

D. HENSON: There is more space to maneuver. Very few things can be traced to some conference in Geneva, but the end of the Cold War, as a construction, I think, interestingly parallels the rise of a grassroots environmental movement—that's talking about control of the means of production. It takes alliances like the National Toxics Campaign [NTC] to actually shift the language, redefine what we are talking about. Communities are starting to organize independently, but with lots of cross-fertilization—that's what makes it a movement, rather than just an odd assortment of events, an aggregation. Basically, people are saying that they want more involvement in the decisions that will affect their lives, their health, their local economy. That plant just can't up and leave. Plant closure laws used to be perceived as socialist-labor vs. capitalist-government. Now it's not the labor unions that are at that vanguard, but community organizations—like at Yellow Creek—Eastern Kentucky, Appalachia. Screwed over bad in an event that lasted many years. The most unlikely people to start talking about taking control over what's done, asking where the money is going.

The companies can't just leave. That's the first thing. Close the back door. GATT [General Agreement on Tariffs and Trade], of course, completely altered that. It's hard not to see it as a response. The Eastern Bloc collapses, opening a space for people to start talking about control over the means of production. Shortly thereafter, just as we're getting a grip on it, GATT takes the back wall out. Paves the exit. The long-standing way we have been able to organize in the civic sector is to win lawsuits, legislative battles, elections, consumer boycotts. The carpet has been pulled out from underneath us.

K. FORTUN: And the National Toxics Campaign has fallen apart. At a time when its importance is increasingly evident. Organizing regional and national coalitions of grassroots groups is one way to engage the structural problems GATT represents. What happened?

D. HENSON: The founder was a brilliant guy. But a big lesson: he couldn't get over a founder's syndrome, and being a particular kind of man, in this culture. Not just him, others too. But he had the vision to say that there are fires out there; communities are bubbling up. Things are happening all at once. They seem unrelated. I'm not a naturalist, and think that conditions are behind everything, causing all the bubbling to be happening at once. And that we just jump into the hot spots, issue organizing like we always do. We had to ask how we could unify. How do we build people power to actually do something very political, very strategic, very radical? We couldn't just walk into communities and start talking. We had to organize. The model was to start a national vehicle for these grassroots groups to have a voice. So much was happening at the national level: RCRA, Superfund. These grassroots people should have been at the table.

But, it all started with the typical line: let's found a national group, not based on community experience. It wasn't the people founding a pyramid, which would result in their projection into national issues. It was a city-based, intellectually influenced group of activists who could see the topography, see what was coming down, conceive of a strategy, figure it out. You can't see that from some county in Georgia facing off a waste management company. Ingeniously, the founders of NTC could see that this bubbling was going on—and needed to be interconnected.

At first, NTC was just five people; a board, funders, East coast philosophers and elitists, your typical city organization. And they transformed. Through incredible pain and struggle. Into an organization with thirty-five people on the board, from all over the country. Crazy, on one hand. Economically and logistically. But every person on the board was a grassroots leader. And really was.

But, when you pick people out of some small town in Idaho, or Georgia, and have them fly into Boston for meetings, it screws up their local identity, and group structure. Constructed from the top down, rather than from the bottom up. Because the bottom just can't move in this way. There are barriers that prevent the kind of organizing you see in other countries. Revolutionary movements. People are placated. So the task is to get people to the table, then turn it over to them. Turn over the process, turn over the money. Not without a struggle. And make it multiracial. Make it a majority of people of color. A majority of women. And have an explicit set of discussions about what is being done, about what needs to be done.

With NTC, the founders continued to manipulate things but it got so democratic that the founder no longer had a real role. This never got resolved—reliance on the founder as the visionary. It destroyed the organization. But this model, this vision that never transpired, it was only a third of the way there, was to develop an Oklahoma toxics network, and an Arkansas toxics network, and so on. We had offices in ten states. Real grassroots groups becoming nationally catalyzed. The Louisiana Environmental Action Network was one that was working on its own. In a Congress structure with democratic accountability: we are the representatives of our local groups, sitting around a table. Built-in reporting. And the people aren't staff. They are grassroots people. And their way is paid. They aren't expected to pay.

K. FORTUN: Paid by whom?

D. HENSON: Grant raising, by the Veatch Foundation. NTC raised a lot of money. People loved this idea. I think it's the right model. ?why?

K. FORTUN: This model that you are talking about. First, it's a representative structure. To what extent is the placating capacity of liberal democracy emergent from this representative structure?

D. HENSON: I've never seen it work, and I've worked with so many organizations. I've tried to be a catalyst, to be an enzyme, to foster a culture that will remember and try to exert democratic control. At Highlander, at NTC, at EPOCA, at every organization I've worked with. Trying to recreate that. And it's always up against the same kind of barriers. But we've got to start somewhere. We've got to keep popping up, and then, eventually, maybe we can make it happen. I think we've got to keep plowing ahead. That it doesn't matter that it doesn't work. It's the best thing we can do. Then I think, what about other models? NTC kind of started out toward this.

Some people up in Boston have an idea, in a back room, but with an organizing director that was an old Alinskyist. He had this oddball organizing model that did not fit with the grassroots. That's another reason it collapsed.

?

The organizing model was to fly someone down to some town we've never been to and break the balls of the local group. Alinskyan machismo. If they are not with us, they are against us, even if they are our most natural allies in the political spectrum. They are the ones who are the most dangerous. We've got to break them first, rein in their power. Then move forward and take over the state. Have a press conference then fly out, just before the thugs come in to beat up the locals. An exaggerated version, but not far from the truth. The founder of NTC never understood that. He was a cowboy, a macho man. It's a classic thing. People are good on some things, and just miss the boat on others. He was really good at some things. No I shouldn't credit him with that—grassroots leaders from around the country forced him to cede control of his organization, to embody his own rhetoric: it's blacks who actually control the organization. Complete bullshit to the very end. It's actually women. . . . The worst kind of . . .

K. FORTUN: patronizing . . .

D. HENSON: It was unbelievable. I've never worked with anyone that bad. I'm sure it's rife in military, government and corporate culture. But here it is in the Left. People from the grassroots seized power from him. They had other business to attend to so it took time. And he went along with it. But when community groups actually controlled the organization, the process by which they could envision and implement strategy was stunted constantly by the peculiarity of them being plucked out of their town and flown into Boston. Personal class issues became important. Where you stay, what you eat. The people left at home are jealous. I think it happened in almost every single community. They were called national leaders, and the founder exploited that—by finding the most victimized women around the country to put on the board. They had lost a child, due to chemical exposure, and he took them around, got them on the podium with his arm around them, so they could cry. The worst kind of manipulation. Then he would slip them money on the side for the kid's education, because he's got a wife with money. A classic problem, that our labor unions have, and so many Left organizations. But had that not happened, I still think there was a barrier to these democratic structures actualizing the vision. Politically. To develop a national organizing strategy based in a grassroots movement. A very good vision. For several reasons. Race. Blew it up. Gender politics. Exploded. Class politics. Fiasco.

We decided to shut down NTC. It was better not happening, although it was a great thing. It was replicating the things we're fighting against. But now there's this huge hole. We're almost worse off than if it never existed. People are scarred; they're burned; they're fried. Nobody is going to fund this kind of thing anymore. For years, we won't be able to have anything national. Not until all those people have disappeared. One of the big costs is that a lot of foundations have gotten very wary. They were infatuated with the grassroots, for a while.

K. FORTUN: Tell me more about the after-effects.

D. HENSON: Some of the after-effects are positive. We learned a lot from the process of quitting, putting together our "post mortem." We met five or six times, and it was a majority of people of color. And it was the majority

of the Board: all the people of color voted to close it, except for two. Most
of the white people voted for it to stay open. We decided to write that paper
to analyze what had happened, and that's why it has so much on race and
gender. But mostly on race because that was the primary experience that
people had. And we went through some incredible meetings together. It
was one of the most important processes for me, ever.

Here are high-powered people from different organizations around the
country, getting together for weekends and just plowing through. We broke
down the entire organization's history, its organizing model, its structure, its
power, everything. And afterwards we were drained. It took over a year. We
held out for a long time, hoping that we could salvage this incredible multi-
racial process. In the end, we had a great analysis of one of the more recent
attempts at multi-racial, national revolutionary organizing, within the mainstream culture. And people just wanted to go away and get back to work
at home, see what comes up next. There was a sort of sinking feeling that
white people and black people, if they're really based in their communities,
can't just start working together at the national level. A lot more has to tran-
spire before that.

K. FORTUN: Are you then saying "separate but equal"?

D. HENSON: I would say separate but equal—to a point. I think the obligation for white organizations is to educate themselves and be engaged with
issues of white racism, out of necessity to the historical situation and what
it's going to take to win in this country, and to make a commitment to engage with groups of color on the terms or the agenda of groups of color. It's
a kind of affirmative action approach. Because otherwise the dynamic, in
fact, is that the white agenda dominates.

I think that modeling, being who we are, is as good as we can do. People
pick up, learn, or they don't. Creating environments where people can learn
from each other, like Highlander. You can facilitate, catalyze. You can't tell
somebody, unless they are asking you to tell them.

The work done by citizens in Yellow Creek, Kentucky, illustrates how
a Highlander approach to grassroots organizing comes together, exempli-
fying what Henson calls a "whole systems approach" to organizing—a
"watershed approach" that recognizes everyone's place in the grand scheme
of things.[12]

Larry and Sheila Wilson's hogs died.[13] So did their other animals. Thirty-
six hogs. Thirteen registered dairy goats. Seven cows. Two hundred chick-
ens. Dead within ninety days after the Wilsons started watering them from
Yellow Creek. It was the summer of 1980. There was a drought. The Wilsons' well had gone dry. They called the city to make sure that the water
in the creek was okay and were assured that is was. The creek was running
a brownish purple color—the color of leather. And there was a tannery
upstream.

After their animals died, the Wilsons started worrying about their children. So did their neighbors. So they formed a citizens' group that called itself Yellow Creek Concerned Citizens (YCCC). Fifteen years later YCCC won a lawsuit against the tannery. The story of how YCCC was organized and sustained is truly extraordinary.

Larry and Sheila Wilson and other residents of Yellow Creek had heard of the Highlander Center. They thought they were probably a bunch of communists, but they decided to give it a shot. They needed help, so they piled into a pickup truck and headed across the mountain. They learned many things. That their distrust of experts was eminently reasonable. That good relations among themselves would not come naturally. That they had a lot to learn, about the Freedom of Information Act, about what to expect from the EPA, about the connections between federal assistance to Appalachia and local corruption, about chemicals, water, and health disorders of many kinds. Most important, however, was what citizens of Yellow Creek learned about themselves. The Highlander Center taught them that they already knew a great deal. Enough to conduct their own health survey, which would provide a great opportunity for talking about their concerns with their neighbors, while generating data to think with. Enough to run for city council. Enough to begin writing for local newspapers.

Highlander's emphasis on the way communities organize themselves had a particularly strong impact in Yellow Creek. It is a small community where most everyone knows each other and many are related. But workshops at Highlander emphasized the need for specific and purposeful efforts to keep their organization democratic—because outside forces will do everything possible to divide and rule them, because benevolent offers of help from the outside often lead to dependence, because even the best people have been socialized in a culture of self-interest.

The accomplishments of YCCC have been multiple. They won a $15.1 million judgment against Middlesboro Tanning Company, most of which was allocated for health monitoring, rather than being distributed to individual claimants. Four executives were held responsible and ordered to pay punitive damages, setting an important precedent for corporate accountability. A story about keeping experts in their place was performed and continues to be retold. And a community was organized in a way that doesn't reproduce the social forms that legitimated uneven distributions of toxic risks.

Dave Henson and figures like him helped people in Yellow Creek accomplish what they set out to do—helping them shape their practice in ways that carried their critiques. Other communities—and coalitions like

CCC—have also been catalyzed by their influence. Henson describes capitalism as an organism that sucks everything into its logic unless it is intentionally stymied. Liberal democracy is capitalism's friendly face. Citizens are invited to tea with EPA officials and are drained of any imagination for another way to do things. Often progressive organizations are not much better. They are run by visionaries who tout fashionable commitments at odds with their practice, reproducing racism, sexism, and other egoisms more predictable elsewhere. The work of grassroots organization is therefore far from "natural." At its best, it is rigorously reflective and continuously cross-fertilized—building on what communities already know, but also on what they learn to understand about collectivity and authority from progressive organizations like the Highlander Center. Community "experience" is recognized as both a resource and a liability.

Eleven

Green Consulting

(11.1) Stakeholder Skepticism

The Bhopal disaster provoked profound changes in the way people perceive and engage with the chemical industry. The chemical industry itself has changed as well. A film made by the Oil, Chemical and Atomic Workers Union (OCAW) in 1992 is illustrative. The title: *Out of Control: The Story of Corporate Recklessness in the Petrochemical Industry* (Bedford 1992). Unlike most OCAW films, it is not directed at workers. It is directed at people outside the plant, who need to know what is happening inside.

The film begins with drive-by footage along the fence line of an enormous chemical plant complex. Smoke stacks come into view. Rail cars and crosscutting tracks. Intricate entanglements of pipe. What look like watchtowers. Pickup trucks. Men in hard hats. The voice has a workaday, but ominous tone: "You're driving by that chemical plant just like you do every day. Your kids ask you what they make in there. And you answer that you're not really sure. And you realize that you should be." The film then cuts to home videos. Black smoke of inky density billows into a cloud that covers half a city block. People are running. Fire trucks. Ambulances. Shots of men strapped to stretchers. The fire in the next one is red, with a white core. It turns to a dense steel gray. Then it explodes. The cloud is mushroom-like. We're told that in the last six years the U.S. petrochemical industry has had over ninety major accidents—160 killed, 1,200 injured, thousands evacuated.

Melanie Masih tells of watching a yellow-gray cloud travel toward her home from Union Carbide's Seadrift plant in March 1991. Gloria Chaplin of Baytown, Texas, near a major Exxon facility, says that she had not planned to get involved: "I'm a quiet person. I'm a wife and mother. But I was scared we were going to die." Residents are not the only ones expressing a new fear. B. T. Walker, a Phillips worker, says everything has changed with increasing reliance on contract labor in the plants: "You can't help feeling, thinking. You see a cloud and the first thing you wonder: what is it? Is it steam? Is it gas? It's an everyday thing." AMOCO worker Jon Foutz tells of how he "smelled a really raw, foul smell and started to hear a rumble. I didn't even take a chance. I just started running. Didn't look around or nothing. I'd taken about three steps when I heard the detonation." Later Foutz describes how it felt like his skin was ripped off his body, pulled over his head like a shirt. Sandy Davis says that she knew that "dealing with chemicals is never completely safe, but I also thought that the chemical industry was taking all the precautions they could—that they were running safe plants. I now know it's not." Davis's husband was killed in an explosion at ARCO in Channelview, Texas, in July 1990.

The film then cuts to Bhopal, where the story is said to have begun, a decade previously. The camera pans rows of bodies covered in white sheets. We're told of controversy over the number of dead. And we're told that "whatever the number, 'Bhopal' has entered our vocabulary forever." In the years since the gas leak, "a movement already mobilized to protest the siting of hazardous waste facilities turned its attention to the whole production cycle." What it found was "an industry operating outside of public scrutiny." People also learned that Bhopal could happen here—in the United States.

We're told of Henry Waxman's investigation following Bhopal, in which he found that facilities in other countries are modeled on plants here. We're told about right-to-know legislation, about the flow of toxic chemicals into the environment revealed in Toxic Release Inventory (TRI) data and about industry public relations campaigns to downplay the significance. And we're told about an "epidemic of plant disasters" that could not be covered over by accountants. There is footage of a horrific fire. Live commentary in the rhythm of an air traffic controller tells us that a huge cloud of hydrocarbons is spinning from Reactor 6. And that it has exploded.

The explosion at the Phillips 66 plant in Pasadena, Texas, on October 23, 1989, was the worst industrial disaster in the United States in forty years. Twenty-three workers were killed. Two hundred and thirty were injured. Two-thirds of the Phillips complex was destroyed. Debris and chemicals were blasted up to six miles away. Phillips was cited for 577 willful violations of health and safety regulations and fined $5.7 million. Later the fine was reduced to $4 million, and all willful violations were erased from

Phillips' record. Phillips also issued a statement confirming that it did not admit to any wrongdoing.

C. T. Roberts, a Phillips worker, was part of OCAW's own investigation. He was most disturbed by the removal of all reference to willful violations from Phillips' record. In his assessment, the disaster was due to reliance on contract labor for plant maintenance—which was becoming an industry-wide norm. So was disaster. In the year and a half following the Phillips explosion, there were two dozen accidents in U.S. petrochemical plants. Sixty-nine workers were killed. Over 90 percent of these accidents involved temporary, nonunion contract workers.

The Occupational Safety and Health Administration (OSHA) released a study on the problems posed by use of contract workers in the petrochemical industry in January 1991. The "John Gray Report," named after the institution commissioned to carry it out, found that contract workers comprised up to 54 percent of the petrochemical industry's workforce; they tended to be much younger than regular employees, with less education, less safety and health training, less time with a particular employer, and less time with the petrochemical industry. In November 1990, *Engineering News and Record* reported that the John Gray Report was delayed at every turn by the Bush administration, and by the Office of Management and Budget in particular.

Workers interviewed for the OCAW film describe what reliance on contract labor means in practice: Weld inspectors are replaced by contractors with no welding experience. There is a never-ending cycle of replacements and shifting assignments. "One day they're a pipe fitter. Next day they're a welder. Next day they're sweeping the floor." "People don't have enough time even in 12 hours to perform their assigned duties." Managers come in on "cost cutting binges, with something to prove." "Not only do they defer maintenance but they're pushing the plants so hard that the margin of safety starts going away. If someone makes a mistake . . ."

Congressional hearings addressed the problems associated with contract labor and "near misses" in the petrochemical industry in November 1991. Gene Busler, an AMOCO worker, provides useful clarification: "They call 'em near misses. I call 'em near hits." In one unit of one AMOCO refinery, OCAW documented over 120 separate near misses over a two-year period. In a Government Operations Subcommittee hearing on October 21, 1991, the focus was on prevention:

> *Representative Tom Lantos:* If this is a very dangerous industry, and I take it we all agree it is, and there are large numbers of near misses, would it not be common sense to expect these data to be publicly available? If we came close to a Bhopal on seven occasions and the good Lord was with us and we

avoided it, would it not be helpful to know what were all of the factors that resulted in that?

Eugene McBrayer, President, Exxon Chemical: The problem, sir, in my opinion is that these unwanted conditions are usually so site specific that it is very difficult to translate that context to another plant in another part of the country.

Representative Lantos: I have trouble believing that given the degree of standardization in the petrochemical industry that everything is so site specific. I just have trouble buying that. Because I think you have a tremendous degree of standardization of machinery, processes and training. And I have trouble understanding what is clearly sort of unwillingness or reluctance with what, to me, is simple logic.

Mr. McBrayer: It's the best first step for OSHA to encourage this kind of open dialogue at the local level, and make sure it's being done.

Representative Lantos: You take upon yourself a hell of a responsibility when you encourage open dialogue when, if we make it mandatory, we might save lives. It's not a leisurely discussion about a new recipe about the bean soup that we try and see how we like it. We are killing lots of people in your industry with the best efforts on your part and everybody's part. Now, we all share the goal of minimizing the people killed . . .

Mr. McBrayer: We share that completely.

Representative Lantos: If we share that goal why are you gentlemen resisting sharing information on near miss situations?

The OCAW film cuts back to McBrayer for a long moment, but nothing is said. Then workers speak again about how we keep reliving history, having learned little. There is brief footage of Texas City in 1947, just following the explosion of a freighter loaded with ammonium nitrate. A chain reaction ripped through the chemical plants that surrounded the city—560 people killed, thousands injured, over 3,000 homes destroyed. A police officer on the scene described it as like what he had seen on Okinawa, in the city of Naha—"after the Marines had gotten through with it." The OCAW film then cuts to an interview with Texas City's current fire chief, who insists that a similar disaster could not happen today. Because ships are built differently. Because people are better prepared. The film suggests that even the fire chief is unreliable. Then the film tells us about the "almost Bhopal" in Texas City in October 1987. A nonunion contract worker dropped a compressor on a tank of hydrofluoric acid at Marathon Oil— 1,000 injured, 3,000 evacuated. Catastrophe is pictured as predictable, and preventable.

The message of the OCAW film is blunt, capturing the political economy of hazardous chemical production. Fundamental shifts in the way people imagine work and everyday life are hinted at, but remain below the surface. For the most part, an industry "out of control" is described in the controlled tones of people who know what they are doing and how to solve the problems before them. The answer is unionization. OCAW President Robert Wages explains that while one worker does not know if an employer is telling the truth—to shareholders or to the public—300 workers do. Together workers know "what they produce, how they produce it, what goes into it, what comes out, and what is done with waste products." Wages insists that OCAW has a "tremendous knowledge that could be put to use in making an employer a good corporate citizen." His bottom line is straightforward: "the union is an advocate for responsibility." And "everybody has to have checks."

Off-camera, OCAW members have elaborated on Wages' final statement, pointing out that what counts as responsibility and as "checks" on knowledge is what remains to be figured out. What they know for sure is that the glut of information produced thus far by environmental regulations has not reduced the risks of hazardous production. So OCAW members have formed alliances with environmental and community groups to develop alternative strategies—creating pressure on chemical companies from both outside and within. Chemical companies have responded with aggressive initiatives to manage risk perception. They must deal with publics that have heard too much, but still want to know more. Publics ready to question everything they are told.

By the late 1980s, possibilities and desire for information were increasing at an extraordinary rate. This alone would have challenged the chemical industry. But what the chemical industry faced was even more complicated: citizens, workers, environmentalists, and even some public officials wanted more information, but they also were ready to question the credibility of any information they were given. Controlling the skepticism became a formidable challenge. Definitive proclamation, no matter how slick or substantiated, no longer worked. So the chemical industry committed to new strategies, claiming to involve all "stakeholders" in the information production process itself. Risk communication experts—whom I call "green consultants"—have been called in to direct the performance.

(11.2) Going Green

Social roles for "green consultants" have consolidated since the 1960s, to help institutionalize corporate environmentalism. Bruce Harrison entered the fray early. An account of his career and contributions is provided in his book *Going Green: How to Communicate Your Company's Environmental*

Commitment (1993). The book's jacket explains that the E. Bruce Harrison Company is "a professional consulting firm specializing in environmental public policy, which works with more than 80 of the Fortune 500." The book articulates Harrison's version of the history of contemporary environmentalism and a sense of the proper place of historical perspective in corporate efforts to "go green." It also tells corporate actors who and where they are in the environmental storm and how they can hope to survive.

Harrison's story begins in the chemical industry's days of grace.[1] The whole world could be viewed through a benzene ring. Bumper crops of babies were being born, calling for pesticides to be just as productive. The perfect lawns of suburbia were growing, but the weeds were not. It was a time of command and control. And everyone was a customer for chemicals. Faith in science and technology was extraordinarily high. In a 1957 report by the National Association of Science Writers, nearly 90 percent of the public agreed that "the world was better off because of science." An equal proportion could not cite a single negative consequence of science (Hoffman 1997, 47). Then came *Silent Spring*.

In the summer of 1962, Bruce Harrison was working at the Manufacturing Chemists Association, which later would become the Chemical Manufacturers Association. He had just been named manager of community relations, since he was in charge of "good neighbor programs" in industrial plant communities, which included an annual observance of Chemical Progress Week. Then the chemical industry was "hit," and Harrison's career took off. The problem: "Rachel Carson's thesis was not only that pesticides would wreak unintentional harm—the hypothetical spring when birds would not sing—but also that because it *knew* of the potential damages of chemicals to wildlife and humans the industry was *evil*. She used the image of the Borgias, cooking up the poison" (Harrison 1993, xiv).

Harrison got a new title, invented especially for the situation—"Manager of Environmental Information, perhaps the first such title in corporate PR." He nonetheless recalls being somewhat naïve regarding the significance of the Carson attack. His mentor, Alan Settle, corrected the picture through a comparison with the Japanese attack on Pearl Harbor. Harrison recalls the lesson: "We weren't ready for this, but we've got to come back fast. If we don't take charge now, we're going to be buried, pure and simple." The lesson stuck; over thirty years later Harrison's strategy is equally aggressive, beginning with knowledge that "taking charge is the only winning move after an attack" (1993, xiv–xv).

The more appropriate analogy, however, may be the Vietnam War, evoking images of environmentalists as guerrilla communists threatening a domino effect that would destroy the foundation of American society. Over

far as your eye can see...larger better equipped, more efficient farms. Today's farm is 70%
than twenty years ago . . . raises much more food per acre. As farmers feed their land more soil-
...ing minerals, every year, the land feeds us better. IMC—world's largest independent miner and refiner
...l-producing minerals—IMC grows by helping the world feed itself, here and in 56 foreign countries.

INTERNATIONAL MINERALS & CHEMICAL CORPORATION

FIGURE 11.1 THE WORLD THROUGH A BENZENE RING   One of a series of IMC advertisements from the early 1960s that view the entire world through a benzene ring. The world
viewed is inevitably well controlled. Nature—and agricultural production, in particular—is
domesticated.

FIGURE 11.2 BIGGEST CROP OF BABIES   An advertisement for fertilizers produced by Lion Oil Company to feed the "bumper crops of babies" born in 1954. The ad ran in the September 9, 1955, issue of *U.S. News and World Report.*

Eleven to One *but the odds now favor you*

Now modern lawn owners have a "great equalizer." Their weapon—the modern pesticide dieldrin! Not one of the eleven leading members of Grass Eaters Incorporated is immune to dieldrin—many other lawn insects also succumb.

Safe and easy to use, dieldrin can be applied as a dust, spray, or as granules. One "shot" of dieldrin kills insects above the ground for many weeks, knocks them out underground for years.

Dieldrin is used throughout the world with outstanding results against insects that feed on agri-

cultural crops. It is equally effective in protecting lawns against such pests as ants, white grubs, cutworms, Japanese beetle grubs and others.

Dieldrin has shifted the odds to favor man—a species greatly outnumbered by the many varieties of insect life. Helping lawn owners control insects with more potent pesticides is another way Shell Chemical helps make modern living more comfortable and pleasant.

Shell Chemical Corporation
Chemical Partner of Industry and Agriculture
NEW YORK

FIGURE 11.3 SHELL: THE ODDS NOW FAVOR YOU   An advertisement illustrating the battles Shell Oil could help middle-class Americans win, from the inside cover of the July 19, 1957, issue of *U.S. News and World Report.*

FIGURE 11.4 EVERYONE IS A CUSTOMER FOR CHEMICALS An advertisement that illustrates the importance of consumer enrollment in the promise of chemicals and the chemical industry. Women, for example, can go shopping within a chemical plant. The ad was run in the May 13, 1957, issue of *U.S. News and World Report*.

the years, the guerrillas have been tamed. Counterattacks were carefully strategized, the terrain defoliated, and the enemy forced into the open. By the early 1990s, environmentalists were coming to meetings in business suits, with their own PowerPoint presentations.

In Harrison's account, Carson's attack on pesticides ushered in the "attack mode" of American environmentalism. The war dragged on through the 1960s and 1970s, reaching a low point with the election of Jimmy Carter in 1976. But Carter's election had ironic consequences. Environmental activists found their way into government bureaucracy. And they built environmental organizations that were "managed like business." By 1990 and the passage of the Clean Air Act Amendments, "environmentalists and business people, as types, began to resemble each other a little" (Harrison 1993, 8).

Then the real watershed: the Second World Industry Conference on Environmental Management (WICEM II), held in Rotterdam in April 1991, which forged a Business Charter for Sustainable Development.[2] Governments and business leaders from around the world came together: Frank Popoff, CEO of Dow; Edgar Woolard, Jr., CEO of DuPont; Pete Silas, CEO of Phillips Petroleum; Robert Kennedy, CEO of Union Carbide.

Visioning is what got them there, most likely. Harrison tells us that visioning is learned from athletes and their coaches. It means thinking things through in advance. Like Olympic medalist Bruce Jenner, who visualized every part of his competitions beforehand, imaging every move and muscle required. The last stage was "seeing himself crossing the finish line in that long race, circling the track in the stadium, hearing the cheers of the crowd." Robert Kennedy is Harrison's example of how visioning helps people "go green," though he admits he can't prove it. Kennedy's company "suffered through the great tragedy of Bhopal," but "resolved to rise from this tragedy and to gain new respect as an environmentally aggressive company" (Harrison 1993, 23–24). Kennedy may well have visualized one of the scenarios Harrison suggests that we start with.

In one scenario, we visualize ourselves sitting alongside firefighters, city officials, and community neighbors as they draft the final stages of a local environmental emergency response plan. Another scenario visualizes hearing a verdict after a jury has been convinced that the environmental class action suit brought against our firm was unfounded and inconsistent with the firm's performance and commitment. Another visualizes a time when we can point to profits in key operations as the direct result of continuous environmental process improvement.

The scenario Kennedy is credited with realizing visualizes the presentation of a paper at a world conference on environmental management, leading to a global green charter. Harrison bets that "Kennedy visualized an outcome like this, and that he's already set his sights on the next positive green milestone." Like any golfer, Kennedy must have known that mental preparation helps the game. The strategy is "really more than foresight. It's a sort of *advance hindsight.* . . . The important point is to break the habit of seeing a contentious, costly, long, or losing condition when the subject is green" (Harrison 1993, 24).

The subject is green because participants in WICEM II created what they called a Green World. This was accomplished by pledging support for "sustainable development" because it has "something for everybody."

The concept of "sustainable development" began circulating in the early 1980s, in efforts to build bridges between the development community and conservationists. The United Nations Environment Program (UNEP) played a leading role, emphasizing nontraditional economic criteria in evaluating cost effectiveness, community self-reliance, and the need for "people-centered initiatives" (Lele 1995 [1991], 230).[3] By 1987, "sustainable development" had gone mainstream, with the publication of *Our Common Future,* a report from a special UN commission chaired by Norwegian Prime Minister Gro Harlem Brundtland. The commission was convened to help adjudicate the divergent perspectives of developed and developing countries on environmental issues. The commission adopted the concept of "sustainable development" to articulate the need for and possibility of reconciling environmental and economic priorities. Indira Gandhi's argument that "poverty is the greatest pollution," made at the Stockholm Conference on the Environment in 1972, provided the basic logic. The Brundtland report posited that underdevelopment is a major cause of environmental degradation. Whether in the Third World or the former Soviet bloc, the environment cannot be preserved without economic growth and the satisfaction of basic human needs. Sustainable development is the answer, even if it cannot yet be fully defined. The working definition of "sustainable" provided by the Brundtland report is "development that meets the need of the present without compromising the ability of future generations to meet their own needs"(World Commission on Environment and Development 1987, 43).

"Sustainable development" has been called an oxymoron. And it has been criticized for veiling the continuing difference of perspective between the First and Third Worlds. Others applaud sustainable development precisely *because* it veils differences, providing terminology for building new partnerships between rich and poor societies. Many see promise in the concept of sustainable development, so long as it is not solely defined by industry.[4]

Business historian Andrew Hoffman, for example, argues that it is crucial to pluralize participation in the definition of sustainable development, to "create the tension necessary to foster institutional change" (Hoffman 1997, 182). Without pluralization, sustainable development will go the way of corporate environmentalism in the 1960s, emerging from an equally insular field, based on institutional and cognitive structures already in place within corporations. The possibility of "revolutionary change" will depend on the entrance of new parties to the sustainable development debate or on the realignment of power relationships among those already at the table. Hoffman argues that revolutionary change involving the restructuring of institutions is important, but that it will depend on the interjection of external events and their interpretation. One possibility is for peripheral actors to force their way to the table through a catalyzing, legitimating event like Earth Day. Another possibility is that "opposing concepts of sustainable development could be thrust onto the organization field by the force of an unpredictable event, one that highlights the inconsistencies in the existing institutional framework, such as the Bhopal incident in 1984" (Hoffman 1997, 195).

Hoffman argues that history remains to be built and that events like the Bhopal disaster could make a difference. Harrison argues that history has already been determined. There is no need for further discussion with peripheral actors. There is no chance that an external event like Bhopal will disrupt the program—because business signed the charter for sustainable development at WICEM II. Like visioning, signing the charter was an act of "advance hindsight." It signed off on the past and wrote in a future in which greening and growth are no longer in opposition. Period.[5]

Harrison confirms the success of the WICEM II Charter through his reading of the United Nations Conference on Environment and Development (UNCED), held in Rio de Janeiro in 1992. By all accounts, the Earth Summit was a landmark event because it initiated cooperative work toward sustainable development—but also because it brought more people to the table than ever before in a UN forum.[6] Nations were not the only invitees. Grassroots representatives were invited, as were representatives from nonprofit organizations—including chambers of commerce. The goal was to transform not only what the UN did, but also how it did it. Pluralization of participation in global environmental politics was an explicit goal.

Harrison attended as a member of the International Chamber of Commerce's Environmental Commission. His contingent was prepared. They came *as* signatories to sustainable development. They preempted the summit's mission. The work of advocacy was to assimilate others into an already codified program. Some claim that business "captured" Rio by playing the upper hand in defining the content of "sustainable development." Even more important, however, was business influence over how "sustainable development" would be troped.

Harrison's account of the coup at the Earth Summit suggests how powerful a business approach can be. The "attack mode" of American environmentalism was, in his view, dead on arrival. The eight thousand journalists in attendance "missed the story," but it was clear to Harrison. As he saw it, the bottom line was that "greening has become the lubricant to grease the machinery of government-sanctioned commerce" (1993, 4). In sum: environmentalism today is "owned by business."

At WICEM II, business visualized a Green World. At Rio, business confirmed it. All things became possible for committed corporations: it was a "new ball game." Business executives could carry the card Harrison provides at his counseling sessions. The card is small enough to be nestled in the palm of an executive's hand so that he can look down at it during press conferences. The message on the little card is bold: "WE ARE THE GOOD GUYS."

The great historical transformation described by Harrison has been accomplished by providing corporations a new language for communicating with the public and for solidifying their own identity. "Sustainable development" is the anchor holding everything in place, allowing basic concepts to be reconfigured. According to Harrison, "[T]he chemical industry has the world's toughest job in public relations. It has to deal with public fears that are ignitable at will. While people absolutely require chemicals to live (every product is touched by chemistry; we ourselves are composed of chemicals), there is no end to the association of *chemical* to *danger,* and therefore no end to the communication challenge" (1993, xv). The challenge for Harrison is to help corporations disassociate *chemical* from *danger.*

COMMUNICATIONS TRAINING:
A STRATEGIC ASSET FOR EXECUTIVES

Change-ability has become the criterion by which successful companies are and will be judged. This resiliency is important if companies are to survive the transformation of the business climate. Tele-communications, public involvement, the globalization of the marketplace, and increasing environmental regulations have forever changed the post-war industrial world. . . . Old top-down leadership styles do not fit the shifting business paradigm. Executives are no longer managing from the peak of the hierarchy, but are operating instead in the middle of everything. The change can be taken quite literally—no longer sequestered in the boardroom, executives have taken to management by walking around.

This interactive approach is in part a response to the public demand for accountability. The *Fortune* magazine headline, *"The King is Dead,"* says it all. The article illustrates how the public and the shareholders are permanently *re*-writing the rules on leadership and corporate responsibility—particularly in the areas of environmental management and communications. The consequence is the undeniable need for executive re-training. . . . Communications training produces the biggest payoff for the effort. . . .

Communication greases the wheels of decision-making in the executive world today, and making choices is growing tougher in a world where issues aren't black and white. Today's issues are more often "dilemmas"—scenarios that pose several equally compelling options or present confounding trade-offs, regardless of outcome.

Nowadays, an executive is more likely to encounter a dilemma than a problem. The word "problem" implies that somewhere, a clear "solution" is available, and has only to be invented or discovered. Much to the chagrin of the decision-maker, what characterizes a dilemma most often is ambiguity. This is the form in which most environmental issues present themselves. Often the key determinants for a dilemma are intangible ideas like *what is just?* or *fair? What will benefit the most people?* and *To whom do I owe allegiance—my community? or my shareholders? What needs should have the highest priority? Whose preferences trump when preferences compete?*

Accordingly, executives act much like a judge in a court of law. Deciding on these dilemmas is at the heart of their job. A critical skill for handling these environmental and other business dilemmas is effective communication. Previously, executives had relied on concrete data supplied by managers, the CFO and numbers to make the choice. But in a context where the best decision is unclear—and dependent at times on abstract ideas—communication can be the only real way of obtaining the kind of information needed to achieve resolution. . . .

At any moment, *Sixty Minutes* or your local newspaper can come knocking at the door. At any moment, a potential crisis can disrupt normal operations. In an instant, executives can find themselves in front of news cameras. Just ask the president of Exxon, United Airlines, Union Carbide, Amtrak and scores of others. . . .

(Grant 1994, 57)

Companies need "envirocomm," a process and term coined by Harrison "to signal that this is a special and in many ways unique form of communication and public relations" (Harrison 1993, xii). The uniqueness of envirocomm, codified into a seven-step management model, lies in recognizing that "going green is a habit, not a hormone." To begin, companies need to become keenly aware of contemporary conditions, recognizing major trends that make environmental public relations imperative. Next, companies need a "winning green attitude" that visualizes success and exorcises "hang-ups"—"pollutants in the channels of green communication" that obstruct "mental greening." Perhaps most important, companies need to stop dropping the ball because, like Charlie Brown's Lucy, "the past got in her eyes." Times have changed. Lucy can now catch that ball, secure in believing that "the green ball is in the court of corporate America" (Harrison 1993, 5).

As bearer of envirocomm, Harrison is part messiah, part therapist. His task is to deliver the People to the Corporation, then to deliver the Corporation from the People. First, he must help draw the People into the Corporation's domain of influence so that they can be treated for perception disorder. Next, he must help solidify the Corporation's own identity by matching what the Corporation thinks it is to external reality—now defined by the promise of sustainable development, which Harrison himself has helped realize. People will learn to trust business as business learns to trust itself.

Envirocomm is a three-part program. It works to reconstitute the corporation by providing a means to deal with what is outside it. The corporation must by harmonized—within, with the world it moves around in, and with the perceptions of those who observe it. Harmonization is to be accomplished through the discourse of sustainable development, which operates on a number of registers. Knowledge that business has signed a global charter on sustainable development allows corporations to recognize their own integrity. The very existence of the charter also proves that the corporation is in sync with the world and is an expression of shared values. This reality is the basis for sustainable relationships between corporations and various "customer-publics," to be built through sustainable communication.

But, first, the media must be dealt with. The media should not be considered a customer-public because you cannot talk to the media (unless one happens to own the newspaper or network). Rather, you talk *through* the media. And it's important to remember this: editors, like nature, abhor a vacuum. This means that, "like it or not, your turn will come to make the news. The great, insatiable news machine will end up in your face" (Harrison 1993, 95). It also means that there will be a story, whether you participate or not. Your choice is only whether or not you will help determine the content. And your model should be the television commercial. "They have everything. They expose and resolve conflict six times a half hour in a nice, neat 30 seconds" (Harrison 1993, 95). The goal is to be a "3" or a "7"—staking a position near the middle of the "News Source Ruler," not so far at either end to sound extreme, but also enough off center not to sound wishy-washy. This will give reporters what they want: conflict and simplicity. If you are a "7," the news story can oscillate between what you say and what is said by whoever takes the "3" position. This, in Harrison's view, is how journalists provide what they consider an objective account.

You should imagine yourself as a teacher or coach—because you should assume that most journalists have science phobia. After all, most probably went into journalism because they hated physics and chemistry. You must be sensitive to this. You must be very clear and abbreviated in all technical explanations, knowing that if you don't provide them, reporters will make them up on their own back in the newsroom. In all, take charge of the interaction. And get personal. Know reporters' names. Indicate that you understand that they have a deadline. Imagine the media as a way *through* to the specific individuals or groups of individuals that comprise your customer publics.

Customer-publics have to be created, and they should be quantified. Don't assume that you *already have* publics, waiting for you like an audience in an auditorium. Customer-publics have to be made by identifying the needs of particular groups and how you can fulfill them. The first goal is *not* to be understood, but to understand. You need to know "what they want," so you have to listen—"aggressively." General George Patton has good advice. "I never start with a strategy. I start by going to the front and looking around." What business needs now, says Harrison, is not spokespersons, but *hear*persons.

Harrison's QUALITY model is a seven-step process. "Q" is for *quantifying* your publics. "U" is for *understanding* their points of view. "A" is for *asking* questions that show your interest and draw out *their* questions. "L" is for *listening.* "I" is for *interpretation* of what you hear so that you can initiate appropriate action. Now is the "time for action—or, more accurately, interaction. The Q, U, A, L, and I steps have brought you to the *creation of public relationships*—the heart of envirocomm. Now you and your organization can establish the sustainable communication process that makes these relationships beneficial" (Harrison 1993, 44–45). This is the "T" step, where you *take charge.* Harrison explains:

> Envirocomm is an active process on three levels. It is *proactive,* in that it
> takes the initiative in relationship construction and maintenance. It is *reactive,* in that it responds to and comes to terms with the action (and attitudes
> that presage actions) of customer-publics. And perhaps most important, it is
> *interactive.* This means that the communication climate is open and information can flow freely in all directions. You go beyond dialogue—one-to-one communication—to *multilogue,* or communication in which a number
> of two-way transactions are handled simultaneously, with the goal of harmonizing perspectives and organizing partnerships. An example of this process
> is the Responsible Care program pioneered by the U.S. Chemical Industry, in
> which local groups of people from all walks of life come together to discuss

environmental conditions and problems, to plan cooperatively for handling environmental emergencies, and to implement agreed-upon programs. . . . Your green communication must now be made evergreen—a continuous operation that is constantly improving and self-enriching. (1993, 45)

The last step is "Y," for "You, Yes, You." Because "the 'radio station' you should be most tuned into is WIIFM—What's In It For Me?" (Harrison 1993, 45).

Your own stakes are important. You need to *believe* that an effective envirocomm process will pay off. Charting potential benefits can be useful, remembering that "publics" have to be "recast as customers":

"Public"	As Potential "Customer"	They May:
Employees	Buy in to your value system	Build and support your green performance and reputation
Suppliers	Buy in to your value system	Give you resources and spread your green reputation "upstream"
Politicians, Regulators	Buy in to your perspective	Provide a more level playing field for your green performance
Media	Buy in to your perspective	Give you a fair representation of your green record, commitment and views
Neighbors	Buy in to your values and perspective	Support & partner in your green performance and build your green reputation
Transporters, Distributors, Retailers	Buy in to your values and commitment	Extend your green influence and reputation "downstream"
Activists	Buy into your record and commitment	Support & partner—or give you a fair opportunity with your other "customer-publics" (Harrison 1993, 32)

Quantifying publics helps us remember that "person-recognition" is crucial. Beware "the old notion that anybody can ever again stand on the mountaintop and hand down chiseled stone tablets to the masses. In the new green world, this is a waste of time, tablets and money" (Harrison 1993, 38). Instead, quantify, customize, and continually adjust your communication for sustainability. Hierarchical, top-down communication won't do. You must talk *with* people, never down to them. The most important thing to *demonstrate* is *commitment*.

Older executives have some relearning to do. They must learn to seek out feedback, recognizing that "the best corporate green policies are grown *organically*" (Harrison 1993, 126). They must cultivate connections, reaching out for alliances, even with environmental groups. And they must be good at showing concern. The most vital question that executives must respond to is basic: do you care?

The goal is not to manipulate emotions. The goal is to manage expectations. Steven Covey, author of *The Seven Habits of Highly Effective People,* offers useful insight: "Instead of projecting your own autobiography and assuming thoughts, feelings, motives, and interpretation, you're dealing with the reality inside another person's head and heart" (quoted in Harrison 1993, 132). The challenge is to be able to give back to people what they already know, affirming that you *mean business* when it comes to the environment.

An attitude of openness and evidence of commitment will go a long way. Executives do, however, need to be prepared for the occasional attack. Union Carbide CEO Bob Kennedy is an authority on such matters. He expects activists to show up at annual shareholders' meetings, so he tries to meet with them in advance—to make sure the company knows the agenda. Then there are periodic meetings with a variety of coalitions. Harrison explains how they should be understood and handled:

> Kennedy says that a number of widely known environmental groups have gotten together and sponsored local grassroots coalitions, known as "Citizens Concerned About (fill in the blank)," which can be marketed virtually anywhere, on virtually any issue. Again, Carbide (and Kennedy, when requested) agrees to meet, but sets ground rules: no press, no tape recorders, no post-meeting statements and about two hours of dialogue. At one such meeting, a "Concerned Citizens . . ." group asked about their involvement in the company's policy making. Carbide pointed to their Community Advisory Panels (CAPs), composed of community teachers, fire chiefs, etc., but the group still felt left out. Carbide agreed to let the group draft a protocol for organizing a CAP. Tactics for setting ground rules are not always successful. One group was critical of a Carbide operating plant in Texas. The company and group discussed demands and a time when the activists could come to the plant to present themselves. When the two couldn't agree on a mutually convenient time, the group came to the plant and staged a mass demonstration. The company was forced to turn them away at the gate, and explained to the media that their suggested date had been turned down. That message was strengthened when Carbide pointed out that the demonstrators weren't even locals, but were out-of-towners who had flown in to stage the protest for the media. Carbide's message was clear and credible: "We don't respond well to demands. We are ready for dialogue." (1993, 187–88)[7]

Harrison offers a few principles of confrontation and guidelines on where activism is headed. One principle is a reminder that the goal of some attacking groups is the conflict itself, not amicable resolution. Another principle reminds us that "key disarming tactics are openness to dialogue and commitment to public-interest principles." But business should not get too comfortable. Precisely *because* business is now leading environmentalism, extreme groups need to "differentiate themselves to their members." They will continue to do this with marketing techniques "which rule out business co-opting environmentalism." The writing is on the wall: "expect [activists] to be less polite" (Harrison 1993, 189).

(11.3) Moving beyond Blame

Union Carbide has accepted "moral responsibility, but no liability" for the Bhopal disaster. Out of broader contexts, the opposition seems oxymoronic. Perception of logical contradiction is, however, culturally constituted and socially situated. Carbide's opposition to responsibility and liability is therefore fully rational, "made comfortable to reason" through the work of new cadres of professionals who insist that environmentalism does matter and that it does not contradict established profit accrual strategies. They insist that "going green" is possible, desirable, and in the process of being realized. The rhetorical strategies for circumventing the contradictions are multifaceted, but one persistent theme provides grounding: moving beyond blame.

The logic of being "beyond blame" insists that environmental devastation is a shared concern and a shared responsibility, generated by a complex of social processes that disallow designation of a single culprit.[8] Corporate environmental excellence is in everyone's interest and should not require litigation to propel it forward. The task of the consultant is to make this mutuality visible, proving that "one of the distinguishing features of the environmental movement . . . is its inclusiveness. Men and women, rich and poor, far and near, are all beginning to ask the same set of questions. As they take off their masks and discover that the battle to save the environment is a journey that cuts across all societal divisions, we discover how universal the concerns for environmental health and safety have become" (Piasecki and Asmus 1990, 170).

A major landmark in this journey "beyond blame" was the Bhopal disaster. According to consultant Bruce Piasecki, the Bhopal disaster engendered an "avalanche of change," signaled by the model Union Carbide has provided the chemical industry in environmental excellence. In Piasecki's reporting, Union Carbide provided such a model before the disaster and has reemerged as a leader since. According to Piasecki, Carbide was rated as "possibly the best" corporate citizen among the multinational companies surveyed in an (unfortunately never published) MIT study based on the company's environmental programs, social responsibility, and responsiveness to host governments before the disaster. Apparently relying on inside information, Piasecki states that "the company had long made safety a priority. Any disabling injury to employees elicited urgent attention at the highest levels of management" (1994, 6).

Piasecki recognizes that the night of December 2, 1984, changed everything, resulting in Carbide's entry into the *Guinness Book of World Records* under the section "Worst Accidents and Disasters in the World." Since then, "the Bhopal legacy has attached itself to Union Carbide like Hawthorne's 'scarlet letter' to the adulteress Hester Prynne, depicting the company as no less immoral" (Piasecki 1994, 7). In response, Carbide is credited with crafting distinctive environmental management initiatives and learning the importance of the search for external credibility. The search for credibility has led Union Carbide to play a leadership role in the Chemical Manufacturers Association's Responsible Care program and in the Global Environmental Management Initiative (GEMI). GEMI, founded on the twentieth anniversary of Earth Day in April 1990, was born of recognition of the need to integrate environmental management into the production process. The goal of GEMI was to develop the tools for benchmarking corporate environmental progress, since this could best be done by corporations themselves.

Piasecki never mentions the health status of gas victims in Bhopal or the role of Union Carbide in litigation regarding asbestos, the Dalkon Shield, and twenty-two Superfund sites (Oil, Chemical and Atomic Workers Union 1990). Instead, he notes that "in fact, history books will show that the Chemical Manufacturers Association's current 'Responsible Care Initiative,' which developed a shared code of environmental management behavior for the industry, evolved directly from Union Carbide's tragedy." In the same paragraph, he notes that "the new consulting divisions of Arthur D. Little, Booz-Allen and Hamilton, and ICF-Kaiser Engineering derive their energy and fees from the fears triggered by the Bhopal disaster" (1994, 8). It was under the guidance of such consultants that Union Carbide responded to the Bhopal disaster. According to Piasecki, the response was systematic. Carbide hired Arthur D. Little. The company held "visioning" sessions of great intensity. Environmental audits became "the buck that no one could pass" (Piasecki 1995, 26).

Piasecki's laudatory analysis of Union Carbide is developed in his book *Corporate Environmental Strategy: The Avalanche of Change since Bhopal* (1995). The overall promise is one of harmonization: strategic leadership can outsmart the opposition between government and industry and between the economy and the environment. Piasecki explains that the genre he has chosen for his account is the dramatic case study because it can provide representative details that help us interpret historic change. Unlike most authors writing about environmental management, who cite hundreds of firms each, he focuses on a few consequential cases. His "selective dramatization" is meant to help us abstract meaning from seemingly isolated examples, providing a "contemplation of cultural change" that has had "fecund momentum" since Bhopal. As Piasecki notes, "When you are dealing with an avalanche of change, it doesn't help much to do a drop-by-drop or flake-by-flake analysis. What follows, then, is a big-picture dramatization of what has happened to the corporate world in general, and the field of environmental management in particular, since Bhopal" (1995, 11).

Piasecki acknowledges that the dramatic case study does have shortcomings, noting that "some feel, for instance, that it moves too rapidly from observation to conclusion, and they miss its somewhat quiet quest for heightened and representative insight. They simply want facts and figures." One risk is that experts will feel disorientation, policy makers will feel uncertainty, and those with ideological motives will feel anger or frustration. His audience, however, is the general manager, "who can benefit from case studies that are not about what should happen, but rather, what is happening, and how it is taking place." Case studies are reliable because they sidestep ideology and political hyperbole, examining both the art and the science of environmental strategy (Piasecki 1995, xii).

How, though, can distinct cases capture what is happening in general, across all companies? "The answer rests in the experience of an avalanche. An avalanche accomplishes two things. First, the massive slide utterly changes the landscape. There are new forms to consider. At first, the newness is disorienting, sometimes downright overwhelming." Next, an avalanche separates different kinds of survivors. "As a downward curtain of terror, an avalanche fans open in a massive sprawl, including a far larger realm than the first boom suggested" (Piasecki 1995, 5). Different companies survive in different ways, evolving strategies that "still resemble the best defense scenarios generated in the silent halls of the Pentagon itself," responding to the specific threats and possibilities that characterize a company's field of action (Piasecki 1995, 4).

Piasecki acknowledges that for the last twenty years corporate executives have responded to environmental regulation with bewilderment and resistance, caught in foreign terrain without the solid footing of conventional, internally focused management tools. He remembers his own experience while climbing the Sierra Divide with friends. They encountered an avalanche. The boom began. Within seconds, miles of snow fanned toward them. Their first reaction was to find the closest rock and hide behind it, like most companies responding to the aftermath of Bhopal, hoping to find a safe haven in the "greenwash" of public relations. The result was false security in a time of severe and irreversible change.

The challenge is to recognize the metaphoric significance of the avalanche, as did Warren Anderson, CEO of Carbide at the time of the Bhopal disaster—the "environmental equivalent of Pearl Harbor, a violent wake-up call that shook many nations and many firms" (Piasecki 1995, 25). According to Piasecki, the press and academia only contributed to Anderson's problem. "Not content to recognize Bhopal as the paramount environmental disaster, scholars and journalists embellished the story with brutality and barbarism as well. Was this not the 'perfect' example of corporate environmental neglect and abuse?" (1995, 23). Warren Anderson responded by insisting that Carbide view the disaster in "moral, not legal terms." A footnote explains that "the Bhopal plant was run entirely by Indian managers and technicians and was in fact partially owned by the Indian government. Yet Carbide's U.S. headquarters aided the victims, cleaned up the site, and investigated the cause of the incident. This was only right and illustrates how an avalanche implicates the entire firm" (Piasecki 1995, 150 fn. 13). Union Carbide's board of directors wasn't as quick as Warren Anderson and only got the "real message" eight months after the Bhopal disaster, following an "incident" in Institute, West Virginia, at another major Carbide facility where methyl isocyanate is produced and stored. Although shocked by Bhopal, the board had assured themselves that "it couldn't happen here." The incident at Institute proved that it could.

Union Carbide's response to the Bhopal disaster included many new initiatives—the development of new classification systems that link regulatory compliance to the legal department's liability-containment strategies, the development of audits on environmental performance that allow executives to compete for capital resources, programs for computer monitoring of "episodic risk," and new management tools that keep senior personnel in touch with liability issues by sentencing them to jail time for environmental crimes (Piasecki 1994, 8). Most important, however, has been the development of a new language of corporate environmentalism by enlivening the environmental audit program. Prior to the Bhopal disaster, environmental audits at Union Carbide were based on complex standards, were heavily oriented to U.S. laws and regulations, and differed "widely in scale of application from actual requirements to nonimplementable ideals. The full set took 14 inches of office shelf space, and that's where they often remained, unread" (Piasecki 1995, 28). According to Piasecki, the standards that emerged after the disaster made their debut boldly, provoking many in the company to note that "it was as if Ross Perot was their chief editor" (1995, 28). The new standards "were only ½ inch thick, simple and focused, readable, and, most important, one could audit against them." The simplicity standard was not met by telling people "how to" meet the grade, but by "expressing each higher standard as an executive expectation." Today, "standards are no longer a topic requiring on-the-shelf research. They have become conversational, the topic of everyday life for executive and line professionals alike" (Piasecki 1995, 28).

Corporate environmental management has "matured considerably since Bhopal," even while public debates remain "locked in wheel-spinning and blame" (Piasecki 1995, 4). The most advanced maneuvers have acknowledged dramatic change in the landscape, modifying how corporations are imagined, correcting for "the distorting authority of noted popular myths" (Piasecki 1995, 10). The first of these myths suggests that corporations don't know what they're doing regarding the environment, telling a tale of corporations as misaligned, out-of-control machine tools, spitting out sparks and shards in all directions. The second myth suggests that corporations are too secretive regarding the environment, relying on manipulative marketing to keep any disclosure "ziplocked with secrecy." These myths, Piasecki states, are at times obsolete. Instead, one should imagine the corporation through the metaphor of a house, "a large rambling structure with plenty of windows, doors and points of entry." Lawyers can peer in via provisions for freedom of information. Communities have the right to know. Even the Securities and Exchange Commission can look inside, through increasing demands for disclosure of risk generated by environmental as well as financial concerns (Piasecki 1995, 11).

The cover of Piasecki and Peter Asmus's book *In Search of Environmental Excellence* (1990) illustrates the logic. It draws us into a vanishing point, suggesting a future where there is no conflict. To get there, one must move forward without looking back or around. All lines of site conflate into a single focus. There is one proper perspective, and the frame of objectivity is clear. The goal is to "cut through blinding complexities that cloud our understanding," providing a simple vision (Piasecki 1995, xiii).

(11.4) Force of Law

Green consultants like Harrison and Piasecki make few overt references to events like the Bhopal disaster. But events like Bhopal inhabit their texts. They know about Bhopal and about plant disasters in the United States. Their job is to contain them, moving us "beyond blame." There is nonetheless a need to fear the law—because of its capacity to prosecute innocence. Green consultants provide protection against this injustice.

Piasecki reminds us that, since Bhopal, the "Justice Department has indicted 521 individuals and 240 corporations for crimes under six major federal environmental laws"—which means that "374 individuals and 176 corporations have pled guilty or have been convicted, with over $157 million in criminal penalties that were assessed with over 348 years of jail time imposed." This kind of legal information "makes many executives sit up in attention." But that still "doesn't capture what makes Dizzie Gillespie a musical genius, or why Lod Cook is a corporate leader worthy of study and imitation" (Piasecki 1995, 66). The point is that legal and technical competence is not enough, which means that "we cannot neglect the manager, the well-paid decision maker, the one who can cope." The authority referenced here is Machiavelli, used to remind us that "there is nothing more difficult to carry out than to 'initiate a new order of things.' " Piasecki insists that we do need environmental laws. Law is the "starting point" because "without a rule-based score, even Dizzie Gillespie couldn't make music that is comprehensible to so many so consistently. Still, environmental law is limited. It cannot solve all our problems. It is always a set of nails seeking hammers" (Piasecki 1995, 64). So environmental strategy is more about leadership than law—about outpacing regulatory compliance. Piasecki provides the tools of change. The narratives he supplies are driven by the possibility of liability. But they also are a means to control liability. Companies remain in charge. A subheading in Piasecki's chapter on AT&T sums up the promise: "The Future Has No Fear" (1995, 94).

Bruce Harrison paints the legal backdrop of corporate environmentalism with bolder colors. Harrison is careful to describe many reasons that corporations should *Go Green*. The cost of compliance with environmental regulations is emphasized, as is the connection among greening and the willingness of banks to lend money and the willingness of insurance companies to underwrite risky production. Rising stockholder activism is predicted, as is an increase in customer attention to the greenness of the products they buy. The "AMP" syndrome is also explained. Created by the synergy of Activists + Media + Politicians, AMP is a vicious feedback circuit in which "public scares" spiral out of control and "box" companies into discourses peopled by victims and "villains (very often, business interests)" (Harrison 1993, 277). According to Harrison, "[No] one is unplugged from greenism" due to new modes of communication and computer databases harboring deep pools of environmental data.

But the threat of going to jail is what is most repeated. A full chapter is devoted to "How It All Starts: A Guide to Green Crime's Humble Origins" (Harrison 1993, 281–91). The motif: refusing *envirocomm* could have criminal implications.

Harrison has already introduced us to the criminal threat, telling us that politicians and regulators "have discovered that nothing gets the attention of a senior manager like the threat of doing time for environmental crime." Current sentencing guidelines are said to indicate that first-time environmental law violation will normally lead to jail time. In 1990, the Justice Department's environmental indictments totaled 134, up 30 percent from 1989. Four out of five were against corporations and their executive officers. The department's conviction rate was 95 percent. And "it was the same story at EPA: A record number of environment convictions in 1989. Half of these got jail sentences, and 85 percent of these actually went behind bars." Further, because of increasing media coverage of environmental incidents, "trials in the court of public opinion" can be just as bad—"obviating the need for prosecution—and eroding the option of defense—at the bar of justice" (Harrison 1993, 6).

The process of law is not, however, described as rational. Corporate executives cannot avoid jail simply by trying not to break the law, or even by not—in fact—breaking it. Their own reasonableness will not be sufficient. One manager in Ohio is said to have gone to jail for a year after pumping rainwater collected on top of a waste cell into a creek. It caused no pollution, but he was found to have broken an Ohio law by failing to evaluate the water before discharging it. In another story, managers go to jail because an engineer developed a creative solution to hazardous waste products—dilution—which neither he nor his bosses knew was against the law.

The "Green-Crime ante" is said to have gone up significantly with the passage of the 1990 amendments to the Clean Air Act. Liability has now been extended by defining a facility's "operator" as any member of senior management. According to Harrison, "[Y]ou don't have to be a bad guy to be convicted; managers need not even take part in or agree to a criminal action to be held accountable for it under the doctrine of 'the responsible corporate officer' upheld by the Supreme Court." Harrison also tells us of how Frank Friedman, vice-president for Occidental Petroleum, explained why many environmental managers are leaving the field—because "a facility manager who inadvertently misses a regulatory deadline can be faced with criminal prosecution no different than a member of the Medellin cartel" (1993, 282).

Four trends are said to have turned green-collar crime into a "greater threat to corporate governance than virtually any other area of concern," creating a vulnerability even more intense than the threat of a hostile takeover. One trend involves the prosecution of people several layers above the actual operation where criminal activity is said to have occurred. The law is moving up the chain of command to get at people who should have had compliance systems in place.

Another trend involves the sheer complexity of environmental laws. The rules governing the Resource Conservation and Recovery Act, for example, comprise about 640 pages of the Code of Federal Regulations. The 1990 amendments to the Clean Air Act are even more out of control, shattering the "rule of thumb that a new law generates roughly 12 pages of regulations for every page of statute." The Clean Air Act "weighs in at 700 pages of text, creates 55 wholly new regulations and will take many thousands of pages to express" (Harrison 1993, 282). This means that it is nearly impossible to keep up with the law; the law can be broken and criminal liability incurred simply as a result of the deluge of legal obligations managers now face.

The third trend also involves information. According to Harrison, "[T]he 'paper' generated in connection with day-to-day environmental compliance activities gets people into trouble more than the underlying violations!" The point here is that monitoring and reporting procedures can be more important than the reality for which they claim to account.

The fourth trend involves the prosecution of corporations under the criminal code even when no individual within that corporation acted in willful disregard of the law. The threat here is strict liability for a corporation's total knowledge—the collective knowledge of all employees. If, for example, "one employee knows that material is being sent to a certain area of the plant for disposal (but believes the activity to be perfectly OK), and another employee is aware that the substance is a waste and may be dangerous (but does *not* know how it is being handled) . . . the collective information is enough on which to base a prosecution of the corporation" (Harrison 1993, 286).

The conclusion of Harrison's book returns again to the criminal threat. The tone suggests that we should really think about it: "Are they really putting business people in jail for committing a crime? Yes, and it's not that hard to do. You don't have to be a midnight dumper. We're not talking about plotting a scheme to pollute. We're not even talking about meeting a compliance deadline. Your own compliance audit can do you in. It's rougher than that. A meeting, held in your company, at which environmental matters were discussed, can do you in. You can be done in, and get into deep trouble with the law, because of well-meaning employees and well-meaning consultants" (Harrison 1993, 320). The only answer to the criminal threat is preventive communication.

(11.5) Rhetorical Solutions

Green consultants have developed innovative rhetorical forms for dealing with disaster. Bruce Piasecki's use of the dramatic case study is illustrative.

In his case study of ARCO, for example, Piasecki begins with an important question: "How did ARCO's CEO Lodwrick Cook head off the movement to methanol-based fuels, win the praise of the Natural Resources Defense Council, rewrite auto-emissions regulations, fatten stockholders' portfolios, and introduce the concept of 'green marketing' to the fossilized oil industry?" The answer, Piasecki says, is "complex and worthy of emulation" (1995, 37).

ARCO is said to have proven that "good deeds and good numbers can live together prosperously in this brave new world of corporate environmentalism. At stake are clean air, livable cities, and our commuter lifestyles" (Piasecki 1995, 38). CEO Lod Cook is the hero of the story. He wants us to remain a nation of automobiles run on gasoline. And he knows that "the issue of the environment is persistent" and that "unless we find ways to make gasoline burn more cleanly, gasoline is dead." So "rather than reacting to news about violations, fines, and industrial accidents, ARCO has focused its staff, their reputation and its financial resources on securing safer fuels for the largest commuter markets, in California and four other western states" (Piasecki 1995, 38, 39).

ARCO's new fuel, EC-1, was first available in California in September 1989. By 1990, the payoff was clear. ARCO "enjoyed $19.9 billion in revenues in 1990, nearly four times its 1986 sales. Moreover, the return on stockholder equity in 1991 was 29.3 percent, making ARCO the best performer in the oil industry. The years 1992, 1993, and 1994 were also brilliant for ARCO" (Piasecki 1995, 38).

On July 15, 1990, there was an explosion at an ARCO plant in Channelview, Texas. Sixteen workers were killed. Piasecki's case study format does not have room for this detail. The changing political economy of the oil industry is addressed at length, but the actual operations of ARCO plants remain out of sight. Exclusion of disaster from corporate environmentalism is often more subtle. But such exclusions are what make strategic corporate environmentalism work.

(11.6) Disharmony

One of the most revered images within environmentalism is that of the Earth rising over the Moon, taken from Apollo 8. It is an organically integrated whole; a smooth surface, green against blue, without the sharp divides of nation and power. *Time* magazine chose this image in 1989, in lieu of its usual Man or Woman of the Year, to announce the "endangered earth" as Planet of the Year.

FIGURE 11.5 RHONE-POULENC'S GLOBE The cover of Rhone-Poulenc's *1994 Environment Report,* highlighting the harmonious integration of natural and social worlds, enabled by the Responsible Care program's integration of public values with corporate values.

**UNION CARBIDE OPERATIONS:
COUNTRIES AND PRINCIPAL PRODUCTS**

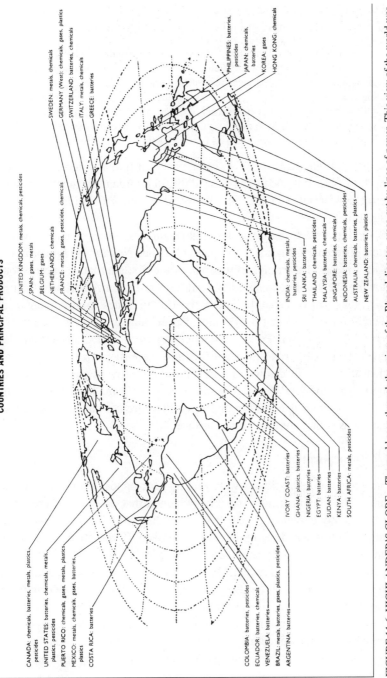

CANADA: chemicals, batteries, metals, plastics, pesticides

UNITED STATES: batteries, chemicals, metals, plastics, pesticides

PUERTO RICO: chemicals, gases, metals, plastics

MEXICO: metals, chemicals, gases, batteries, plastics

COSTA RICA: batteries

COLOMBIA: batteries, pesticides

ECUADOR: batteries, chemicals

VENEZUELA: batteries

BRAZIL: metals, batteries, gases, plastics, pesticides

ARGENTINA: batteries

IVORY COAST: batteries

GHANA: plastics, batteries

NIGERIA: batteries

EGYPT: batteries

SUDAN: batteries

KENYA: batteries

SOUTH AFRICA: metals, pesticides

UNITED KINGDOM: metals, chemicals, pesticides

SPAIN: gases, metals

BELGIUM: gases

NETHERLANDS: chemicals

FRANCE: metals, gases, pesticides, chemicals

GERMANY (West): chemicals, gases, plastics

SWITZERLAND: batteries, chemicals

ITALY: metals, chemicals

GREECE: batteries

PHILIPPINES: batteries, pesticides

JAPAN: chemicals, batteries

KOREA: gases

HONG KONG: chemicals

INDIA: chemicals, metals, batteries, pesticides

SRI LANKA: batteries

THAILAND: chemicals, pesticides

MALAYSIA: batteries, chemicals

SINGAPORE: batteries, chemicals

INDONESIA: batteries, chemicals, pesticides

AUSTRALIA: chemicals, batteries, plastics

NEW ZEALAND: batteries, plastics

FIGURE 11.6 HIGHLANDER'S GLOBE   The world seen through the prism of the Bhopal disaster, cross-cut by lines of power. This view of the world was depicted in the pamphlet titled *No Place to Run: Local Realities and Global Issues of the Bhopal Disaster*, published jointly by the Highlander Research and Education Center (United States), the Center for Science and the Environment (India), and the Society for Participatory Research in Asia (India) in 1985.

The Earthrise image was also central to the vision of the seminal Brundt-land Report, "Our Common Future." I quote from this report:

> In the middle of the 20th century, we saw our planet from space for the first time. Historians may eventually find that this vision had a greater impact on thought than did the Copernican revolution of the 16th century. . . . From space we see a small and fragile ball, dominated not by human activity and edifice, but a pattern of clouds, greenery, oceans and soil. Humanity's inability to fit its activities into that pattern is changing boundary systems fundamentally. (World Commission on Environment and Development 1987)

Casting environmentalism as an endeavor to fit humans into natural patterns doesn't accommodate the Bhopal disaster. Organicist images are also ripe for appropriation. Changing the world becomes a simple matter of a child's brushstroke.

Bhopal has produced other, less lulling images—images that draw out critical differences. An image in a pamphlet titled *No Place to Run* provides a place to start. The pamphlet was one of the first alternative publications following the 1984 gas leak in Bhopal, a joint Indian-American effort. It tells a story that connects India and the United States, environmentalism and labor issues, multinational companies and human rights. It is an image cut by lines of power. Points of connection are Union Carbide facilities around the globe. Trade in chemicals and their risks provide the logic. Rather than force fit Bhopal into the unifying images of conventional environmentalism, *No Place to Run* runs the risk of something different: the New World Order is not harmonized. Its lines are not natural; the cuts of society and politics remain on the surface.

(11.7) Incorporation

During the 1960s, pesticides were cast in stark opposition to disease and famine. The chemical industry acknowledged that pesticides were "poison," patiently explaining that "that's why they work." By the 1990s, the chemical industry had reconfigured its approach. Corporate brochures and public statements carry a new message. Science has a less prominent role. Expertise is no longer flaunted. Most important, *everything* is said to be made of chemicals—our bodies, the food we eat, the water we drink. We are told that even water is dangerous if you stick your head under it and that even potatoes can kill you if you eat too many of them. Glossy brochures drive the message home. Wildlife plays on the covers. Images of the Earth shot from Apollo 8 run alongside photographs of baby seals and soaring eagles.

FIGURE 11.7 CARBIDE'S ENVIRONMENTAL EXCELLENCE   Union's Carbide's promise, building on the title of a seminal book titled *In Search of Excellence,* first published in 1982. The book provides case studies of stellar corporations to show what it takes to succeed. This ad ran in the September 1989 issue of *Scientific American.*

FIGURE 11.8 OXYCHEM'S HANDS   An interpretation of the cupped hands symbolizing the Chemical Manufacturers Association's Responsible Care program—by OxyChem, the chemical operations component of Occidental Petroleum, a major player in the Love Canal disaster. This image is the cover of a pamphlet that describes OxyChem's Responsible Care commitment. The centerfold is titled "Our Survival Depends On It." Against a black background, the words curl gracefully around a vibrantly colored butterfly, fish, bird, and lizard. Only the panda bear lacks pink and yellow tones.

The new emblem of the chemical industry, developed by the Chemical Manufacturers Association in the late 1980s, is an image of two hands, cupped as though they are praying for the world. In the standardized image, three spheres—presumably molecules—float above the cupped hands. The suggestion is that there is nothing heavyhanded about chemical production. When lifted onto the brochures of individual companies, a butterfly can substitute for the spheres, and the hands can become those of a child.

The dramatic style of corporate environmentalism in the 1990s was recognizably different from that of earlier periods. Many people applauded the "take charge" approach called "strategic environmentalism," linking it to the strategic management styles associated with Total Quality Management. The posture was proactive, cooperative, and expressly interested in what the public had to say. Corporate environmentalism promises to help us clean up the past and manage future risks, while continuing to provide "better living through chemistry."

Corporate environmentalism can be difficult to critique because it promises good things. Bluer skies. Cleaner water. Vibrant democracy. It does not seem appropriate, or effective, to condemn it completely. Because corporations do need to demonstrate environmental excellence. Because "the problem" with corporate environmentalism is more in its mode than in its overtly articulated promise.

Like other contemporary processes of political economy and culture, corporate environmentalism operates globally, linking issues, people, and institutions that only a few years ago were not imagined as related. Corporate environmentalism incorporates the differences. There are confrontations between opposing elements. But they are resolved to create a new unity. Steady progress is promised, in clean developmental narratives that move us "beyond blame." All opposition is overcome. The past is settled.

Epilogue

Corrosions

Gas victims learned of disaster in Bhopal running in the dark. They woke late in the night thinking neighbors were burning chili peppers. Soon breathing became painfully difficult. Most fled. Many headed for Bhopal's New Market area, where the Old City overlaps with offices of the modern state, new businesses, and middle-class homes. Because there was no evacuation plan or even any information on wind direction, they ran into the gas, alongside its destructive currents. Some told me their logic, stressing the faith they once had that whoever had "turned on" operations of industrial modernity could also turn them off. So they ran toward Bhopal's New Market, the hub of a modern retrofit of an old city. Agency and the possibility of change were located there; hospitals lined the route.

My understanding of the Bhopal disaster has followed the disastrous path of gas victims, running through the dark. Stumbling, without clear direction. Sickened by the corruption. Witness to extraordinary displays of charity. Witness to callous abandon and aggressive pursuits of self-interest. Knowing only that we had to keep going—that obligation was out of joint with expertise conventionally conceived, that one could never be prepared for disaster.

My work in Bhopal was shaped by both too much and too little information. I engaged what information I had with limited expertise and an enduring sense of potentially tragic results. Even the smallest move of politics seemed my responsibility, even if a matter of law, medicine, or technology that begged for expert analysis I could not offer. Focused inquiry was deferred, and I simply accepted the barrage of data that characterizes disaster at the grassroots.

Data come from all directions, but one still wants to hear more. Everything seems methodologically scrambled, contingent, and at odds with any desire for functional efficacy. At first, one can only listen. The racket threatens to deafen and to mute all possibility of response. What can one say when medical data produced by different institutions so differ that comparative analysis seems all but impossible? What is the reply when one set of lab results finds no toxins in the soil or water, while another set of lab results finds dichlorobenzene? How can one find words when rehabilitation schemes turn into the Bhopal Beautification Plan, legitimating the forced relocation of slum dwellers who are said to pose threats to public health? All one hears resembles noise, discordant sound lacking any syntax, disturbances that interfere with the reception of signals and useful information.

Advocacy became a way of knowing, a prosthetics for making sense of a world asphyxiated with meaning. Advocacy directed my gaze and provided protocols for evaluating validity. My posture was not that of a neutral observer. Nor was it the posture of radical opposition. Established systems of state, law, and science deserved vehement criticism, but I could not stand outside them. I spoke within the force fields of established systems, tragically aware of their insufficiency. The bodies of gas victims were force fit into the diagnostic taxonomies of biomedicine. Everyday life was written into proposals for bureaucratic schemes. Justice was sought within the law.

In Bhopal, I worked alongside destructive currents, relying on the institutions and concepts of modernity even while critiquing them. Like gas victims, I had to hope that whatever had turned on the disaster could also help turn it off. Irresolvable contradiction became matter of fact; proficiency depended on an ability to work within constraint, within paradox, within disaster.

With time, I have learned that advocacy can and must take diverse forms. In Bhopal, I spoke and wrote in the language of those I hoped to challenge. My ethnographic writing needs to be different, creating a space for idioms that cannot be said in the courtroom or even in pamphlets prepared for political supporters. Ethnography must take responsibility for questioning and transfiguring the very concepts and rhetorics I deployed in Bhopal—not because they were wrong, but because more remains to be said.

But ethnography, too, works alongside destructive currents. Ethnography is part of the pluralist project. It is a project worth pursuing; it is a project requiring pervasive critique. Recolonization of marginal perspectives is difficult to avoid. Juxtaposition of diverse truth claims can be read as aggregation—working, not unlike the law, to resolve differences into single judgments.

Aggregation is a modernist trope, relied on to raise questions about the reliability of any single truth claim. But it also can affirm the perspective of he who stands outside competing claims, surveying and assessing the evidence to arrive at a final synthesis. An ethnography of Bhopal should not work toward final synthesis. The case should remain unsettled; continuing liability should be assumed. The challenge is to engage Bhopal as disaster—as a call to respond to a new order of things—an order so disorderly that established ways of imagining "the whole" must be rethought.

Ethnography, like political activism, must aspire to address the "whole problem." Local realities must be understood within global systems. Interpretations must be situated within political-economic trends. Discourses of law and science must be understood as translatable—into legal proof or therapeutic programs, into means of distributing equality, or into mechanisms of exclusion. The result can never be comprehensive. Expertise itself becomes a paradox, as does ethics. One is always confronted with more to understand and more to address than is possible. One must choose a focus, knowing that responding well to one problem ignores another.

The choices that must be made cannot be directed by established codes of good conduct. Values embedded in such codes may persist. But they must be imagined in an emergent future that resists analytic simplification or hierarchical organization of priorities. Asking questions about what is *most* valued won't work. Nor will detached evaluation. Experts must work within the systems they evaluate, moving between competing modes of knowledge, without a God's eye view. A dizzying oscillation is engendered by any act of choice; even the smallest move has the potential to set off a runaway reaction. But one must move, without fully understanding the complex systems in which one works. Ethics happen within such movement. Ethics play out in ways that cannot be controlled.

Advocacy, conceived of as a performance of ethics in anticipation of a future, is more than calculation. Responsibility is assumed for consequences that cannot be fully predicted. Liability is recognized as a debt that is never fully paid. What seems good and possible emerges from complex interactions that produce more than the sum of their parts. The many "stakeholders" in these interactions do not remain the same over time. The moves made by stakeholders in one locale resonate with those made by stakeholders in other locales, transforming the systems in which they operate, repositioning everyone within them. Advocacy must negotiate the change.

Advocacy, as I have come to understand it, requires continual interpretation. One must watch the world, listening for what escapes explanation by science, law, and other established discourses. Accounting for what established systems discount as noise.

Disaster accelerates the pace. No time seems like the right time for conceptual critique when people have desperate practical needs and basic functions of law remain to be accomplished. Meanwhile, the basic functions of law legitimate continued disaster—provoking chronic awareness of the need to rethink what "counts" as knowledge, as adequate representation, as a justifiable plan of action. Bhopal is this double bind.

Whacher is what she was.
She whached God and humans and moor wind and open night.
She whached eyes, stars, inside, outside, actual weather.

She whached the bars of time, which broke.
She whached the poor core of the world,
wide open.

To be a whacher is not a choice.
There is nowhere to get away from it,
no ledge to climb up to—like a swimmer

who walks out of the water at sunset
shaking the drops off, it just flies open.
To be a whacher is not in itself sad or happy,

although she uses these words in her verse
as she uses the emotions of sexual union in her novel,
grazing with euphemism the work of whaching.

But it has no name.
It is transparent.
Sometimes she calls it Thou.

(Carson 1995 [1992], 4–5)

Futures

"Remember Bhopal." A slogan repeated in the streets of Delhi; of Institute, West Virginia; of Geismer, Louisiana. A slogan reminiscent of other disasters: Hiroshima, Auschwitz, Vietnam. "Remember Bhopal." What does this slogan demand? What does it promise?

I have pursued this question as an ethnographer, and as a political activist, in India and the United States. The convergence of scholarship and activism in my work was neither planned nor ever fully explicated. Disaster is unsettling. Established logics and agendas of action no longer seem sensible; urgency directs one's gaze and delimits one's choices. Thus, I have pursued remembrance of Bhopal—far from perfectly, at odds with claims that the time of responsibility comes after full understanding, when full knowledge can rationalize one's roles and positions, when it is clear what must be done to fulfill one's obligations. Remembrance, I have come to know, is difficult to observe. But its enactment is inevitable.

There was a time when I understood remembrance quite differently. Remembrance was opposed to forgetting. Progressives choose to remember Bhopal. Union Carbide and the government of India chose to forget. Now my understanding of how remembrance happens is more complicated— and even more insidious. Remembrance, as I have come to understand it, is not only a positive term, easily opposed to forgetting. Remembrance and forgetting cannot be bifurcated—they happen simultaneously and ubiquitously. In other words, remembrance is more than a purposeful act. It is something that happens all the time by each and every one of us. It is carried out by individuals and institutions. It is imbricated in legal decisions and scientific claims. It is harbored within the most mundane daily acts as well as within grand and official statements. And it is something that cannot be punctuated with a full and final stop. Remembrance is not an identifiable act or statement that is accomplished once and for all. Every move of remembrance sets up the next. Remembrance is implicitly promissory. Liability, then, is also transfigured. Obligations—to remember, in particular—are never accomplished once and for all.

My thoughts about remembrance and liability have been shaped by theoretical articulations of the way language works as well as by political-economic theories that highlight the multiple determinants of change. As important, however, has been the materiality and sociality of Bhopal. Gas victims and those working alongside them have reached for ways to articulate what they want in the wake of disaster. They want rehabilitation. They want legal judgments of wrongdoing. And they want measures taken that would prevent future Bhopals. Often these demands are articulated in seemingly simple terms—the phrase "We want justice" recurs in street rallies, in pamphlets, and in the graffiti painted throughout Bhopal. The seeming simplicity of the call for justice is, however, deceptive. Gas victims and their allies seem to know this—often emphasizing their awareness that justice itself will always be deferred, that there is no way to fully rehabilitate Bhopal. Gas victims' demand for "continuing liability" articulates this understanding.

Continuing liability calls upon us to question the order of things—to refigure how law works—to rethink how the past should be built into the future. Advocacy becomes a historical project. A project that contests the origins and order of things. A project that is interminably recursive, running back over history again and again, reaching for ways to figure the future differently.

The Bhopal disaster can be remembered as a singular catastrophic event that killed many, injured many more, and thrust people around the world into new understandings of the way risk percolates through modern society. December 3, 1984, is the day to mourn. But there are also other possible origins of the disaster, both before and after 1984: 1600, the year the East India Company was chartered in London, establishing an enduring relationship between India and the West. 1934, the year the Ever Ready Company (I) Ltd. was incorporated in India to assemble dry cell batteries. 1959, the year the Ever Ready Company became Union Carbide, to signal its diversification into chemicals and plastics. 1969, the year Union Carbide began pesticide formulation in Bhopal—to enter the market created by the Indian government's "green revolution" campaign to modernize agriculture. 1974, the year Union Carbide India was granted an industrial license by the government of India to manufacture as well as formulate pesticides in Bhopal. 1980, the year pesticide use in India declined dramatically—because pesticides did not work like they were supposed to, because indebted farmers had to buy seeds and fertilizers first. 1982, the year cost-cutting initiatives at Union Carbide's Bhopal plant accelerated, affecting both maintenance regimes and worker training. 1984, when BASF, a German multinational chemical company, locked out its workers at its plant in Geismer, Louisiana—jumpstarting a decade of union busting in the U.S. chemical industry. 1984, when the first enumeration of destruction and suffering in Bhopal was done, foreshadowing how much it would mean when some things and people count, and others do not.

Later dates can also be spoken of as origins. 1985, the year of an "almost Bhopal" at Union Carbide's plant in Institute, West Virginia—a plant considered the "sister plant" of the plant in Bhopal because built on the same design. 1986, the year U.S. courts denied jurisdiction in the Bhopal case, claiming that important decisions leading to the disaster were made in India and that the case did not reflect the public interest of the United States. 1988, the year Union Carbide formally presented its "sabotage theory" to explain the disaster, blaming a single "disgruntled employee" said to have put water into the methyl isocyanate storage tank, knowing that it would cause it to explode. February 1989, when the Indian Supreme Court ordered Union Carbide to pay $470 million in damages, as "full and final settlement" for the Bhopal case.

Each of these origins executes the future. Together they fold the past into the way reality presents itself, in ways that figure what will come. The future is anteriorized.

The future anterior is not unlike determinism, but dispersed, rather than operationalized through linear causality. The future inhabits the present, yet it also has not yet come—rather like the way toxics inhabit the bodies of those exposed, setting up the future, but not yet manifest as disease, or even as an origin from which a specific and known disease will come. Toxics, like the future anterior, call upon us to think about determinism, but without the straightforward directives of teleology. Toxics, like the future anterior, iterate rather than duplicate the past.[1]

The future anterior is anticipatory. What is said and done now draws the fullness of its meaning from what will be said and done in the future. But the future can be folded back in different ways. The future can be folded back as teleology, mandating continuation of established systems and ideals. Or the future can be folded back differently—opened up, to things and people that didn't otherwise count.[2]

What once seemed settled can be unsettled. The past can be pulled into significance in different ways. History can become an altercation.

What is promised by remembering differently? The change mobilized by remembrance can be difficult to see. But it operates, reshaping what we think is possible, necessary, and good. A disaster can become a prism—a means of refracting New World Orders.

The Bhopal disaster comes to operate like the cumulative effect of toxics—inhabiting us, reconfiguring us in ways that science cannot register. Responsibility comes to be understood as kaleidoscopic, rather than direct. The relationship between knowledge and ethics gets tangled. The figure of the advocate shifts—from modernist hero to something different, and more complex—to deal with double bind; to deal with disaster.

All related

Appendix

QUESTIONS
→→→→

↓ CHAPTERS ↓ ↓	What double binds call advocates to speak?	How have advocates strategized and configured their own role in disaster?	What is the Bhopal disaster said to mean?
Prologue	The times		
Introduction		The fieldwork	The analysis
1 Plaintive Response	1.1 Standing tall before the world	1.2 Oscillating roles	1.3 Justice before, and after, the law
2 Happening Here	2.1 Cynicism and contradiction	2.2 Safer systems	2.3 Rights to know
3 Union Carbide: Having a Hand in Things	3.1 Things fall apart	3.2 Assuming moral responsibility	3.3 Bhopal is history now
4 Working Perspectives	4.1 Knowing that you don't know	4.2 Culpable by comparison	4.3 Sabotaging sabotage
5 States of India	5.1 Constitutional contradictions	5.2 Science as reason of state	5.3 Categorical imperatives
6 Situational Particularities	6.1 Matter out of place	6.2 From confusion to categorization	6.3 Statutory violence
7 Opposing India	7.1 In the middling	7.2 To be an environmentalist	7.3 Social problems, technical solutions
8 Women's Movements	8.1 Beyond their means	8.2 Breaking in	8.3 Continuing liability
9 Anarchism and Its Discontents	9.1 Speaking of . . .	9.2 Powers of practice	9.3 It isn't famine
10 Communities Concerned About Corporations	10.1 Constitutional constraints	10.2 Language learning	10.3 Bottoms up
11 Green Consultants	11.1 Stakeholder skepticism	11.2 Going green	11.3 Moving beyond blame
Epilogue			

What focusing devices have advocates used to understand the disaster and connect it to other things?	Through what organizational, rhetorical, and textual forms has advocacy been disseminated?	What corrodes or undercuts this advocacy?	What catalyzes this advocacy?
The focus	The text		
1.4	1.5	1.6	1.7
Material effects	De bonis non	Idealic images	Aesthetics and expertise
2.4	2.5	2.6	2.7
Sheltering in place	New vehicles of meaning	Covering events	Re-writing company towns
3.4	3.5	3.6	3.7
Psychosis?	Responsible care?	Designing disaster	Old dogs, new tricks
4.4	4.5	4.6	4.7
Systems theories	Demonizing technology	Be-laboring environmentalism	Repetitions
5.4	5.5	5.6	5.7
India, Inc.	Remaking India	From the study of women	And yet it moves
6.4	6.5	6.6	6.7
Depression and paralysis	Refusals	Maelstroms	Flagrante delicto
7.4	7.5	7.6	7.7
Imagining disaster	Encroaching on civil rights	In search of famine	Catalytics
8.4	8.5	8.6	8.7
Ritual openings	Law of the grassroots	Endemics	Interruptions
9.4	9.5	9.6	9.7
Rebuilding as subversion	Writing alternatives	At odds	Toward a politics of disaster
10.4	10.5	10.6	10.7
Intimate connections	Just cause	Structural problems	The many faces of the new expert
11.4	11.5	11.6	11.7
Force of law	Rhetorical solutions	Disharmony	Incorporation
		Corrosions	Futures

NOTES

Prologue

1. The *Wall Street Journal*'s description of the fall of the Berlin wall is telling: "With Lenin and Marx little more than icons on the Kremlin wall, the U.S. heads into a new era, hoping to demonstrate again that free markets, free institutions and free workers are the answer to future prosperity" (Morris, Robinson, and Kroll 1990, 210, 216).

2. For a review of the Gulf War as a "hygienic project" mobilized by fear of contamination, see Avital Ronell's "Activist Supplement: Notes on the Gulf War." Ronell wrote:

> In the name of symbolic health (a unity of world that sees its image in wholesomeness and the project of renewal), we have waged war on what was repeatedly represented as a degenerate, sickly "something" that carried the threat of contagion. . . . The hygienic project has everything to do with establishing a new world order that consists in nothing less than purifying an imaginary—and real—territory, guaranteeing that it be proper and that it espouse values of propriety and property. The invisibility of the enemy inhabits this logic, which is essentially viral. (1994, 293, 301–2)

3. For an interpretation of the Los Angeles riots as a index of the United States in the early 1990s, see John Agnew's "Global Hegemony versus National Economy: The United States in the New World Order." Agnew argues that the Los Angeles riots

> captured the paradoxical position of the United States in the contemporary world. Although it is acquiring an apparent military and cultural hegemony at the global scale, its national economy is suffering from stagnation, increased income inequality between regions and within metropolitan areas, and growing social polarization. . . . What emerges beyond the statistics of the riots—2,383 injured, 5,383 fires, 16,291 arrests, $785 million in damages—is a picture of a rigidly segregated and socioeconomically divided city. . . . My central claim is that the New World Order being made under American auspices is a "transnational liberal economic order" that now detracts from, rather than augments, the material conditions of life for both the "average" and the poorest people within the territorial boundaries of the United States. (1994, 269–70)

4. The Montreal Protocol on Substances That Deplete the Ozone Level was negotiated with an unprecedented level of corporate involvement in international diplomacy. DuPont, the inventor of chlorofluorocarbons (CFCs), was a particularly prominent ally for environmentalists. DuPont had also developed the alternatives. India and China, on whom the success of the agreement was said

to depend (Makhijani, Bickel, and Makhijani 1990), became party to the agreement a few years later—after negotiating both a "grace period" for reducing use of CFCs and the establishment of an international fund to help Third World nations gain access to technologies necessary to implement CFC reductions and substitutions. The international fund established by the "London Accord" set the stage for a reorganization of international aid in terms of global environmental problems, rather than in terms of the Cold War. By the 1992 Earth Summit in Rio, the environment was overtly referred to as a "bargaining chip" in efforts to secure international transfers of resources. The Montreal Protocol has also become an important example of preventative initiative "before the science is in." The particular interests of DuPont and other companies that stood to profit from the transition complicate the story.

Introduction

1. Quotations without citations throughout this book are drawn from my field notes. Many of the primary documents included in the book—particularly those produced by the Bhopal Group for Information and Action and the Bhopal Gas Affected Working Women's Union—were also part of my field notes, since I co-authored them.

2. Jacques Derrida acknowledges that the "future anterior could turn out to be—and this resemblance is irreducible—the time of Hegelian teleology. Indeed, that is how the properly philosophical intelligence is usually administered, in accord with what I called above the dominant interpretation of language—in which the philosophical interpretation precisely consists" (1991, 36–37). Derrida also suggests other possibilities, describing how the concept of the future anterior could have

> drawn us toward an eschatology without philosophical teleology, beyond it in any case, otherwise than it. It will have engulfed the future anterior in the bottomless bottom of a past anterior to any past, to all present past, toward that past of the trace that has never been present. Its future anteriority will have been *irreducible* to ontology. An ontology, moreover, made in order to attempt this impossible reduction. This reduction is the finality of ontological movement, its power but also its fatality of defeat: what it attempts to reduce is its own condition. (1991, 37)

Derrida writes of the future anterior to think about what both demands and makes possible a future that would not be merely a continuation of the present. The future anterior puts in place a "lace of obligation" that both binds and unbinds the ethical actor. The possibility of pursuing justice beyond the determinations of law is one important effect (Derrida 1990, 329).

Michel Foucault provides an alternative articulation, more directly laying out the writing yet to be done:

> A task is thereby set for thought: that of contesting the origins of things, but of contesting it in order to give it a foundation, by rediscovering the mode upon which the possibility of time is constituted—that origin without origin or beginning, on the basis of which everything is able to come into being. Such a task implies the calling into question of everything that pertains to time, everything that has formed within it, everything that resides within its mobile element, in such a way as to make visible that rent, devoid of chronology and history, from which time issued. . . . In that case the origin is that which is returning, the repetition toward which thought is moving, the return of that which has already always begun. . . . (1973, 332)

There is, then, an injunction: thought must "keep watch in front of itself, on the ever receding line of its horizon" so that origins can be reordered, and time can proceed on a different axis (Foucault 1973, 332).

Foucault's teacher, Georges Canguilhem, explains the future anterior in very specific terms, suggesting the specific relevance of the concept to my consideration of how facts, exclusion, and effacement have worked within the Bhopal disaster:

> The abnormal, as ab-normal, comes after the definition of the normal, it is its logical negation. However, it is the historical anteriority of the future abnormal which gives

rise to a normative intention. The normal is the effect obtained by the execution of the normative project, it is the norm exhibited in the fact. In the relationship of the fact there is then a relationship of exclusion between the normal and the abnormal. But this negation is subordinated to the operation of negation, to the correction summoned up by the abnormality. Consequently it is not paradoxical to say that the abnormal, while logically second, is existentially first. (1991 [1966], 243)

Reference to Jacques Lacan's conceptualization of retroaction, anticipation, and punctuation is also useful, particularly for thinking about how the future anterior is actualized through the practice of advocacy. Retroaction, also translated as "deferred action," is Lacan's *apres coup* and Freud's *Nachtraglichkeit*. Retroaction refers to the way the present affects past events a posteriori—either entrenching or dislodging habituated structures of memory. What concerns the analyst (and the advocate) is how the past exists for the patient now, as evident in the way the patient reports it (in the courtroom, for example). Lacan shows how discourse is structured by retroaction: only when the last word of the sentence is uttered do the initial words acquire full meaning. Anticipation refers to the way the future affects the present. Like retroaction, anticipation marks the structure of speech; the first words of a sentence are ordered in anticipation of the words to come. Punctuation allows the analyst (and the advocate) to intervene in the patient's construction of past and future. By punctuating the patient's discourse in an unexpected way, the analyst (and the advocate) can retroactively alter seemingly fixed (settled) meaning. Punctuation can also be used to draw out items of signification that seem like random externalities. What the analyst (and the advocate) does is insert herself in time, reworking the past so that the future is anteriorized— differently. There is a series of confrontations between opposing constructions of meaning. And the confrontations do produce movement. But the movement is not toward final synthesis—because of the irreducibility of the unconscious (like the irreducibility of disaster), analysis (and liability) is interminable (Lacan 1977 [1966], 292–325). The future anterior is where the future is worked out, now.

3. Michael Fischer was my Ph.D. advisor. George Marcus was a member of my committee and the chair of the Department of Anthropology at Rice University throughout the time I studied there (1987–93).

4. Akhil Gupta and James Ferguson have argued that "cultural critique," as proposed by Marcus and Fischer, "assumes an already existing world of many different, distinct 'cultures' and an unproblematic distinction between 'our own society' and an 'other' society" (Gupta and Ferguson 1997a, 42). In their account, thinking in terms of "different societies" spatializes cultural difference in familiar ways, supporting a conceptualization of the world as a pretty mosaic, justified by a liberal political model. As should be evident throughout this book, I share Gupta and Ferguson's criticism of liberal pluralism. Bhopal implodes habitual opposition between "us and them," as between here and there. And we need political models and forms of responsibility that acknowledge this. But I don't think this means that we can't talk in terms of social and cultural difference. It's just harder. Rather than assuming difference a priori, we have to describe it into visibility so that it can interrupt us, so that we don't forget starkly uneven distributions of risk and reward. Part of the problem is textual. As is no doubt clear here, representing difference without evoking pluralism poses serious challenges. Whatever one says operates in a recursive structure that predisposes a pluralist effect, no matter how much one writes against it. Experimentation with different textual models, as proposed by Marcus and Fischer, seems crucial. So does attention to our spatialization devices—as in the volume edited by Gupta and Ferguson titled *Culture, Power, Place: Explorations in Critical Anthropology* (Gupta and Ferguson 1997b).

5. For an evocative rendering of the dilemma, see Maurice Blanchot's *The Writing of the Disaster* (1986).

6. I use the term "enunciatory" to refer to theories of significance and subjectivity developed by Foucault and Lacan in particular. In linguistic theory generally, enunciations are differentiated from statements. Statements are understood in terms of abstract grammar, divorced from the specific context of articulation. Enunciations are made by specific individuals, in specific times and places. Lacan builds on this differentiation to highlight the unconscious dimensions of

speech and subjectivity; the conscious dimension of speech takes the form of a statement, while the unconscious is signaled in enunciation (Lacan 1981 [1973], 138–42; 1977 [1966], 298–316). Unconscious dimensions of articulation are important here, in my effort to track both intended and unintended enactments of advocacy—and the way the past shapes what advocacy becomes.

Foucault problematizes the distinctions between statements and enunciation by describing the impossibility of finding structural criteria of unity for the statement because the statement "is not in itself a unit, but a function that cuts across a domain of structures and possible unities, and which reveals them, with concrete contents, in space and time." Foucault's project was to describe the function of statements—what he calls the "enunciative function" in actual practice: "its conditions, the rules that govern it, the field in which it operates" (1972 [1969], 87). My project has been to describe the function of the Bhopal disaster in the articulations of various "enunciatory communities" that have joined related political debates.

Foucault is careful to argue that the

> statement, then, must not be treated as an event that occurred in a particular time and place, and that the most one can do is recall it—and celebrate it from far off—in an act of memory. But neither is it an ideal form that can be actualized in any body, at any time, in any circumstances, and in any material conditions. Too repeatable to be entirely identifiable with the spatio-temporal coordinates of its birth (it is more than the place and date of its appearance), too bound up with what surrounds it and supports it to be as free as a pure form (it is more than a law of construction governing a group of elements), it is endowed with a certain modifiable heaviness that allows of various uses, a temporal permanence that does not have the inertia of a mere trace or mark, and which does not sleep on its own past. Whereas an enunciation may be *begun again* or *re-evoked,* and a (linguistic or logical) form may be *reactualized,* the statement may be *repeated*—but always in strict conditions. This repeatable materiality that characterizes the enunciative function reveals the statement as a specific and paradoxical object, but also as one of those objects that men produce, manipulate, use, transform, exchange, combine, decompose and recompose—and possibly destroy. Instead of something being said once and for all—and lost in the past like the result of a battle, a geological catastrophe, or the death of a king—the statement as it emerges in its materiality, appears with a status, enters various networks and various fields of use, is subjected to transferences or modifications, is integrated into operations and strategies in which its identity is maintained or effaced. Thus the statement circulates, is used, disappears, allows or prevents the realization of a desire, serves or resists various interests, participates in challenge and struggle, and becomes a theme of appropriation or rivalry. (1972 [1969], 88)

Foucault's theorization is important here for a number of reasons. First, Foucault unsettles the opposition between particularity and generality and the related opposition between that which can (legitimately) be forgotten and that which is universal and thus worthy of remembrance. And he does this by showing how things that were once thought unified and isolatable are crosscutting. Second, his approach is not "anthropological" in the conventional sense. Rather than analyzing the assertions of individuals, he describes the places within which individuals become speaking subjects. What they say is understood through force fields that distribute status, techniques, and material resources. Third, Foucault examines how the places within which subjects speak consolidate the subject and establish identity through opposition with what the subject is not. But he is even more interested in how enunciation divides and disperses the subject. He explains, for example, that the analysis he proposes "would not try to isolate small islands of coherence in order to describe their internal structure; it would not try to suspect and to reveal latent conflicts; it would study forms of division. Or again; instead of reconstituting *chains of inference* (as one often does in the history of the sciences or philosophy), instead of drawing up *tables of difference* (as the linguists do), it would describe *systems of dispersion*" (Foucault 1972 [1969], 37). Fourth, Foucault emphasizes how enunciation manifests continuity—it is something to which subjects return and begin again. But he also emphasizes the strategies through which enunciation transforms and even disposes of given ways of speaking.

7. Gregory Bateson's ethnographic study of the Naven ceremony of the Iatmul of New Guinea foreshadowed his work on double-bind theory in important ways (1958 [1936]). The "Epilogue" to the 1958 reprint discusses the book in terms of the problems posed for an ethnographer when confronted by messages of different logical types. Sharon Traweek provides a more recent example of work with the concept-metaphor of double bind, even if with little overt reference, which is itself instructive. Her book *Beamtimes and Lifetimes: The World of High Energy Physics* was written to reach diverse readers, including both physicists and social scientists. By her own accounts, work toward this task caught Traweek within a residing double bind: to talk about physics in new ways, to different kinds of readers, she needed to produce a text that was both "accessible" and very carefully textualized. One strategy Traweek used mapped Bateson's theories of different ways of learning across the tropes of metaphor, metonymy, synecdoche, and irony, to provide an "unsayable" structure to the book's form and argument. The textual grid created by this mapping exercise guided Traweek's efforts to nest multiple layers of description within passages that read as "just stories" (Traweek 1988).

8. For discussion about how double binds operate and how researchers have tried to understand them, see the essays collected in *Double-Bind: The Foundation of the Communicational Approach to the Family,* edited by Carlos Sluzki and Donald Ransom (1976). The essay by Anthony Wilden and Tim Wilson, "The Double Bind: Logic, Magic and Economics," provides an important clarification:

> A true double bind is not simply an awkward situation in which "we are damned if we do and damned if we don't" for this usually amounts to no more than a choice between the lesser of two evils. Neither is it simply a binary opposition or contradiction, for here it is possible to make a stable choice between one side or the other, and the two sides may differ in real or apparent value. Nor is a double bind equivalent to the "horns of a dilemma," where one is presented with a choice between conflicting alternatives, both of which are similarly unfavorable. A true double bind—or a situation set up or perceived as one—requires a choice between two states which are equally valued and so equally insufficient that a self-perpetuating oscillation is engendered by any act of choice between them. . . . It is the result of the fact that one must choose, and moreover choose between incompatible alternatives. (1976, 276)

9. Enunciatory communities should also be differentiated from "epistemic communities," a unit of analysis developed by Ernst Haas in *When Knowledge Is Power* (1990). Epistemic communities are transnational networks of experts held together by common values and by ways of testing the truth, understanding causality, and approaching the policy enterprise. They are recognizable by their shared command of potentially instrumental technical knowledge; consensual knowledge explains their particular way of nesting problems and solutions. In Ernst Haas's account, epistemic communities are responsible for articulating the knowledge relevant to the functioning of international institutions. Peter Haas illustrates the importance of epistemic communities in environmental politics in his book *Saving the Mediterranean* (1990). In his account, a global ecological epistemic community emerged in the 1980s around shared belief about how causality operates in tightly coupled systems. Many in this community were trained in the relatively new science of systems ecology, informed by holistic beliefs about the nature of social and physical systems.

Nor are enunciatory communities the "interpretive communities" described by Stanley Fish, despite his insistence that they posit "language centered identities." Interpretive communities are groups of people who "share a way of organizing experience" including "categories of understanding and stipulations of relevance" (Fish 1989, 41).

10. Enunciatory communities should be understood as subject to what Foucault calls the microphysics of power—power that operates insidiously rather than overtly, and cumulatively rather than in one blow. Gilles Deleuze provides important clarification:

> These power-relations, which are simultaneously local, unstable and diffuse, do not

emanate from a central point or unique locus of sovereignty, but at each moment move "from one point to another" in a field of forces, marking inflections, resistances, twists and turns, when one changes direction, or retraces one's steps. This is why they are not "localized" at any given moment. They constitute a strategy, an exercise of the non-stratified, and these "anonymous strategies" are almost mute and blind, since they evade all stable forms of the visible and the articulable. Strategies differ from stratifications, as diagrams differ from archives. It is the instability of power-relations which defines a strategic or non-stratified environment. Power relations are therefore not *known*. Here again Foucault somewhat resembles Kant, in whom a purely practical determination is irreducible to any theoretical determination or knowledge [connaissance]. It is true that, in Foucault, everything is practical; but the practice of power remains irreducible to any practice of knowledge [savoir]. To mark this difference in nature, Foucault will say that power refers back to a "microphysics." But we must not take "micro" to mean a simple miniaturization of visible and articulable forms; instead it signifies another domain, a new type of relation, a dimension of thought that is irreducible to knowledge. "Micro" therefore means mobile and non-localizable connections. (Deleuze 1988 [1986], 73–74)

11. Gary Downey's early work on conflicts over nuclear power in the United States showed the importance of accounting for different fields of reference when describing how people negotiate risk (Downey 1986a, 1986b). His later work on engineering education backtracks to explore how different fields of reference are created. Both moves are important here, particularly in working to understand how the conditions that led to the Bhopal disaster made sense in an everyday way to engineers at Union Carbide (Downey and Lucena 1995).

12. Evelyn Fox Keller has provided particularly evocative descriptions of how "objects of concern" are configured and of the blindness generated in the process. The discourse of gene action, for example, configures "genes as active agents, capable not only of animating the organism but of enacting its construction." Keller explains that this configuration was "immensely productive for geneticists, both technically and politically . . . it permitted the framing of their question in terms that allowed for a remarkably prolific research program." But the discourse of gene action also had its costs; it worked so well partly because of what it excluded. Keller explains that

the very glow of the geneticists' spotlight cast a deep and debilitating shadow on the questions, on the methods, indeed on the very subject of embryology. It allowed neither time nor space in which the rest of the organism, the surplus economy of the soma, could exert its effect. What is specifically eclipsed in the discourse of gene action is the cytoplasmic body, marked simultaneously by gender, by international conflict, and by disciplinary politics. (Keller 1995, xv)

Keller also reminds us that the story is not only about recuperating embryology and supplementing the discourse of gene action with the more adequate discourse of gene activation. Things are much more complicated than this, partly due to the ways restoration of embryogenesis to biology depended on and extended notions of distributed power drawn from information theory. Adding embryology didn't merely "flesh out" biology's object, but also made the object into a very different animal, so to speak. Discourses of disaster evidence similar moves of spotlighting and occlusion. The discourse of risk management, for example, manages disaster by configuring the object of its concern as an aggregation of all potential problems and groups of people potentially affected. A risk management plan aggregates atomized parts of the problem to form an operationally whole solution. The discourse of risk management is productive because it claims to include everything and everyone. But it works because "the rest of the organism" (political-economic context of production, employee reward systems, the possibility total system failure, etc.) is discounted. Like in the discourse of gene action, "the surplus energy of the soma" (here, the potential irrationalities of disaster) is not given time or space to exert its effect. If it were, a risk manager's object of concern not only would be more complete, but also would be a different concern, productive of a very different mode of advocacy.

13. My conception of advocacy as a social practice that takes many forms draws on the example of others who have shown the importance of using ethnography to disrupt monolithic constructs. John and Jean Comaroff provide a particularly evocative explanation, with multiple levels of relevance to this study:

> The image of colonialism as a coherent, monolithic process seems, at last, to be wearing thin. That is why we are concerned here with the tensions of empire, not merely its triumphs; with the contradictions of colonialism, not just its crushing progress. This is not to diminish the brute domination suffered by the colonized peoples of the modern world, or to deny the Orwellian logic on which imperial projects have been founded. Nor is it to deconstruct colonialism as a global movement. It is, instead, to broaden our analytic compass; to take in its moments of incoherence and inchoateness, its internal contortions and complexities. Above all, it is to treat as problematic the *making* of both colonizers and colonized in order to understand better the forces that, over time, have drawn them into an extraordinarily intricate web of relations. (1992, 183)

14. Charts running comparisons of "major industrial disasters of the twentieth century" are a familiar component of discussions about Bhopal. The incidence of industrial disaster is often compared to the incidence of natural disasters (Bogard 1989, 41; Shrivastava 1987). Comparisons of settlement amounts are also common (Cassels 1993, 225). The Bhopal figures here are drawn from accounts by Claude Alvares (1994) and Indira Jaising (1994).

15. Kwame Anthony Appiah indicts liberal pluralism and associated assumptions that identity is an essence rather than an effect in *In My Father's House: Africa in the Philosophy of Culture* (1992). Appiah reminds us that to assume essential commonality is to tread into racist territory because it often invokes biology by default. His point is that identity is a result of interpretation, often driven by instrumentalism. My challenge here is to create narrative strategies that embody Appiah's critique.

16. Much can be learned from critical race theory on the significance of "verbing." Zora Neal Hurston, Aime Cesaire, Baraka—they all wrote about and with verbs. Nathaniel Mackey emphasizes the significance of verbing with relation to swing: when appropriated by white musicians, swing became a noun, and it all started to sound the same. Verbed, swing was extraordinarily generative, becoming bebop, hard bop, cool. Working on different registers, improvising, refusing to be contained by the demands of commodification, providing black artists a stage. Ethnography can borrow the style, noting how difference is brought about and maintained through wide-ranging techniques—of measurement, of timing, of turning and sliding between things once thought of as distinct.

17. Sally Merry (1992) outlines the importance of tracking these interactions in a review of efforts to account for transnational processes in the analysis of local legal phenomena. Merry argues for reengagement with concepts of legal pluralism, despite the fact that previous reliance on pluralism as an analytic category failed to explore the interaction between different legal systems and thus failed to account for power inequalities among them. In Merry's account, pluralism can be used as a heuristic for exploring the mutually constitutive nature of different legal systems and for extending "what counts" as legitimate foci of study—to include a range of informal normative ordering systems. She goes on to emphasize how theories of unequal but mutually constitutive legal systems lead to new questions: How do these systems interact and reshape one another? To what extent is the dominant system able to control the subordinate? How do subordinate systems subvert or evade the dominant system? In what ways do the disputation strategies of subordinate uses reshape the dominant system?

18. Bateson's *Naven* (1958 [1936]) provides an important example of an ethnographic text organized to repeatedly come back to the same "object" of study with a different set of questions—to illustrate how different questions produced different descriptions, but also to provide a cumulatively more nuanced description of the Naven ceremony.

19. Michael Fischer and Mehdi Abedi inspire this effort toward a hopscotched text in *Debating Muslims: Cultural Dialogues in Postmodernity and Tradition* (1990), as does Julio Cortazar, in his

novel *Hopscotch*. Cortazar, too, is concerned with representing both cultural difference and the way worlds blend together. The first section of his book, "The Other Side," is "about" Europe. The second section, "This Side," is about Latin America. The third section, "From Diverse Sides" (or "Expendable Chapters"), disintegrates into an intertextual mélange of "messages" of different logical types—including missing parts of the story already told, excerpts from other texts, and commentary on the nature of the novel. Cortazar's protagonist, Oliveria, is no cheerleader for the masala. She obsesses over the falsity of contact between "isolated orbs" and over the "unbridgeable space between you and me," insisting that "all endearment is an ontological clawing" (Cortazar 1967 [1963], 326–27). My textual strategy also draws from the orthogonal arrangement of Levi-Strauss's *Mythologiques* (Levi-Strauss 1969 [1964], 1973 [1966], 1978 [1968], 1981 [1971]). If deconstruction had to traverse phenomenology, I presume that ethnography will have to traverse Levi-Strauss. The challenge, it seems, is to push structuralism so far that it tips over into something else.

20. I borrow the metaphor of "corrosion" from Bruce Lincoln's *Authority: Construction and Corrosion* (1994)—which provides a beautiful analysis of how authority is an effect rather than an entity, dependent on particular combinations of place, time, rhetoric, props, audience, and speaker.

21. These questions are intended to catalyze a number of different reactions or chains of signification. Questions move across chapters along a synchronic, substitutive axis, driven by metaphor. Movement through each chapter follows a diachronic, combinatorial axis that operates metonymically. The two axes are meant to interrupt rather than confirm each other. The interruption works because the metonymic axis (the domain of the signifier) is itself tropologically produced. This nesting of tropes within tropes unsettles binary opposition between inside and outside, as between signifier and signified—which means that neither chapters nor the entire text will ever "settle down" into one conclusive statement: their full meaning will always be deferred because every move of signification will call for another. There will always be more to say because things are out of joint and there is no stable center.

Jacques Derrida shows how the absence of a center motivates rather than paralyzes articulation, and this has been important in my thinking about the possibility of a noncentered, nontotalizing advocacy—embodied in ethnography as well as in other political forms. The logic is not one of insufficiency. Productive nontotalization

> can also be determined in another way: no longer from the standpoint of finitude as relegation to the empirical, but from the standpoint of the concept of *play*. If totalization no longer has any meaning, it is not because the infiniteness of a field cannot be covered by a finite glance or a finite discourse, but because the nature of the field—that is, language and a finite language—excludes totalization. This field is in effect that of *play*, that is to say, a field of infinite substitutions only because it is finite, that is to say, because instead of being an inexhaustible field, as in the classical hypothesis, instead of being too large, there is something missing from it: a center that arrests and grounds the play of substitutions. One could say—rigorously using that word whose scandalous signification is always obliterated in French—that this movement of play, permitted by the lack or absence of a center or origin, is the movement of *supplementarity*. One cannot determine the center and exhaust totalization because the sign which replaces the center, which supplements it, taking the center's place in its absence—this sign is added, occurs as a surplus, as a *supplement*. The movement of signification adds something, which results in the fact that there is always more, but this addition is a floating one because it comes to perform a vicarious function, to supplement a lack on the part of the signified. (Derrida 1978, 289)

22. The arguments made via the form of the text are meant to reiterate its surface arguments. Most concretely, the text is structured to provide a forum for a hearing of the Bhopal case as a still unsettled dispute, a dispute that cannot be localized, a dispute that challenges representations of the late-twentieth-century world as harmonically interconnected, a dispute that recalls what environmentalism must be to assume "continuing liability" for the Bhopal disaster.

The text borrows many of the strategies of structuralism to show how things connect and operate relationally. But there is a difference. Structuralism privileges commensurability in order to demonstrate systemic integration. Collating disaster cannot be so clean. To advocate Bhopal, a text must trace intricate connections. But it also must recognize what gets lost in translation, what escapes codification despite expert design, and what exceeds adjudication by formal law. A text of disaster may, then, be imagined as a semiotic grid that has exploded—its pure formality contaminated by corrosive elements. Runaway reaction sets in; the grid's contents are released, dispersed, become asphyxiants. Disaster manifest.

23. Recall Italo Calvino's *If on a Winter Night a Traveler* (1981), with its novels within the novel, films within the novel, still photographs within the novel. The trope is not aggregative; the different narrative levels interrupt rather than confirm one another. Calvino develops his perspectives on competing levels of reality within (and in the production of) literary texts in *The Uses of Literature* (1986)—comparing what happens when different levels of reality remain separate to what happens when they meld. Separation is maintained in Shakespeare's *A Midsummer Night's Dream;* the world of the aristocratic court, the world of the supernatural characters, and the rustic world of Bottom and his friends (which merges with the animal kingdom) coexist, but with little mutual determination. The effect is Comic. The fusion of different worlds is the story of Hamlet, which Calvino describes as a whirlpool that sucks in all levels of reality. There is a play within the play. There is the realistic, day-to-day narrative. There is Hamlet's reflection. There is Hamlet's self-presentation, as mad. They congeal, and the effect is Tragic.

24. George Marcus has described how such movement recurs in research attempting to link local and global processes, suggesting that the conundrums I describe are not unique to those instances when politicization of scholarship is manifest and purposeful. His concept of "circumstantial activism" describes how contemporary research is often caught within multiple, competing interpretations, disallowing any possibility of detached observation or articulation. Politics and scholarship collide not only when affiliation with a particular movement or cause is overt, but also in the very process of research engagement in a world system wherein language games do much of the work of securing power differentials. From the start, contemporary research is embroiled in interpretive multiplicity, inflecting any and all articulation with issues of power (Marcus 1995).

Chapter One

1. Opinion and Order, In Re: Union Carbide Corporation Gas Plant Disaster at Bhopal India in December 1984, 846 F. Supp. 842 (S.D.N.Y. 1986).

2. Advocacy can be thought of in terms of translation. Gayatri Spivak suggests how, articulating in the process the links among risk, safety, and systems that are major concerns of environmentalism:

> How does the translator attend to the specificity of the language she translates? There is a way in which the rhetorical nature of every language disrupts its logical systematicity. If we emphasize the logical at the expense of the rhetorical interferences, we remain safe. "Safety" is the appropriate term here, because we are talking of risk, of violence to the translating medium. . . . Logic allows us to jump from word to word by means of clearly indicated connections. Rhetoric must work in the silence between and around the words in order to see what works and how much. The jagged relationship between rhetoric and logic, condition and effect of knowing, is a relationship by which a world is made for the agent, so that the agent can act in an ethical way, a political way, a day-to-day way; so that the agent can be alive, in a human way, in the world. Unless one can construct a model of this for the other language, there is no real translation. . . . The relationship between logic and rhetoric, between grammar and rhetoric, is also a relationship between social logic, social reasonableness and the disruptiveness of figuration in social practice. These are the first two parts of our three-part model. But, then, rhetoric points at the possibility of randomness, of contingency as such, dissemination, the falling apart of language, the possibility that things might

not always be semiotically organized. . . . The dominant groups' way of handling the three-part ontology has to be learned well—if the subordinate ways of rusing with rhetoric are to be disclosed. (Spivak 1993, 180–81, 186–87)

Rayna Rapp's description of genetic counseling as a "translation industry" suggests how exploring translation between different discourses—particularly involving technoscience—illuminates cultural, class, and power differentials. Rapp also provides particularly acute analyses of how diagnostic "information" accrues legitimacy and becomes the grounds for far-from-neutral categorization schemes (Rapp 1991, 1992, 1994, 1995).

3. Spivak elaborates on ways narrative forms affect an author's ability to tell "the whole truth." She does not argue that either first- or third-person narration should be privileged, that microlevel analysis obviates the need for macro theorization, or that literature can do the work of politics. Instead, she reminds how oscillation between forms carries the critical project. No one form can tell the whole truth, but forms cannot be merely aggregated either. Instead, we must bounce between forms, allowing a return to the seemingly comprehensive narration of the third person "with its ground mined under." The challenge, according to Spivak, is to work within the exigencies of a given form, within an effort to "strive moment by moment to practice a taxonomy of different forms of understanding, different forms of change, dependent perhaps on resemblance and seeming substitutability—figuration—rather than on the self-identical category of truth" (Spivak 1987, 88). Spivak is elaborating on Margaret Drabble's *The Waterfall,* a story about Jane's love affair with James, her cousin Lucy's husband. Rivalry between women is not the theme, though it does create one of the story's many double binds. Instead, Drabble explores the "conditions of production and determination of microstructural heterosexual attitudes within her chosen enclosure." The question, in short, is "Why does love happen?" The literal enclosure is a middle-class home where Jane delivers a baby, without the father's presence, by choice. Lucy and James care for Jane, and something new begins. James is given "the problem of relating to the birthing woman through the birth of 'another man's child.' " The relationship cannot be legalized, or defined in terms of James's "possessive ardor toward the product of his own body," since the child is not "his." Jane is left to narrate the story, which cracks as it shifts from third- to first-person narration. Jane has to admit that her third-person narration hasn't really told the truth, or hasn't told enough, and now must acknowledge that the qualities she has staked out "are interchangeable: vice, virtue: redemption, corruption: courage, weakness: and hence the confusion of abstraction, the proliferation of aphorism and paradox." Propelled by the ill logic and double binds of her affair with James, Jane has to shift between first- and third-person narration to make her story hold. But, in the end, Jane gives up, failing to engage the inadequacy of conventional rules and modes of description as an opportunity to create new understandings of what is sensible and virtuous. According to Spivak, "The risk of first person narration proves too much for Drabble's fictive Jane. She wants to plot her narrative in terms of the paradoxical category—'pure-corrupted love'—that allows her to *make* a fiction rather than try, *in* fiction, to report on the unreliability of categories." Had Jane persisted, dwelling within the double binds rather than trying to avoid them, she could, perhaps, have forged a narrative form particularly appropriate for representing the unconventional social collaboration in which she was entangled. Spivak points to the implications: "To return us to the detached and macrostructural third person narrative after exposing its limits could be an aesthetic allegory of deconstructive practice" (1987, 89).

4. What I describe here is not unlike what interdisciplinarians confront within academia. In Science and Technology Studies, for example, protocols for evaluating rigor differ, often dramatically. Modes of constructing meaning are vehemently contested. Like the ethnographer-in-advocacy, interdisciplinarians often feel that their expertise has become so dispersed as to become almost useless. They find themselves embedded within and thus responsible for multiple discourses, none of which can be handled as thoroughly as specialization would demand. Sensibilities of competence and accomplishment seem forever forestalled, ad infinitum.

5. Recognizing that we have no universal rule of judgment that can legitimately adjudicate between diverse genres of discourse, Jean-Francois Lyotard asks that we "bear witness to the differend," developing language technologies that acknowledge "1) the impossibility of avoiding

conflicts (the impossibility of indifference) and 2) the absence of a universal genre of discourse to regulate them (or, if you prefer, the inevitable partiality of the judge): to find, if not what can legitimate judgment (the 'good' linkage), then at least to save the honor of thinking" (Lyotard 1988 [1983], xii).

6. In an active voice, the advocate would bear gifts. In the middle voice, the advocate would distribute gifts she herself had received. Derrida explains the philosophical context:

And we will see why that which lets itself be designated *differánce* is neither simply active nor simply passive, announcing or rather recalling something like the middle-voice, saying an operation that is not an operation, an operation that cannot be conceived either as passion or as the action of a subject on an object, or on the basis of the categories of agent or patient, neither on the basis of nor moving toward any of these *terms*. For the middle voice, a certain nontransitivity, may be what philosophy, at its outset, distributed into an active and a passive voice, thereby constituting itself by means of this repression. (Derrida 1986 [1972], 9)

7. Derrida, praising Marcel Mauss for actually performing *The Gift* and thus an ethics configured by involvement (indebtedness) rather than by the assertion of a guaranteed truth (Capital), articulates these elements of the conventional view of advocacy:

Which implies that by good ethical standards—and here the good ethical standards of scientific discourse—one must not take sides unless one is able to do so neither in the dark, nor at random, nor by making allowance for chance, that is, for what cannot be thoroughly anticipated or controlled. One should only take sides rationally, one should not get involved beyond what analysis can justify and beyond what can accredit or legitimate the discourse in which the taking of sides, the *parti pris* or bias is stated. Otherwise, one pays with words or *on se paie de mots,* as we say in French [one gets paid in words, i.e., one talks a lot of hot air], by which one understands that words in this case are simulacra, money without value—devalued or counterfeit—that is, without gold reserves, or without the correspondent accrediting value. (Derrida 1992 [1991], 61)

8. For an excellent review of the exclusionary effects of essentialism, see Angela Harris's "Race and Essentialism in Feminist Legal Theory" (1991).

9. Hayden White provides a succinct description of the organicist impulse and its construction of the micro-macro relationship:

Organicist world hypotheses and their corresponding theories of truth and argument are relatively more "integrative" and hence more reductive in their operations. The Organicist attempts to depict the particulars discerned in the historical field as components of synthetic processes. At the heart of the Organicist strategy is a metaphysical commitment to the paradigm of the microcosmic-macrocosmic relationship; and the Organicist historian will tend to be governed by the desire to see individual entities as components of the processes which aggregate into wholes that are greater than, or qualitatively different from, the sum of their parts. Historians who work within this strategy of explanation . . . tend to structure their narratives in such a way as to depict the consolidation or crystallization, out of a set of apparently dispersed events, of some integrated entity whose importance is greater than that of any of the individual entities analyzed or descried in the course of the narrative. (White 1973, 15)

Note, however, Donna Haraway's reminder that organicism, while a recognizable mode of putting things together, is not one thing—as evident in the work of different scientists who have relied on it as a conceptual tool (Haraway 1976, 39–40).

10. In his deconstruction of structuralism (particularly in anthropology), Derrida elaborates an alternative possibility:

Perhaps something has occurred in the history of the concept of structure that could be called an event . . . its exterior form would be that of a *rupture* and a redoubling. Nevertheless, up to the event which I wish to mark out and define, structure—or rather the structurality of the structure—although it has always been at work, has always been neutralized or reduced, and this by a process of giving it a center or of referring it to a point of presence, a fixed origin. The function of this center was not only to orient, balance and organize the structure—one cannot in fact conceive of an unorganized structure—but above all to make sure that the organizing principle of the structure would limit what we might call the *play* of the structure. By orienting and organizing the coherence of the system, the center of the structure permits the play of its elements inside the total form. And even today the notion of a structure lacking any center represents the unthinkable itself.

Nevertheless, the center also closes off the play which it opens up and makes possible. As center, it is the point at which the substitution of contents, elements or terms is no longer possible. At the center, the permutation or the transformation of elements (which may of course be structures enclosed within a structure) is forbidden. At least this permutation has always remained *interdicted* (and I am using this word deliberately). Thus it has always been thought that the center, which is by definition unique, constituted that very thing within a structure which while governing the structure, escapes structurality. This is why classical thought concerning structure would say that the center is, paradoxically, *within* the structure and *outside* it. The center is at the center of the totality, and yet, since the center does not belong to the totality (is not part of the totality), the totality *has its center elsewhere*. The center is not the center. The concept of centered structure—although it represents coherence itself, the condition of the *episteme* as philosophy or science—is contradictorily coherent. And as always, coherence in contradiction expresses the force of a desire. The concept of centered structure is in fact the concept of a play based on a fundamental ground, a play constituted on the basis of a fundamental immobility and a reassuring certitude, which itself is beyond the reach of play. And on the basis of this certitude anxiety can be mastered, for anxiety is invariably the result of a certain mode of being, implicated in the game, of being caught by the game, or being as it were at stake in the game from the outset.

If this is so, the entire history of the concept of structure, before the rupture of which we are speaking, must be thought of as a series of substitutions of center for center, as a linked chain of determinations of the center. Successively, and in a regulated fashion, the center receives different forms or names. The history of metaphysics, like the history of the West, is the history of these metaphors and metonymies. Its matrix—if you will pardon me for demonstrating so little and for being so elliptical in order to come more quickly to my principal theme—is the determination of Being as *presence* in all sense of this word. (Derrida 1978, 278–79) (emphasis in original)

11. For critical readings of the figure of Gandhi, see Shahid Amin's "Gandhi as Mahatma" (1988) and Gyanendra Pandey's "Peasant Revolt and Indian Nationalism" (Pandey 1988).

12. A classic feminist articulation of the paradox here is Donna Haraway's: "Cyborg politics is the struggle for language and the struggle against perfect communication, against the one code that translates all meaning perfectly, the central dogma of phallogentrism. That is why cyborg politics insist on noise and advocate pollution" (Haraway 1991 [1985], 176).

13. Note, for example, Kevin Kelly's description of the infamous model of the global environment produced by the Club of Rome:

The Limits to Growth model treats the world as uniformly polluted, uniformly populated, and uniformly endowed with resources. This homogenization simplifies and uncomplicates the world enough to model it sanely. But in the end it undermines the purpose of the model because the locality and regionalism of the plant are some of its most striking and important features. Furthermore, the hierarchy of dynamics that arise out of differing local dynamics provides some of the key phenomena of Earth. The Limits to Growth modelers recognized the power of subloops—which is, in fact, the chief virtue of the Forrester's system dynamics underpinning the software. But the model entirely ignores the paramount subloop of the world: geography. A planetary model without geography is . . . not a world. Not only must learning be distributed throughout a simulation; *all* functions must be. It is the failure to mirror the distributed nature—the swarm nature—of life on Earth that is this model's greatest failure. (1994, 445)

14. The double bind I describe here is theorized in Derrida's argument that to critique Reason one cannot merely oppose it, which would be a call for an even higher rationalism. One must work alongside Reason's destructive currents, strategizing "something more to say when all is said and done." The double game is possible because of the duplicities within Reason (here, official settlements) itself. One reason for making this connection is to suggest how the reading practices taught by poststructuralism can be embodied in ethnographic work. Another point is more specific, relating to ways advocacy as a social practice can be understood as the kind of double game Derrida describes, wherein struggles for change work right up against what most constrains change: disassembly from within, so to speak. Derrida's own articulation is particularly relevant: "Since the revolution against reason, from the moment it is articulated, can operate only *within* reason, it always has the limited scope of what is called, precisely in the language of a department of *internal* affairs, a disturbance" (1978, 36).

15. Feminist legal theorists have explicated law's demand for equivalencies with particular force. See Cornell's *Transformations: Recollective Imagination and Sexual Difference* (1993).

16. Similarly competing demands recur within feminism, suggesting how the central problematics of this book are feminist problematics. The oppositions described here recur with particular antagonism in debates between "cultural feminists" and "post-structural feminists." Many insist that the paradoxes I describe must be resolved if feminism is to secure political force. I take a different tack, suggesting that temporalizing ethicopolitical judgment allows one to work productively within paradox. My point is not to enshrine paradox, but to acknowledge that paradox often cannot be "man-handled" away. This is not to say that political work within paradox is straightforward. To the contrary. Old blueprints and metanarratives have little explanatory force. Interpretation must be continual, and continually interrupted by "situational particularities." And one must find ways of being accountable for the performative effect of one's interpretive technique. For a clear articulation of the need to resolve the paradox, see Linda Alcoff's "Cultural Feminism versus Post-Structuralism: The Identity Crisis in Feminist Theory" (1994, 107).

17. My emphasis on situating ethicopolitical judgment in time responds to Derrida's critique of the metaphysics of presence. Drucilla Cornell provides a succinct articulation: "Differánce can be understood as the 'truth' that being is only represented in time; therefore, there can be no all-encompassing ontology that claims to tell us the truth of all that is. . . . Differánce, to use Derrida's word, temporizes. It breaks up the so-called claim to fullness of any given reality, social or otherwise, because reality only 'presents' itself in intervals so that, to return to Luhmann, there can no longer be sufficient continuity between each present with other presents" (1992, 128).

18. Note that the root of *virtue* is the same as the root of *virile:* pertaining to, characteristic of, or befitting a man. By implication, feminine access to virtue is only by way of the negative: women can lose their virtue by losing their virginity—which suggests that women have virtue only to the extent they are not connected to the rest of the world. Virtue, it seems, needs to be retroped—to provide access to a morality that works within the world, rather than only from a detached distance.

19. Note Spivak's displacement of the conventional assumption that pleasure can be equated with ease by describing how the womb, conceived as workshop, becomes a place where pain is a "normal" site of creative production: "I would like to suggest that in the womb, a tangible place of production, there is the possibility that pain exists within the concepts of normality and productivity. (This is not to sentimentalize the pain of childbirth.) The problematizing of the phenomenal identity of pleasure and unpleasure would not be operated only through the logic of repression. The opposition of pleasure-pain is questioned in the physiological 'normality' of woman" (1987, 80).

Chapter Two

1. In 1976, the Occupational Safety and Health Administration (OSHA) revealed that workers at Velsicol's Bayport, Texas, plant had developed serious central nervous system disorders due to

exposure to Phosvel. Other workers referred to those visibly injured as "Phosvel Zombies" because they lost their coordination and their ability to think and speak clearly (Weir and Schapiro 1981, 23). In 1977, workers at an Occidental plant in California learned that many of them were sterile due to exposure to DBCP—1,2,-Dibromo-3-chloropropane—used to kill soil-dwelling worms that attack bananas (Weir and Schapiro 1981, 20). Congress held hearings and issued a report on export products banned by U.S. regulatory agencies. The General Accounting Office (GAO) documented how banned pesticides end up back on tables in the United States. The GAO report said that 10 percent of all food imports were officially contaminated with pesticides and that half of that 10 percent was nonetheless distributed to consumers without any warning. The GAO also found that the U.S. pesticide market was nearing saturation, compelling ever more interest in export potential. During the decade ending in 1984, Africa was expected to quintuple its pesticide use. In Africa, as in India, the threat of hunger and famine provided justification (U.S. General Accounting Office 1979).

2. OSHA's citation of Carbide included 130 instances of 8 willful violations and 28 serious violations. The first willful violation was failing to provide respiratory protective equipment to employees who were required to determine the source of phosgene gas leaks by sense of smell. The exposure standard for phosgene, a chemical warfare agent used during World War I, is 0.1 ppm over eight hours. The odor threshold for phosgene is between .125 and 1.0 ppm. The remaining 129 instances of willful violations were for failing to record occupational injuries and illness in an OSHA-designated log. Company officials insisted that in all cited cases the injuries had been recorded in company files. The twenty-eight serious violations related to "the company's ability to detect leaks of carbon monoxide, chlorine and phosgene; design deficiencies or work practices that contributed to the probability of major releases of chemicals; failure to provide adequate backup systems in the event of a major chemical spill or release; and potential fire and explosion hazards." Specific citations included "corrosion and deterioration of nuts and bolts on components of a chlorine tank; failure to plug an open-ended valve on an acrolein line to prevent failure or accidental valve opening; and metals susceptible to fracture and incompatible with ethylene oxide used in the casing and nuts of transfer pumps for two oxide storage tanks" (Dembo, Morehouse, and Wykle 1990, 84).

3. All quotes from the August 1985 meeting at West Virginia State College and all quotes from the pro-Carbide rally are drawn from a video titled *Chemical Valley,* produced in 1991 by Appalshop (Pickering and Johnson 1991). My own fieldwork in West Virginia was limited to three brief trips—in the summer of 1994, the fall of 1994, and the summer of 1995.

4. SARA is an amendment to the Comprehensive Environmental Response, Compensation, and Liability Act (CERCLA), originally passed in 1980 as a response to disasters such as Love Canal. CERCLA empowered the EPA to order parties responsible for hazardous sites to undertake necessary cleanup measures. Alternatively, the EPA would undertake the cleanup and then bill responsible parties. The "Superfund," partially underwritten by taxes on the petrochemical industry, was intended to cover initial expenses; funds recovered from responsible parties were supposed to be returned to it.

5. The more general political-economic context has not helped either. Michael Kraft and Norman Vig explain:

> In constant dollars (that is, adjusting for inflation), the total authorized by the federal government for natural resource and environmental programs actually declined slightly (2.7%) between 1980 and 1994. The declines were much steeper in some areas, particularly pollution control, where spending fell by 21 percent for the same period; the budget had fallen by a far larger amount before recovering somewhat under the Bush and Clinton administrations. Nonetheless, it is surprising to see that the federal government was actually authorizing considerably *less* for pollution control in 1994 than it had fourteen years earlier. (Kraft and Vig 1997, 18)

6. Worst-case scenarios are required by the Risk Management Programs (RMP) rule, section 112[r] of the Clean Air Act of 1990. The 1990 act was actually an amendment that built on previous environmental legislation. The Air Quality Act of 1967 introduced the regional approach to

air-quality standards that is still relied on today. The 1967 act was significantly rewritten in 1970 and then amended again in 1977. The 1970 Clean Air Act favored a health-based, technology-forcing approach that allowed emissions standards to be set at levels deemed necessary to protect health regardless of the current availability and cost of control technology. New sources of pollution, however, were to be regulated with "best available technology." Repeated extension of deadlines for meeting mandated standards further eroded the technology-forcing approach. The 1990 amendments made the shift to a technology-driving approach complete. A new idiom was created for evaluating toxic production, signaled by acronyms like MACT (maximum available control technology, required at all existing sources with the goal of reducing toxic air emissions by 74 percent by the year 2000), RACT (reasonably available control technology, required in "nonattainment" areas), and LAER (lowest achievable emission rate, a strict standard to be applied to new and modified pollution sources).

The 1990 amendments also mandated that new and modified pollution sources "offset" their emissions with reductions from other sources or acquire pollution credits from companies that had reduced their emissions below legally permitted levels. Localized markets for trading emissions had existed previously; the 1990 amendments created a national market. The first auction of EPA pollution allowances was run by the Chicago Board of Trade on March 29, 1993. Each allowance unit sold permitted the discharge of one ton of sulfur dioxide per year. More than 150,000 allowances were sold the first day of trading, at prices ranging from $122 to $450 per allowance (Bryner 1996; Melnick 1983; Valente and Valente 1995).

Selling pollution on the world's largest commodity market has not pleased everyone. Some oppose market mechanisms because they do not reduce the overall amount of pollution permitted by law. Others object to emissions trading because it signifies the *right* to pollute. Toxic emissions are treated like property rights that can be bought, sold, and traded; pollution becomes an entitlement. The 1990 amendments do specify that emissions allowance provisions do not amount to a property right that could be constitutionally protected, should the legislation be repealed at a future point (Valente and Valente 1995, 340). But this doesn't change the way toxic pollution has been renormalized.

Nor does it change what people refer to as the "moral overtones" of market incentives. Market incentives, as many interpret them, imply that production decisions should remain in the private sector. They also put negotiations over risk into the hands of experts, now including financial as well as technical experts. And they cannot guarantee that overall pollution loads will be reduced. "Trading under a cap" establishes limits, but does not ensure that particularly dirty areas will ever be clean. And the catastrophic risk that accompanies "routine emissions" is not addressed at all. The possibility of disaster in regions like the Kanawha Valley remains out of the loop, offset entirely.

7. A call for shelter in place was made by Rhone-Poulenc in February 1996. Afterward residents filed a class action lawsuit alleging that Rhone-Poulenc had a negligent attitude toward safety and that faulty maintenance had caused the leak. Local newspapers covered the public meeting that followed. Hal Foster, Rhone-Poulenc's plant manager, was quoted as having told the audience: "If every time a shelter-in-place is called, we're going to get hit with a lawsuit, we're going to be very hesitant to call a shelter-in-place. . . . [The class action lawsuit gives Rhone-Poulenc] a strong financial incentive not to call a shelter-in-place" (Messina 1998). Foster later denied these statements, under oath. In August 1998, Ken Ward, a reporter for the *Charleston Gazette,* was subpoenaed to testify regarding his coverage of the meeting. Controversies over Ward's First Amendment rights have added still more complexity to environmental problems in West Virginia (Messina 1998).

8. The *Houston Chronicle* reported that Brown & Root, Houston's largest private employer, was importing workers from Alabama, Texas, and Virginia. Brown & Root said that about 60 percent of the workers on the three job sites in the Kanawha Valley were from West Virginia (Sixel 1993, 10-A).

9. In April 1994, the National Labor Relations Board found that Brown & Root had discriminated against workers who listed union associations on job applications. Brown & Root was or-

dered to post a notice at all of its work sites in West Virginia stating that the company will not refuse to hire union activists (Ward 1994b).

10. The mortality figure I have seen most frequently in descriptions of Hawk's Nest is 700. David Dembo, Ward Morehouse, and Lucinda Wykle explain part of the instability of the figure:

> Union Carbide obtained the right to all documents from the plaintiffs' attorneys as part of the out-of-court settlement agreement, including figures on the number of men who died. They are still the only ones with any accurate figures on the death toll of workers. Estimates range from 65 (the figure given by the President of Rinehart & Dennis [the construction firm contracted to do the work]) to 2,000 (an opinion given by Senator Holt at the Congressional Hearings in 1936). Union Carbide has only admitted to 109 total deaths from the tunnel project, 66 from respiratory disease, four times the expected number of respiratory disease death rate in Fayette County at that time. (1990, 29; see also Cherniack 1986, 90–91)

11. The economic boom brought to West Virginia by World War I is what drew one side of my own family there, after emigrating from Lebanon. Druze immigrants in particular settled both in Charleston and in coal and chemical production towns in the hills. Most owned small grocery stores, restaurants, bars, or similar small businesses. My Druze grandfather married a "girl from the coalfields" ("an American") whose family had moved to Charleston after her father was killed in the mines. Many members of that side of the family worked for Union Carbide. I spent a few weeks in West Virginia every summer as I was growing up. The connection between this history and this research project was not purposeful. As I describe elsewhere, I did not plan to study in Bhopal, much less study Union Carbide in particular. This history does, however, inflect my sense of the continuing tragedy of West Virginia. The ironies of viewing the world through the eyes of a company town (and union) are familiar, threaded with memories of the terrible stink of chemicals and of clothes that blackened as they hung on the line. Another fortuitous connection between this research and personal history is the Gulf Coast of Texas, where I grew up—living in Houston, driving each weekend to Baytown (where my paternal grandparents lived), through the beautiful lights (and terrible stink) of the world's largest petrochemical corridor.

Chapter Three

1. For a discussion of how the problem of relating parts to wholes recurs in ethnography, see John and Jean Comaroff's essay "Ethnography and the Historical Imagination" (Comaroff and Comaroff 1992, 17).

2. The explosion and fire at Seadrift were caused by overpressurization of an ethylene oxide production unit. When the oxide unit column blew, a large piece of shrapnel hit the pipe rack and ruptured lines containing methane and other products.

3. The figure here for the proposed fine was drawn from George Draffan's research compendium on Union Carbide (Draffan 1994); Louis Ember reports that OSHA announced proposed penalties against Union Carbide of $2,817,500 (Ember 1995). OSHA levied these fines under its egregious policy, which allows $25,000 for each violation. In 1986, Union Carbide was the first facility cited by OSHA under its egregious policy for violations at their facility in Institute, West Virginia.

4. Ember also notes that a check with EPA's Region 6 office found no air, water, or land (hazardous wastes) violations over a ten-year period for the facility. The plant was fined $8,000 in 1987 by the Texas Natural Resource Conservation Commission (TNRCC, formerly the Texas Water Commission) for unauthorized hazardous discharges and an inadequate groundwater monitoring system at its North Landfill. TNRCC spokesman Terry Hadley said that the company is now following an extensive compliance plan. In January of 1991, the plant was fined $1,000 for violating the Emergency Planning Community Right-to-Know Act. The law requires companies to report within twenty-four hours any release over 1,000 pounds. Carbide reported, after the fact, a release of 7,400 pounds of cyclohexane over a five-day period. In addition to paying the fine,

the company donated $1,000 in equipment to the Calhoun County Local Emergency Planning Committee.

5. One of the interviews with Munoz was made into a video by environmental activist Josh Karliner, titled *Setting the Record Straight* (1994). I interviewed Munoz in November 1995 at his home in San Francisco.

Chapter Four

1. I was a typist, organizer, and editor of Chouhan's book. Excerpts from the book included here are from the last draft I submitted to Apex Press, which published the book. Ward Morehouse and others at Apex Press did the final edits prior to publication.

2. Industrial workers are certainly not the "the poorest of the poor" in India. In urban areas, this name goes to the masses of workers in the "informal sector." The informal sector has extraordinary needs for support. And middle-class activists have a well-developed rhetorical and social style for providing it. By the 1980s, public culture supported the habit. Workers were coming to be viewed as more privileged than ever as a "new agrarianism" began to receive national attention. The deceleration of the 1980s also led to talk about an "informalization of employment," whereby cheap labor could be exploited, while circumventing restrictive laws.

Lloyd Rudolph and Susanne Rudolph elaborate:

> If some workers in the organized sector seem more privileged than others, the issue of whether workers in the organized economy as a whole were more privileged than those in the unorganized economy had become by 1980 a central issue in national politics. . . . Rural-urban difference did not emerge as a live national issue until the second half of the seventies, when the politics of the new agrarianism began to take shape. Farmers' agitations in Maharashtra, Tamil Nadu, Karnataka, Gujarat, and Punjab and nation kisan rallies in the early eighties made it plain that the issue of the organized sector's privileged position had come of age and would be central to the domestic politics of that decade. (1987, 265–67)

Chapter Five

1. The settlement was challenged from within the government of India itself. When the settlement was announced, opposition members of parliament went on strike. Then the Congress (I) lost the elections, and the National Front came to power under the leadership of V. P. Singh. In January 1990, one month after the elections, the *New York Times* carried a quote from Singh denouncing the settlement with the argument that "no one has the right to bargain with the corpses of people" (Hazarika 1990, A9). His critique of the settlement was based on the insufficiency of the settlement amount and the injury categorization scheme through which the amount was said to have been calculated. This critique focused on both the process and the outcome of law, but not its institutional structure.

The new attorney general, Soli Sorabji, had a new role to play, despite his involvement in the case from the outset. Earlier, as a private lawyer, Sorabji had argued the victims' case in challenging the Bhopal Act, challenging, in the process, the presumption that the Indian government could and must act as legal guardian of victims. In his role as attorney general in the National Front government, Sorabji *assumed* the role of guardian.

The effects of changes in Sorabji's argument, and structural position, were soon clear: interim relief finally became available to victims, through government of India funds, rather than through funds available through the settlement award. The distribution of interim relief was poorly designed, inefficient, and corrupt. But it helped people survive.

2. It is important to note that the directive principles, while encoded with a socialist impulse, have provided space for a range of coercive laws, including the Preventive Detention Act, the Maintenance of Internal Security Act, and the Unlawful Activities Prevention Act (Thakur 1995, 51).

3. This announcement was covered by the *Indian Express,* February 2, 1989, in an article titled "Bangalore Scientists to Fight Carbide Deal in Court."

4. This subheading borrows the title of an introduction to an important collection of essays

edited by Ashis Nandy and published in 1988, under the title *Science, Hegemony and Violence: A Requiem for Modernity* (1988). The critiques expressed in this volume circulated widely in the voluntary sector during the years I was in India, indicating, among other things, the porosity between academic and "activist" domains of work in India.

5. The study *Against All Odds* found the process of injury assessment followed by the official Directorate of Claims faulty on multiple counts: (1) A claimant is considered "gas-exposed" only if able to produce medical documents for the postexposure period. (2) Categorization depends on an arbitrary scoring system that converts signs and symptoms into numbers, which are imposed on a hierarchical ranking of body systems. No attempt is made to diagnosis the disease from which the claimant is suffering, rendering the records useless for the purpose of prognosis and rehabilitation. (3) Categorizations are based on minimal investigations, which partially measure respiratory function and ignore other body systems. (4) X-rays, pulmonary function tests, and exercise tolerance tests were done on very few people due to the government's argument that "it is just not practical to subject every claimant to these time consuming investigations in mass operations like this." (5) Translation of "injury" into "disability" is arbitrary and subjective due to the failure to include data on occupation. (6) A person would be categorized as "permanently injured" only if his or her health was worse than in the period immediately following exposure. *Against All Odds: The Health Status of Bhopal Survivors* was coordinated by Dr. C. Sathyamala, Dr. Nishith Vohra, and K. Satish with technical help from the Centre for Social Medicine and Community Health, Jawaharlal Nehru University (Centre for Social Medicine and Community Health 1989).

6. For a discussion of abortion in relation to the availability of amniocentesis in India during the 1980s, see M. Kishwar's essay in *Man-Made Women: How Reproductive Technologies Affect Women* (1987). Rayna Rapp's essays on how amniocentesis has been configured in the United States provide critical insight on the social and cultural specificity of these debates (1991, 1992).

7. Research biases against women have been a consistent problem in research conducted by the Indian Council of Medical Research. Dr. Sathyamala's report states that "in the processing of claims, gynecological problems have not been given adequate weightage. ICMR has in fact categorically stated that they do not have any evidence to show that an increase in dysfunctional uterine bleeding, chronic cervicitis, non-specific leucorrhoea and pelvic inflammatory disease is gas related" (Medico Friends Circle 1990). In categorization data produced by the government of Madhya Pradesh, direct neglect of research on gynecological problems has been furthered by an evaluatory approach that hierarchically ranks primary body systems and thus both ranks fertility problems very low and denies the significance of multisystemic indicators. For a review of the literature on the health of women and children in Bhopal, see "The Health Situation of Women and Children in Bhopal: Final Report on the International Medical Commission on Bhopal 1994" (Eckerman 1996).

Chapter Six

1. Sambhavna Trust, *Sambhavna from Bhopal* (October 1996), translated from Urdu by Indra Sinha.

2. In India, "communalism" denotes sectarian bias against members of a religion other than one's own. Most often the term refers to antagonism between Hindus and Muslims, such as that which followed the partition of India and Pakistan in 1947; there are significant exceptions, such as the anti-Sikh communal riots that followed the assassination of Indira Gandhi in 1984 by her Sikh bodyguards.

3. Following the destruction of the Babri Masjid, twenty-one places in Madhya Pradesh were seriously affected by communal unrest. By official accounts, 161 people were killed in the state altogether, 139 of them in Bhopal (People's Union for Democratic Rights and Sanskritik Morcha 1993, 4).

Chapter Seven

1. Shrivastava cites a brochure titled "We Shall Overcome," published by the Directorate of Information and Publicity, Bhopal, November 1985.

2. For a intriguing description of the "middleness" of nineteenth-century reformers, and of the

Calcutta middle class in particular, see Partha Chatterjee's chapter titled "The Nationalist Elite" in *The Nation and Its Fragments* (1993, 35–75).

3. For a useful overview of the crisis described here, see Gail Omvedt's chapter titled "The Crisis of Traditional Politics" in *Reinventing Revolution: New Social Movements and the Socialist Tradition in India* (1993). The figures Omvedt gives to illustrate rising consumerism are as follows: "Refrigerator production rose from 150,000 in 1975 to a projected 900,000 in 1990; motor scooters and motorbikes from 172,000 to two million; television sets from 100,000 to four million; and automobiles from 46,000 in 1985 to 500,000 in 1990" (1993, 187).

4. Rajiv Lochan Sharma, a physician who worked with the Bhopal Group for Information and Action during the years I was in Bhopal, also left Bhopal in 1993—to work as one of four doctors who staff the miners' hospital in Dalli Rajhara. While in Bhopal, Sharma was extremely resistant to practicing allopathic medicine because of the abuses he saw in government hospitals in particular. The alternative context provided by Shaheed Hospital has given Sharma the opportunity to use and develop his biomedical skills.

5. The first year of activist work in Bhopal included the participation of people with widespread political and intellectual experience. A simple breakdown between "activists" and "academics" is impossible. Radha Kumar, for example, was known for her savvy and intense work on the ground in Bhopal. In 1983, Kumar had published a study of women's employment in the Bombay textile industry between 1919 and 1939, linking the declining numbers of women in manufacturing to colonialist discourses that discounted family forms unlike those of the British middle class (1983). Kumar continued to work on the Bhopal case from Delhi during the years I was in Bhopal. In 1993, Kumar published a book titled *A History of Doing: An Illustrated Account of Movements for Women's Rights and Feminism in India, 1800–1990* (1993).

6. Rajni Kothari is among India's most well known political scientists. He became overtly connected to "radical" approaches through his opposition to the State of Emergency imposed in India in 1975.

7. In December 1984, controversy among and over NGOs came to a head in government proposals to change the Foreign Contribution Regulation Act such that any group receiving foreign funds would have to have registered with the Ministry of Home Affairs. Controversy was reignited by a section of the Seventh Plan in 1986 on the role of "voluntary agencies"—which would have legitimated voluntary agencies in antipoverty programs, but also granted the government the right to oversee them—through a proposed "Council on Voluntary Agencies," which would formulate and implement a code of conduct, possibly to include such things as limits on salaries. Some NGO activists attacked those opposed to the proposals for not wanting supervision—and thus accountability. Those opposing the proposals insisted that they were "establishment" efforts to freeze dissent, or at least to paint it as foreign inspired.

8. Gail Omvedt, a political scientists and activist based in Maharashtra, has been critical of Lokayan itself for its elitism. Her critique of Lokayan is at least threefold. In her view, Lokayan's "anti-communism" made it insufficiently attentive to class and to political economy generally— and turned its work into reconstituted Gandhianism. Related to this was Lokayan's faith in "indigenous" alternatives, which Omvedt says suggests that traditional Indian society is relatively healthy and integrated. Third, Omvedt is critical of the Lokayan emphasis on building intermediary institutions—like Lokayan itself—for linking local-level struggles. In her view, Lokayan "tended to systematically overlook the way in which movements and activists had a broader scope and organization and an antisystemtic thrust." Omvedt contests the assumption that dispersed local-level struggles need elite intervention to become sufficiently linked to become a movement (1993, 193).

9. *In Search of Famine* won the Silver Bear Award at the Berlin Film Festival in 1981 and awards in India for the best feature film, best direction, best screenplay, and best editing. The script was reconstructed and translated by Samik Bandyopadhyay and published by Seagull Books (Sen 1985 [1983]).

10. The IMCB convened at the recommendation of the 1992 session of the Permanent People's Tribunal on Industrial and Environmental Hazards and Human Rights, organized by the International Coalition for Justice in Bhopal, a global network of public interest groups working to secure

justice for gas victims and to deter similar environmental and industrial disasters elsewhere. The work of the IMCB is spearheaded by the Bhopal Action Resource Center, a New York–based organization involved in numerous projects to support alliances among grassroots activists, workers, and concerned citizens responding to corporate misconduct.

11. The People's Health and Documentation Clinic, located near the Union Carbide plant, was providing medical care though a combination of allopathy, ayurveda, and yoga—in an attempt to move beyond the symptomatic approach relied on at government hospitals. The clinic was also monitoring gas-related deaths through a technique known as verbal autopsy, developed by the London School of Tropical Medicine. The staff of the clinic included two allopathic doctors, one ayurvedic doctor, and two yogis. For details on the clinic's accomplishments, see a 1998 report titled *The Bhopal Gas Tragedy, 1984 –?* (Sambhavna Trust 1998).

Chapter Eight

1. Imagine a woman of Bhopal standing before the judge of one of these claims courts. She embodies the dis-ease of trying to relate generality to particularity that shapes this book. Jacques Derrida articulates the double bind, in both general and particular terms:

> How are we to reconcile the act of justice that must always concern singularity, individuals, irreplaceable groups and lives, the other or myself *as* other, in a unique situation, with rule, norm, value or the imperatives of justice which necessarily have a general form, even if this generality prescribes a singular application in each case? . . . To address oneself to the other in the language of the other is, it seems, the condition of all possible justice, but apparently, in all rigor, it is not only impossible (since I cannot speak the language of the other except to the extent that I appropriate and assimilate according to the law of an implicit third) but even excluded by justice as law (*droit*), inasmuch as justice as right seems to imply an element of universality, the appeal to a third party who suspends the unilaterality or singularity of the idioms. . . . It is unjust to judge someone who does not understand the language in which the law is inscribed or the judgment pronounced, etc. We could give multiple dramatic examples of violent situations in which a person or group of persons is judged in an idiom they do not understand very well or at all. And however slight or subtle the difference of competence in mastery of the idiom is here, the violence of an injustice has begun when all the members of the community do not share the same idiom throughout. Since in all rigor this ideal situation is never possible, we can perhaps already draw some inferences about what the title of our conference calls "the possibility of justice." (1990, 949, 951)

2. Judge Deo, frustrated by the way negotiations had "bogged down in the din of diverse voices," located his authority to pass the interim relief order in the "inherent jurisdiction" of the court. Judge Deo acknowledged that his move was novel, but insisted that courts often actively work to preserve the rights of parties to litigation as it proceeds. He also referred to the decision by the Indian Supreme Court in the Shriram case, which he said opened up the possibility of an interim relief order by establishing broad principles of strict liability. Judge Deo further noted that Union Carbide had itself relied on the Shriram case in arguments for dismissal of the Bhopal litigation from U.S. courts. Carbide had argued that the Shriram case was evidence that the Indian legal system was sufficiently innovative to handle the complex issues surrounding the Bhopal disaster. Judge Deo insisted that Union Carbide "relied on this source of law in obtaining judgment on *forum-non-conveniens,* and is, therefore, bound by it." While there was no explicit precedent for interim compensation, Judge Deo insisted that the "law must grow to meet the problems raised by such changes . . . including the problems of industrialization." His concluding remarks raised an important question: "Can the gas victims survive till the time all the tangible data with meticulous exactitude is collected and adjusted in fine forensic style for working out the final amount of compensation with precision . . . ?" (Cassels 1993, 199).

3. The Madhya Pradesh High Court upheld the liability of Union Carbide for the Bhopal disaster, but reduced the amount of interim compensation. Justice Seth also delinked the payment

amount from liability, insisting that the order to pay interim relief was not tied to the merits of the case, as was held by Judge Deo. Instead, the interim compensation amount was to be conceived as payment of damages under the substantive law of torts. Referring to a previous judgment by the Indian Supreme Court, Judge Seth argued that the measure of damages payable had to be correlated to the magnitude and capacity of the enterprise.

4. For a thoughtful description of working as a sociologist within the antiwar movement, recollected with the insights of feminism, see Barrie Thorne's "Political Activist as Participant Observer." Thorne describes the epistemological conflict she confronted as she became aware that "the movement's ways of defining and interpreting experience ran counter to the more detached and routinizing perspectives I maintained as a sociological observer" (1983, 225). This does not lead Thorne to argue for a simple separation of politics and research:

> Comparing my experiences in the Resistance and in the feminist movement, I realized that the sociological imagination—the insight that can come from detachment, comparison and systematic analysis—should be distinguished from other components of the research role. Sociological understanding and information can be organized in various ways, including as part of movements for social change. For example, I believe my contributions to discussions of strategies and tactics in the Resistance (e.g. in our long debates about the efficacy of draft card turn-ins and draft counseling) were strengthened by my ability to think sociologically, and by the systematic observations I had made of the movement over time. However, putting these insights into a dissertation and journal articles, geared for a different audience, was less useful for the movement. (Thorne 1983, 233)

Thorne's academic publications have, however, been useful for my movement work, suggesting that while the political relevance of scholarly accounts may be spatially and temporally deferred, it does materialize.

5. My understanding of how ethnicity can become a resource if allowed to operate across multiple fields of reference has been greatly influenced by the teaching of Michael Fischer. See his essay titled "Ethnicity and Postmodern Arts of Memory" (1986).

6. Throughout my fieldwork in India, I was particularly influenced by the work of Gayatri Spivak. Her identification as a "feminist, Marxist, deconstructivist literary critic" appealed to my concern with many different approaches to politics and scholarship, all of which seemed to demand both commitment and critique. See *In Other Worlds: Essays in Cultural Politics* (Spivak 1987).

7. Operation Faith was a government-managed effort to process all methyl isocyanate remaining in the Bhopal plant following the gas leak on December 3. Many recognize Operation Faith as an attempt by the Indian government to resecure its own authority, both scientific and political. Despite repeated assurance that Operation Faith would be an exercise in complete control, at least 300,000 Bhopalis fled the city.

8. Even the stability of the lower figure has wavered, with some commentators insisting that 225,000 lived in gas-affected areas, while others have used a figure of 250,000. For one articulation of the critique described here (which uses the figure of 225,000), see Shiv Visvanathan, with Rajni Kothari, "Bhopal: The Imagination of Disaster" (Visvanathan and Kothari 1985, 52). Claude Alvares's phrasing of his critique suggests later destabilization of the 600,000 figure: "Though the affected population at that time was 250,000, as many as 700,000 fresh ration cards were liberally distributed and a sum of Rs. 2 crore was spent every month in this way on this scheme." Alvares also reports that exchange rates in 1984 were approximately Rs. 14 to U.S. $1 (1994, 116). Paul Shrivastava reports on this period as follows: "In the first six months after the accident, the government distributed about $8 million in free food, for the most part grain and rice, to both affected and unaffected areas. By October 1985, this total had increased to $13 million, but the food distribution ceased by the end of the year." Shrivastava's footnotes indicate that his source for these figures was an article published in the *Madhya Pradesh Chronicle* on December 28, 1985, titled "Call to Ascertain Total Casualties." See Paul Shrivastava, *Bhopal: Anatomy of a Crisis* (1987, 93, 156).

9. Shrivastava reports that as of a year following the gas leak, 85 percent of the people living in the thirty-six wards officially identified as gas affected had not received any financial assistance (1987, 94).

Chapter Nine

1. An extremely useful account of the early days of activism in Bhopal is Ravi Rajan's "Rehabilitation and Voluntarism in Bhopal" (1988). Rajan went to Bhopal to report on the first anniversary of the disaster and stayed for nine months to work with the Bhopal Group for Information and Action (BGIA) and other voluntary organizations. For a brief, but thorough, overview of the first ten years of struggle to respond to the Bhopal disaster, see Claude Alvares's "Afterward" to *Bhopal: The Inside Story,* by T. R. Chouhan (Alvares 1994). Alvares's "Afterward" covers each year following the gas leak, focusing on the effects various developments had on victims and how victims responded. A second "Afterward" by Indira Jaising (Jaising 1994) also provides a brief, but thorough, review of the years leading to the tenth anniversary, focusing on the "Legal Let-Down." The literature on the Bhopal disaster is now quite broad; these few references directly expand upon some of the issues I explore here.

2. Throughout this chapter, I use the collective "we" and "our" to designate the object of my analysis; my intent is not to suggest that all of my former housemates would concur with my description of what "we" were up to.

3. CPI–M refers to the Communist Party of India–Marxist, which split from the Communist Party of India in 1964. For a very useful comparison of the gender politics of CPI–M and Shramik Sangathana, a grassroots organization in Dhulia district, Maharashtra, see Basu (1992). Basu's account complicates the history of the CPI–M I learned in Bhopal and recount here.

4. The study "Evaluation of Some Aspects of Medical Treatment of Bhopal Gas Victims" was conducted June 27–July 3, 1990, by BGIA and the Indore branch of Socially Active Medicos. The study was based on interviews with 522 patients who visited the outpatient departments of Jawaharlal Nehru Hospital and Shakir Ali Khan Hospital during the study period and who had been prescribed medications.

5. The constant need to respond to official proclamations in Bhopal becomes more interpretable when one recalls how forcefully language has operated as an instrument of colonization. Aldon Nielsen articulates this reality well: "Indeed, one seldom acknowledged aspect of the oppressions of slavery has been that for hundreds of years African Americans have had to endure the unceasing discourses of whites, who would never stop talking to them" (1994, 5).

Chapter Ten

1. CCC (Communities Concerned About Corporations). Press release, December 3, 1994, Charleston, West Virginia.

2. For an excellent account of what is invoked by a unitary constitutionalist "we," see the introduction to Aldon Nielsen's *Writing between the Lines: Race and Intertextuality.* An opening passage deserves emphasis:

> American readers, educated under the state sign of oneness out of multiplicity, the sign given currency on our great seal, shape themselves into a people on state occasions represented by a Latin Motto that few of us can parse. The very sign of our unity is spoken over our heads in a language not our own. Yet we have made that sign coin of our realm and identify ourselves in the spending of it. We, the people, know what we mean when we say "we," or at any rate we think we know what is meant when, in the political celebrations of our peoplehood, our leaders say "we" in our place. Yet one has only to look at photographs of such events, the dedication of the Lincoln Memorial is one example, to see how difficult it has been for us to become a people. In aerial photography of the crowd that turned out to honor the Great Emancipator and the savior of our nation's wholeness, one sees a dark line of demarcation through the crowd marking the place where America divided against itself, where one America bordered on becoming another. (Nielsen 1994, 2)

3. This story about Lois Gibbs (without my speculative elaboration) is told by Michael R. Edelstein, in *Contaminated Communities: The Social and Psychological Impacts of Residential Toxic Exposure* (1988, 167).

4. A 1985 appellate decision upheld the award for loss of quality of life, but reversed the awards for emotional distress and medical surveillance. In 1987, the New Jersey Supreme Court upheld the quality of life award, reinstated the medical surveillance award, and confirmed the reversal of the emotional distress award. Michael Edelstein argues that these decisions set important precedents and that the only limiting precedent "involved the court's refusal to grant damages for enhanced health risk, in part, because the risk could not be quantified and declared 'reasonably probable' even if recognized as significant" (1988, 163).

5. The connection between race and toxics was the focus of a government report in 1983 (U.S. General Accounting Office 1983) and of a report by the United Church of Christ in 1987 (United Church of Christ 1987). Academic writing specifically focused on race and toxics also began to be published in the early 1980s. Robert Bullard has been a key figure (Bullard 1983).

6. Texans United had negotiated an agreement with ARCO, requiring the company to drill groundwater-monitoring wells adjacent to the community, to conduct off-site sampling with citizen participation, and to increase the number of chemicals to be monitored and the frequency of monitoring. ARCO refused to agree to an independent audit with citizen participation and to make public its hazard assessments.

7. I'm not sure which year these figures reference. I drew them (as well as the rest of this account) from an OCAW documentary released in 1988 titled *Locked Out: The Story of OCAW Local 4–620* (Bedford 1988).

8. I have heard Foreman's remarks repeated, almost verbatim, by environmental activists in both India and the United States—usually in the process of questioning the complicity of environmentalism with dominant institutions and ideology. Dowie locates Foreman's reference to famine in Ethiopia in an interview by Bill Devall, published in the Australian journal *Simply Living*. The article on AIDS as corrective to excess population was published in the May 1, 1987, edition of the journal *Earth First!*

9. Robert Reich explains:

> Symbolic analysts solve, identify and broker problems by manipulating symbols. They simplify reality into abstract images that can be rearranged, juggled, experimented with, communicated to other specialists, and then, eventually, transformed back into reality. The manipulations are done with analytic tools, sharpened by experience. The tools may be mathematical algorithms, legal arguments, financial gimmicks, scientific principles, psychological insights about how to persuade or to amuse, systems of induction or deduction, or any other set of techniques for doing conceptual puzzles. (Reich 1992, 178)

10. The term "knowledge worker" was invented by management guru Peter Drucker (1993) to describe a social class that has emerged within postcapitalist society.

11. A more extended version of this interview is published as "Citizens, Inc.: Bottom-Up Organizing in Bottom-Line Contexts" (Fortun 1998).

12. Henson's work in Latin America was for the Environmental Project on Central America (EPOCA), organized to link issues usually kept separate—U.S. foreign policy, the environmental impact of militarization, toxic dumping, human rights abuses, and others. Henson's job included grassroots outreach, the production of EPOCA's *Green Paper* series, and the organization of many conferences and speaking tours. Henson also played an important role in planning and managing the closing of EPOCA in 1990 because he thought shifting times called for a shift in focus for progressive work.

In 1991, Henson co-founded the Environment and Democracy Campaign, a joint project of the Highlander Center and the National Toxics Campaign (NTC). Its mission was to internationalize the work of NTC and the Highlander Center by arranging exchanges between activists from different countries. Many activists were brought to the Highlander Center's facility in Tennessee. A delegation of activists toured Eastern Europe to run environmental organizer training sessions. And a grassroots delegation was organized to participate in the 1992 United Nations Conference on Environment and Development (UNCED) in Rio de Janeiro. Henson helped organize an enormous *Grassroots Report for UNCED,* filled with personal testimony from activists around the country that effectively undercut official U.S. claims to have solved many of its environmental problems. He also organized a delegation of sixteen grassroots activists to go to Rio.

In the mid-1990s, Henson began building the Occidental Arts and Ecology Center (OAEC), in Occidental, California, north of San Francisco, into a new resource for grassroots communities. OAEC runs workshops on organic gardening and permaculture, on appropriate technology, and on arts projects with an ecological focus. Henson continues to run training programs for community organizers, with links to Corporations, Law and Democracy—the group that has generated the analysis of corporations so important to CCC.

13. Larry and Sheila Wilson have worked with CCC. I visited them in the summer of 1995, and they helped me conduct interviews with other members of Yellow Creek Concerned Citizens.

Chapter Eleven

1. Harrison actually begins his account in 1962. I elaborate here to provide context.

2. It is easy to lose sight of the many different business organizations working to formulate proposals for sustainable development, since many share the same language (and often labels for themselves). The charter signed at WICEM II was drafted by the International Chamber of Commerce. Similar effort is documented in *Changing Course: A Global Business Perspective on Development and the Environment* (Schmidheiny 1992). This book draws on the work of the Business Council for Sustainable Development, organized by Stephan Schmidheiny after he was asked to become business and industry advisor to Maurice Strong, secretary general of the Earth Summit. The brief case studies provided in *Changing Course* are useful; the tone is much more sober than Harrison's. Other international organizations' efforts in this direction include the Chemical Manufacturers Association's Responsible Care program and the Global Environmental Management Initiative (GEMI). Consultant Bruce Piasecki, whom I discuss in the next section, is a promoter of GEMI. Corporate executives apparently complain that there are now too many programs demanding their allegiance (Hoffman 1997).

3. Lele is quoting a speech by M. K. Tolba to the World Commission on Environment and Development in October 1984. Lele's review of the development, critique, and dilution of the concept of sustainable development is very useful. For an early critique addressing how some issues (for example, urbanization, the international political order) were left out of the equation, see A. Khosla's "Alternative Strategies in Achieving Sustainable Development" (1987).

4. E. Woolard (chairman of DuPont) published his views in "An Industry Approach to Sustainable Development" (1992). Sustainable development was the theme of a series of articles in the *Academy of Management Review* in 1995 (vol. 20). See particularly P. Jennings and P. Zandbergen, "Ecologically Sustainable Organizations: An Institutional Approach." For a critique of sustainable development because it allows government and industry to "embrace environmentalism without commitment," see M. Jacobs, *The Green Economy: Environment, Sustainable Development and the Politics of the Future* (1993).

5. See Arturo Escobar's *Encountering Development: The Making and Unmaking of the Third World* (1995) for an extremely useful explication of how the discursive figuration of "development" has precluded many alternatives, establishing "what counts" in advance, before consideration of local issues and perspectives.

6. The official Rio conference included representatives from 150 nations and 1,400 nongovernmental organizations (NGOs), which were granted standing in the negotiations for the climate and biodiversity treaties. A second Global Forum was held elsewhere in Rio. Participating NGOs had hoped to formally present their national reports at UNCED itself. This didn't happen, nor was a clearinghouse established to facilitate circulation of the reports later. It is interesting to compare UNCED with the Stockholm Conference on the Environment in 1972. Peter Haas, Marc Levy, and Edward Parson provide figures, which they insist should be taken "with a grain of salt," given inconsistent reporting of attendance: More than 100 heads of state attended the Rio Summit, while only 2 went to Stockholm. Of the 1,400 NGOs accredited by UNCED, about one-third were from

the developing world; at Stockholm, only 134 NGOs attended, with only about 10 percent from developing countries (Haas, Levy, and Parson 1995, 162). For an account from the perspective of the U.S. grassroots delegation, see my interview with Dave Henson titled "Citizens, Inc.: Bottom-Up Organizing in Bottom-Line Contexts" (Fortun 1998).

7. Harrison seems to be referring to a series of meetings with members of Communities Concerned About Corporations, which I focus on in chapter 10. If so, they were "critical of a Carbide plant operating in Texas" because it blew up—on March 12, 1991. The "mass demonstration" by outsiders at the Union Carbide plant would have included about twelve people, who were there to provide technical assistance to Diane Wilson, who lives near the Seadrift/Port Lavaca plant. Some of these people did come in from out of town, for a prescheduled meeting that did not take place. Details on this particular exchange between CCC and Union Carbide are included in chapter 3.

8. Many blame environmental activists for contributing to this discourse of shared concern about the environment. By many accounts, this followed Nixon's establishment of the Environmental Protection Agency and the passage of a series of major legislative initiatives that allowed corporations to appear cooperative. Opposition was then pushed to insist that "factories don't pollute; people do," ushering in consumer environmentalism and rhetorics of "every day is Earth Day." See Timothy Luke's "Green Consumerism: Ecology and the Ruse of Recycling" (1993).

Epilogue

1. For further elaboration on the difference between a "graphic of iterability" and the "logic of repetition," see Derrida's *Limited, Inc.* (1977) and Gayatri Spivak's "Revolutions That as Yet Have No Model: Derrida's *Limited Inc.*" (1980). Spivak comments that

> if such a reading [of iterability, in Derrida's reading of his own reading of Descartes] were to be translated to the social text, it would require an extremely sharp eye for "history." Clear-cut oppositions between so-called material and ideological formations would be challenged as persistently as those between literal and allegorical uses of language. The sedimentation of investment of history as political, economic, sexual "constructions" would be seen as irreducible. Material objects, and seemingly non-textual events and phenomena would have to be seen not as self-identical but as the space of dispersion of such "constructions," as the condition or effect of interminable iteration. (1980, 39–40)

2. Drucilla Cornell sketches connections between the future as mere continuation, habits of calculation, and the possibility of something different:

> As with Derrida, Levinas' conception of time has implications for his understanding of justice. For Levinas, Justice is messianic. The "avenir" is not just the limit created by the aporias Derrida indicates, but instead inheres in the otherness of the Other that cannot be encompassed by any present system of ideality. The Other is other to the system. Incorporation into the system is the denial of the Other. Justice is sanctity for her "otherness." Nonencompassable by the system, the Other is also noncalculable. The right of the Other, then, is infinite, meaning that it can never be reduced to a proportional share of an already-established system of ideality, legal or otherwise. It is the Other as other to the present that echoes in the call to justice. The echo breaks up the "present," because the Other is there before the conception of a system of ideality and remains after. For Derrida and Levinas, if for different reasons, the future is distinguished from the present that merely reproduces itself. Justice, in other words, whether as a limit, as echoed in the necessary demand of the Good, or as the call of the Other that cannot be silenced, is the opening of the beyond that makes "true" transformation to the "new" possible. Without this appeal to the beyond, transformation would not be transformation, but only evolution and, in that sense, a continuation. The very concept of continuation as evolution of the system implies the privileging of the present. (Cornell 1992, 137)

BIBLIOGRAPHY

Abu-Lughod, Lila. 1986. *Veiled sentiments: Honor and poetry in a Bedouin society.* Berkeley: University of California Press.

———. 1990. The romance of resistance: Tracing transformations of power through Bedouin women. *American Ethnologist* 17:41–55.

Agnew, John. 1994. Global hegemony versus national economy: The United States in the new world order. In *Reordering the world: Geopolitical perspectives on the 21st century,* edited by D. Demko and W. Wood. Boulder, Colo.: Westview Press.

Alcoff, Linda. 1994. Cultural feminism versus post-structuralism: The identity crisis in feminist theory. In *Culture, power, history: A reader in contemporary social thought,* edited by N. Dirks, G. Eley, and S. Ortner. Princeton, N.J.: Princeton University Press.

Alternative Survey Group. 1996. *Alternative economic survey.* Delhi: Delhi Science Forum.

Alvares, Claude. 1994. Bhopal ten years after. In *Bhopal: The inside story,* edited by T. R. Chouhan. New York: Apex Press.

Amin, Shahid. 1988. Gandhi as mahatma: Gorakhpur District, Eastern UP, 1921–2. In *Selected subaltern studies,* edited by R. Guha and G. C. Spivak. New York: Oxford University Press.

Anzaldua, Gloria. 1987. *Borderlands/la frontera: The new mestiza.* San Francisco: Aunt Lute.

Appadurai, Arjun. 1990. Disjuncture and difference in the global cultural economy. *Public Culture* 2 (2): 1–24.

———. 1991. Global ethnoscapes: Notes and queries for a transnational anthropology. In *Recapturing anthropology: Working in the present,* edited by R. Fox. Santa Fe, N.M.: School of American Research.

———. 1996. *Modernity at large: Cultural dimensions of globalization.* Minneapolis: University of Minnesota Press.

Appiah, Kwame Anthony. 1992. *In my father's house: Africa in the philosophy of culture.* New York: Oxford University Press.

Arendt, Hannah. 1977. *Eichmann in Jerusalem.* New York: Penguin.

Attali, Jacques. 1977 [1996]. *Noise: The political economy of music,* translated by B. Massumi. Minneapolis: University of Minnesota Press.

Bahti, Timothy. 1989. Lessons of remembering and forgetting. In *Reading de Man reading,* edited by L. L. Waters and W. Godzich. Minneapolis: University of Minnesota.

Basu, Amrita. 1992. *Two faces of protest: Contrasting modes of women's activism in India.* Berkeley: University of California Press.

Bateson, Gregory. 1955. A theory of play and fantasy. *Psychology Research Report* 2: 39–51.

———. 1958 [1936]. *Naven.* Stanford, Calif.: Stanford University Press.

———. 1972. *Steps to an ecology of mind.* New York: Balanchine Books.

Bateson, Gregory, Don Jackson, Jay Haley, and John Weakland. 1976 [1956]. Toward a theory of schizophrenia. In *Double-bind: The foundation of the communicational approach to the family,* edited by C. Sluzki and D. Ransom. New York: Grune & Stratton.

Baviskar, A. 1995. *In the belly of the river: Tribal conflicts over development in the Narmada Valley.* Delhi: Oxford University Press.

———. 1996. Reverence is not enough: Ecological Marxism and Indian adivasis. In *Creating the countryside: The politics of rural and environmental discourse,* edited by E. M. DuPuis and P. Vandergeest. Philadelphia: Temple University Press.

Baxi, Upendra, and Amita Dhanda, eds. 1990. *Valiant victims and lethal litigation: The Bhopal case.* Bombay: N. H. Tripathi.

Bedford, Chris. 1988. *Locked out: The story of OCAW Local 4–620.* Boulder, Colo.: Oil, Chemical and Atomic Workers Union. Video.

———. 1992. *Out of control: The story of corporate recklessness in the petrochemical industry.* Boulder, Colo.: Oil, Chemical and Atomic Workers Union. Video.

———. 1996. Anatomy of a petrochemical explosion. *Just Cause* 3 (7): 1–3.

Benjamin, Walter. 1969. *Illuminations,* translated by H. Zohn and edited by H. Arendt. New York: Schocken.

Berliner, Paul. 1994. *Thinking in jazz: The infinite art of improvisation.* Chicago: University of Chicago Press.

Bhabha, Homi, ed. 1990. *Nation and narration.* New York: Routledge.

———. 1994. *The location of culture.* New York: Routledge.

Bhagwati, P. N. 1989. Travesty of justice. *India Today,* 15 March, 45.

Bharucha, Rustom. 1983. *Rehearsals of revolution: The political theatre of Bengal.* Calcutta: Seagull Books.

Bhopal Action Resource Center. 1988. *The Bhopal gas leak: Corporate negligence or sabotage?* New York: Bhopal Action Resource Center.

Bhopal Group for Information and Action. 1990. *Voices of Bhopal.* Bhopal: Bhopal Group for Information and Action.

———. 1992. *Compensation disbursement: Problems and possibilities.* Bhopal: Bhopal Group for Information and Action.

Blanchot, Maurice. 1981 [1943–1969]. *The gaze of Orpheus and other literary essays,* translated by L. Davis. Barrytown, N.Y.: Station Hill.

———. 1986. *The writing of the disaster,* translated by A. Smock. Lincoln: University of Nebraska Press.

———. 1989 [1955]. *The space of literature,* translated by A. Smock. Lincoln: University of Nebraska Press.

Boehmer, Konrad. 1997. Chance as ideology. *October* 82 (Fall): 62–76.

Bogard, William. 1989. *The Bhopal tragedy: Language, logic and politics in the production of a hazard.* Boulder, Colo.: Westview Press.

Boon, James. Further operations of "culture" in anthropology: A synthesis of and for debate. *Social Science Quarterly* 52:221–52.

Boyce, David. 1985. Foreign plaintiffs and forum non conveniens: Going beyond Reno. *Texas Law Review* 64:193.

Browning, Jackson. 1986. Letter to the editor. *New York Times,* 18 August.

Bryner, Gary. 1996. *Blue skies, green politics: The Clean Air Act of 1990.* 2nd ed. Washington, D.C.: Congressional Quarterly.

Bullard, Robert. 1983. Solid waste sitings and the Houston black community. *Sociological Inquiry* 53 (Spring): 273–88.

Cage, John. 1997. Reflections of a progressive composer on a damaged society. *October* 82 (Fall): 77–94.

Cage, John, with Joan Retallack. 1996. *Musicage: Cage muses on words, art, music.* Hanover, N.H.: Wesleyan University Press.

Calton, Jerry M., and Nancy B. Kurland. 1995. A theory of stakeholder enabling: Giving voice to an emerging postmodern practice of organizational discourse. In *Postmodern management and organizational theory,* edited by D. M. Boje, R. P. Gephart, and T. J. Thatchenkery. Thousand Oaks, Calif.: Sage.

Calvino, Italo. 1981. *If on a winter night a traveler,* translated by W. Weaver. New York: Harcourt Brace Jovanovich.

———. 1986. *The uses of literature,* translated by P. Creagh. New York: Harcourt Brace Jovanovich.

Canguilhem, Georges. 1991 [1966]. *The normal and the pathological.* New York: Zone Books.

Carson, Anne. 1995 [1992]. *Glass, irony and God.* New York: New Directions.

Carson, Rachel. 1962. *Silent spring.* Boston: Houghton Mifflin.

Cassels, Jaime. 1993. *The uncertain promise of law: Lessons from Bhopal.* Toronto: University of Toronto Press.

Cefkin, Melissa. 1998. Toward a higher order merger: A middle manager's story. In *Corporate futures: The diffusion of the culturally sensitive corporate form,* edited by G. Marcus. Chicago: University of Chicago Press.

Central Bureau of Investigation—India. 1987. *Bhopal case investigation charge sheet.* New Delhi: Government of India.

Centre for Social Medicine and Community Health. 1985. *An epidemiological and sociological study of the Bhopal tragedy.* Delhi: Jawaharlal Nehru University.

———. 1989. *Against all odds: The health status of Bhopal survivors.* New Delhi: Jawaharlal Nehru University.

Cerrell Associates, Inc. 1984. *Political difficulties facing waste-to-energy conversion plant siting.* Los Angeles: California Waste Management Board.

Charleston Gazette. 1994. Editorial: Bhopal lessons: Did Rhone-Poulenc learn? *Charleston Gazette,* 17 May, 11A.

Chatterjee, Partha. 1986. *Nationalist thought and the colonial world: A derivative discourse.* London: Zed Books.

———. 1993. *The nation and its fragments: Colonial and postcolonial histories.* Princeton, N.J.: Princeton University Press.

Chemical Manufacturers Association. 1994. *Reporting worst case scenarios: Manag-*

ing our risks together. Arlington, Va.: Chemical Manufacturers Association. Video.

Cherniack, Martin. 1986. *The Hawk's Nest incident: America's worst industrial disaster.* New Haven, Conn.: Yale University Press.

Chouhan, T. R. 1994. *Bhopal: The inside story: Carbide workers speak out on the world's worst industrial disaster.* New York: Apex Press.

Chughai, Ismat. 1990. *The quilt and other stories,* translated by T. Naqvi and S. Hameed. Delhi: Kali for Women.

Clarke, Lee, and James Short. 1993. Social organization and risk: Some current controversies. *Annual Review of Sociology* 19:375–99.

Cohen, Bernard. 1987. *An anthropologist among the historians, and other essays.* New Delhi: Oxford University Press.

Comaroff, Jean. 1985. *Body of power, spirit of resistance: The culture and spirit of a South African people.* Chicago: University of Chicago Press.

Comaroff, John, and Jean Comaroff. 1987. The madman and the migrant: Work and labor in the historical consciousness of a South African people. *American Ethnologist* 14 (2): 191–209.

———. 1992. *Ethnography and the historical imagination.* Boulder, Colo.: Westview Press.

Communities Concerned About Corporations. 1994. Press release. 3 December, Charleston, W.V.

Conca, Ken, Michael Alberty, and Geoffrey D. Dabelko, eds. 1995. *Green planet blues: Environmental politics from Stockholm to Rio.* Boulder, Colo.: Westview Press.

Copjec, Joan. 1994. *Read my desire: Lacan against the historicists.* Cambridge: MIT Press.

Cornell, Drucilla. 1992. *The philosophy of the limit.* New York: Routledge.

———. 1993. *Transformations: Recollective imagination and sexual difference.* New York: Routledge.

Cortazar, Julio. 1967 [1963]. *Hopscotch.* New York: Signet.

Covey, Stephen. 1989. *The seven habits of highly effective people.* New York: Simon & Schuster.

Crapanzano, Vincent. 1992. *Hermes' dilemma and Hamlet's desire: On the epistemology of interpretation.* Cambridge: Harvard University Press.

Daly, H., and J. Cobb. 1994. *For the common good.* Boston: Beacon Press.

Das, Veena. 1993. Moral orientations to suffering: Legitimation, power and healing. In *Health and social change in international perspective,* edited by L. C. Chen, A. Kleinman, and N. Ware. Oxford, England: Oxford University Press.

———. 1995. *Critical events: An anthropological perspective on contemporary India.* New York: Oxford University Press.

de Man, Paul. 1983. *Blindness and insight: Essays in the rhetoric of contemporary criticism.* Minneapolis: University of Minnesota Press.

Dehejia, Jay. 1993. Economic reforms: Birth of an Asian tiger. In *India briefing 1993,* edited by P. Oldenburg. Boulder, Colo.: Westview Press.

Deleuze, Gilles. 1988 [1986]. *Foucault,* translated by S. Hand. Minneapolis: University of Minnesota Press.

Deleuze, Gilles, and Felix Guattari. 1986 [1972, 1977]. *Anti-Oedipus: Capitalism and schizophrenia,* translated by R. Hurley, M. Seem, and H. Lane. Minneapolis: University of Minnesota Press.

Dembo, David, Ward Morehouse, and Lucinda Wykle. 1990. *Abuse of power: The case of Union Carbide.* New York: New Horizons Press.

Derrida, Jacques. 1987 [1978]. *The truth in painting,* translated by G. Bennington and I. McLeod. Chicago: University of Chicago Press.

————. 1976 [1967]. *Of grammatology,* translated by G. C. Spivak. Baltimore: Johns Hopkins University Press.

————. 1978. *Writing and difference,* translated by A. Bass. Chicago: University of Chicago Press.

————. 1981 [1972]. *Dissemination.* London: Athlone.

————. 1986 [1972]. *Margins of philosophy,* translated by A. Bass. Chicago: University of Chicago Press.

————. 1990. Force of law: The mystical foundation of authority. *Cardoza Law Review* 11:5–6.

————. 1990 [1974]. *Glas,* translated by J. Leavey Jr. and R. Rand. Lincoln: University of Nebraska Press.

————. 1991. *At this very moment in this work here I am, re-reading Levinas,* translated by R. Berezdivin and edited by R. Bernasconi and S. Critchley. Bloomington: Indiana University Press.

————. 1992 [1991]. *Given time: I. Counterfeit money,* translated by P. Kamuf. Chicago: University of Chicago Press.

————. 1994 [1993]. *Specters of Marx: The state of the debt, the work of mourning and the new international,* translated by P. Kamuf. New York: Routledge.

————. 1995 [1993]. *On the name,* translated by D. Wood, J. P. Leavey Jr., and I. McLeod and edited by T. Dutoit. Stanford, Calif.: Stanford University Press.

Devi, Mahasweta. 1986. *Five plays,* translated by S. Badyopadjyay. Calcutta: Seagull Books.

Dowd, Ann Reilly. 1994. Environmentalists on the run. *Fortune,* 19 September, 91*ff.*

Dowie, Mark. 1995. *Losing ground: American environmentalism at the close of the twentieth century.* Cambridge: MIT Press.

Downey, Gary Lee. 1986a. Risk in culture: The American conflict over nuclear power. *Cultural Anthropology* 1 (4): 388–412.

————. 1986b. Ideology and the clamshell identity: Organizational dilemmas in the anti-nuclear power movement. *Social Problems* 33 (5): 357–73.

Downey, Gary, Joseph Dumit, and Sharon Traweek, eds. 1998. *Cyborgs and citadels: Anthropological interventions on the borderlands of technoscience.* Seattle: University of Washington Press.

Downey, Gary, and J. C. Lucena. 1995. Engineering studies. In *Handbook of science and technology studies,* edited by S. Jasanoff, G. E. Markle, J. C. Petersen, and T. Pinch. Thousand Oaks, Calif.: Sage.

Draffan, George. 1994. *Research compendium on the Union Carbide Corporation.* Seattle: Institute on Trade Policy for Communities Concerned About Corporations.

Dreyfus, Hubert, and Paul Rabinow. 1983 [1982]. *Michel Foucault: Beyond structuralism and hermeneutics.* Chicago: University of Chicago Press.

Dreze, Jean, and Amartya Sen. 1995. *India: Economic development and social opportunity.* Delhi: Oxford India Paperbacks.

Drucker, Peter. 1993. *Postcapitalist society.* New York: Harper Collins.

Dupuis, E. M., and P. Vandergeest, eds. 1996. *Creating the countryside: The politics of rural and environmental discourse.* Philadelphia: Temple University Press.

Duras, Marguerite. 1961. *Hiroshima mon amour,* translated by R. Seaver. New York: Grove Weidenfeld.

Eckerman, Ingrid. 1996. The health situation of women and children in Bhopal: Final report on the International Medical Commission on Bhopal 1994. *International Perspectives in Public Health* 11 & 12:29–36.

Economist. 1997. Survey: India's economy. *Economist,* 22 February, 3–26.

Edelstein, Michael. 1988. *Contaminated communities: The social and psychological impacts of residential toxic exposure.* Boulder, Colo.: Westview Press.

Ember, Lois. 1995. Responsible Care: Chemical makers still counting on it to improve image. *Chemical and Engineering News,* 29 May, 10–18.

Escobar, Arturo. 1995. *Encountering development: The making and unmaking of the Third World.* Princeton, N.J.: Princeton University Press.

Fabian, Johannes. 1983. *Time and the other: How anthropology makes its object.* New York: Columbia University Press.

Felman, Shoshana. 1987. *Jacques Lacan and the adventure of insight: Psychoanalysis and contemporary culture.* Cambridge: Harvard University Press.

Ferguson, James. 1997. Country and city on the Copperbelt. In *Culture, power, place: Explorations in critical anthropology,* edited by A. Gupta and J. Ferguson. Durham, N.C.: Duke University Press.

Ferrante, Joan. 1992. *Sociology: A global perspective.* Belmont, Calif.: Wadsworth.

Fischer, Michael. 1986. Ethnicity and postmodern arts of memory. In *Writing culture: The poetics and politics of ethnography,* edited by J. Clifford and G. Marcus. Berkeley: University of California Press.

Fischer, Michael, and Mehdi Abedi. 1990. *Debating Muslims: Cultural dialogues in postmodernity and tradition.* Madison: University of Wisconsin Press.

Fish, Stanley. 1989. *Doing what comes naturally.* Durham, N.C.: Duke University Press.

Fisher, William, ed. 1995. *Toward sustainable development? Struggling over India's Narmada River.* Armonk, N.Y.: Sharpe.

———. 1997. Doing good? The politics and antipolitics of NGO practises. *Annual Review of Anthropology* 26:439–64.

Fortun, Kim. 1998a. The Bhopal disaster: Advocacy and expertise. *Science as Culture* 7 (2): 193–216.

———. 1998b. Citizens, Inc.: Bottom-up organizing in bottom-line contexts. In *Late editions 5, corporate futures: The diffusion of the culturally sensitive corporate form,* edited by G. Marcus. Chicago: University of Chicago Press.

Fortun, Kim, and Michael Fortun. 2000. The work of markets: Filming within Indian mediascapes, 1997. In *Late editions 7, para/sites: New locales of cultural production,* edited by G. Marcus. Chicago: University of Chicago Press.

Foucault, Michel. 1972 [1969]. *The archaeology of knowledge and the discourse on language,* translated by A. M. Sheridan Smith. New York: Tavistock/Random House.

———. 1973 [1966]. *The order of things: An archaeology of the human sciences.* New York: Vintage/Random House.

———. 1975. *The birth of the clinic: An archaeology of medical perception,* translated by A. M. Sheridan. New York: Vintage/Random House.

———. 1980. *The history of sexuality.* Vol. 1, *An introduction,* translated by R. Hurley. New York: Vintage/Random House.

Franklin, Sarah. 1995. Science as culture, cultures of science. *Annual Review of Anthropology* 24:163–84.

Freeman, R. E. 1984. *Strategic management: A stakeholder approach.* Boston: Pitman.

Fritz, Charles. 1961. Disaster. In *Contemporary social problems,* edited by R. K. Merton and R. A. Nisbet. New York: Harcourt.

Frohmann, Lisa, and Elizabeth Mertz. 1994. Legal reform and social construction: Violence, gender and the law. *Law & Social Inquiry* 19 (4): 829–52.

Frye, Northrop. 1973 [1957]. *Anatomy of criticism: Four essays.* Princeton, N.J.: Princeton University Press.

Fujimura, Joan. 1992. Crafting science: Standardized packages, boundary objects and "translation." In *Science as practise and culture,* edited by A. Pickering. Chicago: University of Chicago Press.

Gadgil, Madhav, and Ramachandra Guha. 1992. *This fissured land: An ecological history of India.* Berkeley: University of California Press.

———. 1995. *Ecology and equity: The use and abuse of nature in contemporary India.* London: Routledge.

Gellner, E. 1985. *Relativism and the social sciences.* Cambridge: Cambridge University Press.

Ginsburg, Faye, and Rayna Rapp, eds. 1995. *Conceiving the new world order: The global politics of reproduction.* Berkeley: University of California Press.

Good, Byron. 1992. A body in pain—The making of a world of chronic pain. In *Pain as human experience: An anthropological perspective,* edited by M.-J. D. Good, P. Brodwin, B. Good, and A. Kleinman. Berkeley: University of California Press.

———. 1994. *Medicine, rationality and experience.* Cambridge: Cambridge University Press.

Good, Mary-Jo DelVecchio, Byron Good, Cynthia Schaffer, and Stuart Lind. 1990. American oncology and the discourse on hope. *Culture, Medicine and Psychiatry* 14:59–70.

Grant, Andrea. 1994. Communications training: A strategic asset for executives. *Corporate Environmental Strategy* 2 (2): 57–60.

Grossman, Richard. 1996. Revoking the corporation. *Just Cause* 3 (8): 1–4.

Grossman, Richard, and Frank Adams. 1993. *Taking care of business: Citizenship and the charter of incorporation.* Cambridge, Mass.: Charter Ink.

Gupta, Akhil. 1992. The song of the nonaligned world: Transnational identities and the reinscription of space in late capitalism. *Cultural Anthropology* 1 (1): 63–79.

Gupta, Akhil, and James Ferguson. 1997a. Beyond "culture": Space, identity and the politics of difference. In *Culture, power, place: Explorations in critical anthropology,* edited by A. Gupta and J. Ferguson. Durham, N.C.: Duke University Press.

———, eds. 1997b. *Culture, power, place: Explorations in critical anthropology.* Durham, N.C.: Duke University Press.

Haas, Ernst. 1990. *When knowledge is power: Three models of change in international organizations.* Berkeley: University of California Press.

Haas, Peter. 1990. *Saving the Mediterranean: The politics of environmental cooperation.* New York: Columbia University Press.

Haas, Peter, Marc Levy, and Edward Parson. 1995. Approaching the Earth Summit: How should we judge UNCED's success? In *Green planet blues: Environmental politics from Stockholm to Rio,* edited by K. Conca, M. Alberty, and G. Dabelko. Boulder, Colo.: Westview Press.

Hadden, Susan. 1994. Citizen participation in environmental policy making. In *Learn-*

ing from disaster: Risk management after Bhopal, edited by S. Jasanoff. Philadelphia: University of Pennsylvania Press.

Haraway, Donna. 1976. *Crystals, fabrics and fields: Metaphors of organicism in twentieth century biology.* New Haven, Conn.: Yale University Press.

——. 1991 [1985]. A manifesto for cyborgs: Science, technology and socialist feminism in the 1980s. In *Simians, cyborgs and women: The reinvention of nature.* New York: Routledge.

——. 1997. *Modest_Witness@Second Millennium. FemaleMan_Meets_Onco-Mouse.* New York: Routledge.

Harding, Sandra. 1986. *The science question in feminism.* Ithaca, N.Y.: Cornell University Press.

Harris, Angela. 1991. Race and essentialism in feminist legal theory [1990]. In *Feminist legal theory: Readings in law and gender,* edited by K. Bartlett and R. Kennedy. Boulder, Colo.: Westview Press.

Harrison, E. Bruce. 1993. *Going green: How to communicate your company's environmental commitment.* Homewood, Ill.: Business One Irwin.

Hart, S. 1995. A natural resource view of the firm. *Academy of Management Review* 20:986–1014.

Harte, John, Cheryl Holden, Richard Schneider, and Christine Shirley. 1991. *Toxics A to Z: A guide to everyday pollution hazards.* Berkeley: University of California Press.

Harvey, David. 1989. *The condition of postmodernity: An inquiry into the origins of culture change.* Oxford, England: Blackwell.

Hasan, M. 1996. Fund raising in name of gas victims continues. *Madhya Pradesh Chronicle,* 28 June.

Hastrup, Kirsten, and Peter Elsass. 1990. Anthropological advocacy: A contradiction in terms? *Current Anthropology* 31 (3): 301–11.

Hazarika, Sanjay. 1990. Bhopal victims still wait for Carbide money. *New York Times,* 30 January, A9.

Herndl, C. G., and S. C. Brown, eds. 1996. *Green culture: Environmental rhetoric in contemporary America.* Madison: University of Wisconsin Press.

Hess, David. 1992. Introduction: The new ethnography and the anthropology of science and technology. *Knowledge and Society* (The Anthropology of Science and Technology Studies) 9:1–26.

Hicks, Jonathan. 1989. After the disaster, Carbide is rebuilt. *New York Times,* 15 February.

Highlander Research and Education Center. 1992. *Environment and development in the USA: A grassroot report for UNCED.* New Market, Tenn.: Community Environmental Health Program.

Hoban, Thomas More, and Richard Oliver Brooks. 1987 [1996]. *Green justice: The environment and the courts.* 2nd ed. Boulder, Colo.: Westview Press.

Hoffman, Andrew. 1997. *From heresy to dogma: An institutional history of corporate environmentalism.* San Francisco: New Lexington Press.

Hong, Peter. 1992. Do two pollutants make you sicker than one? *Business Week,* 28 September, 7–8.

Huber, Peter. 1998. Green alchemy. *Forbes,* 23 March.

Indian Council of Medical Research. 1990. *Bhopal Gas Disaster Research Center: Annual report.* Bhopal, India.

Ivanovich, David. 1993. Clean Air Act choking refineries. *Houston Chronicle,* 20 June.

Iyer, V. R. Krishna. 1991. Bhoposhima: Crime without punishment, case for crisis management jurisprudence. *Economic and Political Weekly,* 23 November, 2705–13.

Jackson, Patricia. 1994. Worst-case scenarios: A communication tool to reduce risk. *Corporate Environmental Strategy* 2 (4): 69–74.

Jacobs, M. 1993. *The green economy: Environment, sustainable development and the politics of the future.* Vancouver, Canada: University of British Columbia Press.

Jaising, Indira. 1994. Legal let-down. In *Bhopal: The inside story,* edited by T. R. Chouhan. New York: Apex Press.

Jakobson, Roman. 1971. The dominant. In *Readings in Russian poetics: Formalist and structuralist views,* edited by L. Matejka and K. Pomorska. Cambridge: MIT Press.

James, Clifford, and George Marcus. 1986. *Writing culture: The politics and poetics of ethnography.* Berkeley: University of California Press.

Jasanoff, Sheila, ed. 1994. *Learning from disaster: Risk management after Bhopal.* Philadelphia: University of Pennsylvania Press.

Jenkins, John. 1991 [1989]. Bhopal and beyond: John Coale and the disaster hustle. In *The litigators: Inside the powerful world of America's high-stakes trial lawyers.* New York: St. Martin's.

Jennings, P., and P. Zandbergen. 1995. Ecologically sustainable organizations: An institutional approach. *Academy of Management Review* 20: 1015–52.

Kalelkar, Ashok. 1988. Investigation of large-magnitude incidents: Bhopal as a case study. Paper presented at the Institution of Chemical Engineers Conference on Preventing Major Chemical Accidents, May 1988, London. Institution of Chemical Engineers Symposium Series No. 110.

Kapaferer, Bruce. 1988. *Legends of people, myths of state.* Washington, D.C.: Smithsonian Institution Press.

Karliner, Josh. 1994. *Setting the record straight.* San Francisco: Josh Karliner. Video.

Kearney, M. 1995. The local and the global: The anthropology of globalization and transnationalism. *Annual Review of Anthropology* 24: 547–65.

Keating, B., and D. Russel. 1992. EPA yesterday and today. *E Magazine,* July–August, 33.

Keenan, Judge John F. 1986. Opinion and Order, In Re: Union Carbide Corporation Gas Plant Disaster at Bhopal India in December 1984, 846 F. Supp. 842 (S.D.N.Y. 1986).

Keenan, Thomas. 1997. *Fables of responsibility: Aberrations and predicaments in ethics and politics.* Stanford, Calif.: Stanford University Press.

Keller, Evelyn Fox. 1992. *Secrets of life: Essays on language, gender and science.* New York: Routledge.

———. 1995. *Refiguring life: Metaphors of twentieth century biology.* New York: Columbia University Press.

Kelly, Kevin. 1994. *Out of control: The new biology of machines, social systems and the economic world.* Reading, Mass.: Addison-Wesley.

Kharbanda, D. P. 1989. Critique of Kalelkar's sabotage theory—Bhopal. Bombay. Manuscript.

Khosla, A. 1987. Alternative strategies in achieving sustainable development. In *Conservation with equity: Strategies for sustainable development,* edited by P. Jacobs and D. A. Munro. Cambridge, England: International Union for Conservation of Nature and Natural Resources.

Killingsworth, M. J., and J. S. Palmer, eds. 1992. *Ecospeak: Rhetoric and environmental politics in America.* Carbondale: Southern Illinois University Press.

Kipling, Rudyard. 1929 [1891]. *Selected stories from Kipling,* edited by W. L. Phelps. New York: Doubleday, Doran.

Kishwar, M. 1987. The continuing deficit of women in India and the impact of amniocentesis. In *Man-made women: How reproductive technologies affect women,* edited by G. Corea, R. D. Klein, J. Hammer, H. B. Holmes, B. Hoskins, M. Kishwar, J. Raymond, R. Rowland, and R. Steinbacher. Bloomington: Indiana University Press.

Kishwar, Madhu, and Ruth Vanita. 1984. *In search of answers: Women's voices from Manushi.* London: Zed.

Kondo, Dorinne. 1990. *Crafting selves: Power, gender and discourses of identity in a Japanese workplace.* Chicago: University of Chicago Press.

Kothari, Rajni. 1989a. *Politics and people: In search of humane India.* New York: New Horizons Press.

———. 1989b. Human rights—A movement in search of a theory. In *Rethinking human rights,* edited by H. Sethi and S. Kothari. Delhi: Lokayan Press.

Kothari, Smitu. 1993. Social movements and the redefinition of democracy. In *India briefing 1993,* edited by P. Oldenburg. Boulder, Colo.: Westview Press.

Kothari, Smitu, and Harsh Sethi, eds. 1989. *Rethinking human rights.* Delhi: Lokayan Press.

Kraft, Michael, and Norman Vig. 1997. Environmental policy from the 1970s to the 1990s: An overview. In *Environmental policy in the 1990s: Reform or reaction?* edited by N. Vig and M. Kraft. Washington, D.C.: Congressional Quarterly.

Kreisberg, Paul. 1993. Foreign policy building in 1992: Building credibility. In *India briefing, 1993,* edited by P. Oldenburg. Boulder, Colo.: Westview Press.

Krishen, Pradip. 1993. Cinema and television. In *India briefing 1993,* edited by P. Oldenburg. Boulder, Colo.: Westview Press.

Kristeva, Julia. 1984. *Revolution in poetic language,* translated by M. Waller. New York: Columbia University Press.

Kumar, Radha. 1983. Family and factory: Women and the Bombay textile industry 1919–1939. *Indian Economic and Social History Review* 20 (1): 81–110.

———. 1989. Contemporary Indian feminism. *Feminist Review* 33 : 20–29.

———. 1993. *A history of doing: An illustrated account of movements for women's rights and feminism in India, 1800–1990.* Delhi: Kali for Women Press.

Kurzman, Dan. 1987. *A killing wind: Inside Union Carbide and the Bhopal catastrophe.* New York: McGraw-Hill.

LaBar, Gregg. 1991. Citizen Carbide? *Occupational Hazards,* November, 33–37.

Lacan, Jacques. 1977 [1966]. *Ecrits: A selection,* translated by A. Sheridan. New York: Norton.

———. 1981 [1973]. *The four fundamental concepts of psychanalysis,* translated by A. Sheridan and edited by J. Alain-Miller. New York: Norton.

———. 1992 [1986]. *The ethics of psychoanalysis: The seminar of Jacques Lacan 1959–1960,* translated by D. Porter and edited by J. Alain-Miller. New York: Norton.

Lele, Sharachandra. 1995 [1991]. Sustainable development: A critical review. In *Green planet blues: Environmental politics from Stockholm to Rio,* edited by K. Conca, M. Alberty, and G. Dabelko. Boulder, Colo.: Westview Press.

Lepkowski, Wil. 1984. Bhopal disaster spotlights chemical hazard issues. *Chemical and Engineering News,* 24 December.

———. 1994. The restructuring of Union Carbide. In *Learning from disaster: Risk management after Bhopal,* edited by S. Jasanoff. Philadelphia: University of Pennsylvania Press.

Leslie, Charles, and Allan Young, eds. 1992. *Paths to Asian medical knowledge.* Berkeley: University of California Press.

Levi-Strauss, Claude. 1969 [1964]. *The raw and the cooked,* translated by J. Weightman and D. Weightman. New York: Harper & Row.

———. 1973 [1966]. *From honey to ashes,* translated by J. Weightman and D. Weightman. New York: Harper & Row.

———. 1978 [1968]. *The origin of table manners,* translated by J. Weightman and D. Weightman. New York: Harper & Row.

———. 1981 [1971]. *The naked man,* translated by J. Weightman and D. Weightman. New York: Harper & Row.

Lincoln, Bruce. 1994. *Authority: Construction and corrosion.* Chicago: University of Chicago Press.

Lochan, Rajiv. 1991. Health damage due to Bhopal gas disaster: Review of medical research. *Economic and Political Weekly,* 25 May.

Lock, Margaret, and Nancy Scheper-Hughes. 1990. A critical interpretive approach in medical anthropology: Rituals and routines of discipline and dissent. In *Medical anthropology: A handbook of theory and method,* edited by T. Johnson and C. Sargent. New York: Greenwood Press.

Lohia, Rammonohar. 1955. *Wheel of history.* Bombay: Sindhu Publications.

———. 1985. *Economics after Marx.* New Delhi: Samta Era Publications.

———. 1987. *Fundamentals of a world mind.* Bombay: Sindhu Publications.

Lokayan Group. 1985. Sixth year report. *Lokayan Bulletin* 5:8.

Long, Janice, and David Hanson. 1985. Bhopal triggers massive response from Congress, the administration. *Chemical and Engineering News,* 11 February.

Luhmann, Niklas. 1987. Closure and openness: On reality in the world of law. In *Autopoietic law: A new approach to law and society,* edited by G. Teubner. Berlin: de Gruyter.

Luke, Timothy. 1993. Green consumerism: Ecology and the ruse of recycling. In *In the nature of things: Language, politics and the environment,* edited by J. Bennett and W. Chaloupka. Minneapolis: University of Minnesota Press.

Lyotard, Jean-Francois. 1988 [1983]. *The differend: Phrases in dispute,* translated by G. Van Den Abbeele. Minneapolis: University of Minnesota Press.

Mackey, Nathaniel. 1992. Other: From noun to verb. *Representations* 39 (Summer): 51–70.

Makhijani, Arjun, Amanda Bickel, and Annie Makhijani. 1990. Beyond the Montreal Protocol: Still working on the ozone hole. *Technology Review,* May/June, 52–59.

Marcus, George. 1989. The debate over parody in copyright law: An experiment in cultural critique. *Yale Journal of Law and the Humanities* 1 (2): 295–316.

———. 1995. Ethnography in/out of the world system: The emergence of multisited ethnography. *Annual Review of Anthropology* 24:95–117.

Marcus, George, and Michael M. J. Fischer. 1986. *Anthropology as cultural critique: An experimental moment in the human sciences.* Chicago: University of Chicago Press.

Martin, Emily. 1994. *Flexible bodies: Tracking immunity in American culture—From the days of polio to the age of AIDS.* Boston: Beacon Press.

Medico Friends Circle. 1985. *The Bhopal disaster aftermath: An epidemiological and socio medical survey.* Bangalore, India: Medico Friends Circle.

————. 1990. *Distorted lives: Women's reproductive health and Bhopal disaster.* Pune, India: Medico Friends Circle.

Melnick, R. Shep. 1983. *Regulation and the courts: The case of the Clean Air Act.* Washington, D.C.: Brookings Institution.

Merry, Sally. 1992. Anthropology, law and transnational processes. *Annual Review of Anthropology* 21:357–79.

Messina, Lawrence. 1998. Reporter must testify about plant leak, fire. *Charleston Gazette,* 4 August, 3A.

Metzger, Heinz-Klaus. 1997. John Cage, of liberated music. *October* 82 (Fall): 49–61.

Monsanto Company. 1962. The desolate year. *Monsanto Magazine,* October.

Morris, Kenneth, Marc Robinson, and Richard Kroll. 1990. *American dreams: One hundred years of business ideas and innovation from the Wall Street Journal.* New York: Lightbulb Press.

Moukheiber, Zina. 1994. Learning from winners. *Forbes,* 14 March.

Munoz, Edward. 1985. Affidavit, In Re: Union Carbide Corp. Gas Plant Disaster at Bhopal India in December 1984, MDL Docket No. 626: Judicial Panel on Multidistrict Litigation.

Murphy, Alexander. 1994. International law and the sovereign state: Challenges to the status quo. In *Reordering the world: Geopolitical perspectives on the 21st century,* edited by G. Demko and W. Wood. Boulder, Colo.: Westview Press.

Nader, Laura. 1990. *Harmony ideology: Justice and control in a Zapotec mountain village.* Stanford, Calif.: Stanford University Press.

————. 1993. Controlling processes in the practice of law: Hierarchy and pacification in the movement to re-form dispute ideology. *Ohio State Journal of Dispute Resolution* 9 (1): 1–25.

————. 1997. Controlling processes: Tracing the dynamic components of power. *Current Anthropology* 38 (5): 711–38.

Nancy, Jean-Luc. 1983. Le Kategoriein de L'exes. In *L'Imperatif Categorique.* Paris: Flammarion.

Nandy, Ashis, ed. 1988. *Science, hegemony and violence: A requiem for modernity.* New Delhi: Oxford University Press.

————. 1989. The political culture of the Indian state. *Daedalus* 118 (4): 1–26.

National Toxics Campaign Fund. 1990. *Union Carbide in Bhopal, India: The lingering legacy.* Boston: Citizen's Environmental Laboratory.

Niebuhr, H. Richard. 1941. *The meaning of revelation.* New York: Macmillan.

Nielsen, Aldon. 1994. *Writing between the lines: Race and intertextuality.* Athens: University of Georgia Press.

Nietzsche, Friedrich. 1956. *The birth of tragedy out of the spirit of music,* translated by F. Golffing. New York: Doubleday.

Oe, Kenzaburo. 1996 [1965]. *Hiroshima notes,* translated by D. L. Swain and T. Yonezawa. New York: Grove Press.

Oil, Chemical and Atomic Workers Union. 1990. Union Carbide Corporation: Financial analysis. Boulder, Colo. Manuscript.

Olsen, Frances. 1991. Statutory rape: A feminist critique of rights analysis [1984]. In *Feminist legal theory: Readings in law and gender,* edited by K. Bartlett and R. Kenney. Boulder, Colo.: Westview Press.

Omvedt, Gail. 1993. *Reinventing revolution: New social movements and the socialist tradition in India.* Armonk, N.Y.: Sharpe.

Ong, Aihwa. 1987. *Spirits of resistance and capitalist discipline: Factory women in Malaysia.* Albany: State University of New York Press.

Paine, Robert. 1998 [1985]. *Advocacy and anthropology: First encounters.* St. John's, Canada: Institute of Social and Economic Research, Memorial University of Newfoundland.

Pandey, Gyanendra. 1988. Peasant revolt and Indian nationalism. In *Selected subaltern studies,* edited by R. Guha and G. C. Spivak. New York: Oxford University Press.

Pandolfi, Mariella. 1990. Boundaries inside the body: Women's sufferings in southern peasant Italy. *Culture, Medicine and Psychiatry* 14:255–73.

People's Union for Democratic Rights and Sanskritik Morcha. 1993. *Report from Bhopal.* Delhi and Bhopal: People's Union for Democratic Rights.

Pepper, Ian. 1997. From the "aesthetics of indifference" to "negative aesthetics": John Cage and Germany 1958–1972. *October* 82 (Fall): 31–48.

Perrow, Charles. 1984. *Normal accidents: Living with high-risk technologies.* New York: Basic Books/Harper-Collins.

Piasecki, Bruce. 1994. Corporate world shows more care for the environment. *Forum for Applied Research and Public Policy* 9 (2): 6–14.

———. 1995. *Corporate environmental strategy: The avalanche of change since Bhopal.* New York: Wiley.

Piasecki, Bruce, and Peter Asmus. 1990. *In search of environmental excellence: Moving beyond blame.* New York: Simon & Schuster.

Pickering, M., and A. Johnson. 1991. *Chemical valley.* Whitesburg, Ky.: Appalshop Video. Video.

Rajan, Ravi. 1988. Rehabilitation and voluntarism in Bhopal. *Lokayan Bulletin* 6:1–2.

Ramaswamy, E. A. 1995. Organized labor and economic reform. In *India briefing: Staying the course,* edited by P. Oldenburg. Armonk, N.Y.: Sharpe.

Rapp, Rayna. 1991. Constructing amniocentesis: Maternal and medical discourses. In *Uncertain terms: Negotiating gender in American culture,* edited by F. Ginsburg and A. Tsing. Boston: Beacon Press.

———. 1992. Reproduction and gender hierarchy: Amniocentesis in contemporary America. In *Gender hierarchies: The anthropological approach,* edited by B. Miller. Cambridge: Cambridge University Press.

———. 1994. Risky business: Genetic counseling in a shifting world. In *Articulating hidden histories,* edited by J. Schneider and R. Rapp. Berkeley: University of California Press.

———. 1995. Accounting for amniocentesis. In *Power and practice: The anthropology of medicine in everyday life,* edited by S. Lindenbaum and M. Lock. Berkeley: University of California Press.

Reich, Robert. 1992. *The work of nations: Preparing for 21st century capitalism.* New York: Vintage Books.

Reisch, Marc. 1990. Carbide's Kennedy sees many challenges for company, industry. *Chemical and Engineering News,* 27 August, 9–13.

Rhone-Poulenc. 1994. *Risk management plan: Institute plant: Methyl isocyanate (MIC).* Institute, W.V.: Rhone-Poulenc.

Ronell, Avital. 1993 [1986]. *Dictations: On haunted writing.* Lincoln: University of Nebraska Press.

————. 1994. *Finitude's score: Essays for the end of the millennium.* Lincoln: University of Nebraska Press.

Rosaldo, Renato. 1989. *Culture and truth: The remaking of social analysis.* Boston: Beacon Press.

Rosenberg, Gerald. 1991. Cleaning house? The courts, the environment and reapportionment. In *The hollow hope: Can courts bring about social change?* PLACE: PUBLISHER.

Rosencranz, Armin, Shyam Divan, and Anthony Scott. 1994. Legal and political repercussions in Bhopal. In *Learning from disaster: Risk management after Bhopal,* edited by S. Jasanoff. Philadelphia: University of Pennsylvania Press.

Rudolph, Lloyd, and Susanne Rudolph. 1987. *In pursuit of Lakshmi: The political economy of the Indian state.* Chicago: University of Chicago Press.

Rukeyser, Muriel. 1994 [1938]. The book of the dead. In *A Muriel Rukeyser reader,* edited by J. H. Levi. New York: Norton.

Sahlins, Marshall. 1976. *Culture and practical reason.* Chicago: University of Chicago Press.

Sambhavna Trust. 1996. *Sambhavna from Bhopal.* Bhopal, India: Sambhavna Trust.

————. 1998. *The Bhopal gas tragedy, 1984–?* Bhopal, India: Sambhavna Trust.

Scarry, Elaine. 1985. *The body in pain: The making and unmaking of the world.* New York: Oxford University Press.

Scheper-Hughes, Nancy. 1992. *Death without weeping.* Berkeley: University of California Press.

Schmidheiny, Stephan, with the Business Council for Sustainable Development. 1992. *Changing course: A global perspective on development and the environment.* Cambridge: MIT Press.

Schneider, Elizabeth M. 1991. The dialectic of rights and politics: Perspectives from the women's movement [1986]. In *Feminist legal theory: Readings in law and gender,* edited by K. Bartlett and R. Kennedy. Boulder, Colo.: Westview Press.

Schumacher, John. 1989. *Human posture: The nature of inquiry.* Albany: State University of New York Press.

Sen, Mrinal. 1985 [1983]. *In search of famine.* Calcutta: Seagull Books.

Sethi, H. 1993. Survival and democracy: Ecological struggles in India. In *New social movements in the south: Empowering the people,* edited by P. Wignaraja. London: Zed.

Sethi, Harsh, and Smitu Kothari. 1989. On categories and interventions. In *Rethinking human rights,* edited by H. Sethi and S. Kothari. Delhi: Lokayan Press.

Sharplin, Arthur. 1989. Union Carbide, Ltd.: The Bhopal gas incident. Paper presented at Center for Business Ethics, at Bentley College.

Shelton, Robert. Hitting the green wall: Why corporate programs get stalled. *Corporate Environmental Strategy* 2 (2): 5–11.

Shrivastava, Paul. 1987. *Bhopal: Anatomy of a crisis.* Cambridge, Mass.: Ballinger.

Shrivastav, P. N., and S. D. Guru. 1989. *Madhya Pradesh District Gazetters: Sehore and Bhopal.* Bhopal: Directorate of Gazetters, Department of Culture, Government of Madhya Pradesh.

Sixel, L. M. 1993. Clash of the titans: Brown & Root fighting union drive in West Virginia. *Houston Chronicle,* 27 July, 10-A, 10-C.

Sluzki, Carlos, and Donald Ransom, eds. 1976. *Double-bind: The foundation of the communicational approach to the family.* New York: Grune & Stratton.

Smith, Merritt Roe, and Gregory Clancey, eds. 1997. *Major problems in the history of American technology: Documents and essays.* Boston: Houghton Mifflin.

Spivak, Gayatri Chakravorty. 1983. Marx after Derrida. In *Philosophical approaches to literature: New essays on nineteenth and twentieth century texts,* edited by W. Cain. Lewisberg, Ohio: Bucknell University Press.

———. 1987. *In other worlds: Essays in cultural politics.* New York: Methuen.

———. 1988. Subaltern studies: Deconstructing historiography. In *Selected subaltern studies,* edited by R. Guha and G. C. Spivak. New York: Oxford University Press.

———. 1989. Imperialism and sexual difference. In *Contemporary literary criticism: Literary and cultural studies,* edited by R. ConDavis and R. Schleifer. New York: Longman.

———. 1993. *Outside in the teaching machine.* New York: Routledge.

State Correspondent. 1993. Carbide comes up with new theory in Bhopal case. *Pioneer,* 27 February.

Steyer, Robert. 1988. Monsanto volunteers to cut toxic emission at all plants. *St. Louis Post-Dispatch,* 1 July, 1.

Stone, Andrea. 1989. Union Carbide up two dollars after Bhopal settlement. *USA Today,* 15 February.

Strathern, Marilyn. 1992a. *After nature: English kinship in the late twentieth century.* Cambridge: Cambridge University Press.

———. 1992b. *Reproducing the future: Anthropology, kinship and the new reproductive technologies.* Manchester, England: Manchester University Press.

Stree Shakti Sanghatana. 1989. *"We were making history": Women in the Telegana people's struggle.* New Delhi: Kali for Women Press.

Subramaniam, Arun. 1985. Why the guilty must be punished. *Business India,* 2–15 December. 42–46.

Suleri, Sara. 1992. *The rhetoric of English India.* Chicago: University of Chicago Press.

Szasz, Andrew. 1994. *EcoPopulism: Toxic waste and the movement for environmental justice.* Minneapolis: University of Minnesota Press.

Taussig, Michael. 1987. *Shamanism, colonialism and the wild man: A study in terror and healing.* Chicago: University of Chicago Press.

Thakur, Ramesh. 1995. *The government and politics of India.* New York: St. Martin's.

Thorne, Barrie. 1983. Political activist as participant observer. In *Contemporary field research,* edited by R. Emerson. Prospect Heights, Ill.: Waveland Press.

Traweek, Sharon. 1988. *Beamtimes and lifetimes: The world of high energy physics.* Cambridge: Harvard University Press.

Tyler, Stephen. 1987. *The unspeakable: Discourse, dialogue, and rhetoric in the postmodern world.* Madison: University of Wisconsin Press.

Ulmer, Gregory. 1985. *Applied grammatology: Post(e)-pedagogy from Jacques Derrida to Joseph Beuys.* Baltimore: Johns Hopkins University Press.

Union Carbide Corp. 1976. *Methyl isocyanate F-41443A.* New York, July.

———. 1989. *Toward environmental excellence: A progress report.* Danbury, Conn.: Union Carbide Corp.

Union Carbide India Limited. 1979. *Operating manual part II: Methyl isocyanate unit.* February.

———. 1984. *Hexagon* (Commemorative Issue, Golden Jubilee).

United Church of Christ, Commission for Racial Justice. 1987. *Toxic wastes and race:*

A national report on the racial and socioeconomic characteristics of communities with hazardous waste sites. New York: United Church of Christ.

U.S. General Accounting Office. 1979. *Better regulation of pesticide exports and pesticide residues in imported food is essential.* Report No. CED-79-43. Washington, D.C.: Government Printing Office.

————. 1983. *Siting hazardous waste landfills and their correlation with the racial and economic status of surrounding communities.* Washington, D.C.: Government Printing Office.

U.S. Senate. 1963. Committee on Government Operations, Subcommittee on Reorganization and International Organizations. *Interagency coordination in environmental hazards (pesticides).* 88th Congress.

Valente, Christina, and William Valente. 1995. *An introduction to environmental law and policy: Protecting the environment through law.* New York: West.

Vig, Norman J. 1997. Presidential leadership and the environment. In *Environmental policy in the 1990s,* edited by N. J. Vig and M. E. Kraft. Washington, D.C.: Congressional Quarterly.

Visvanathan, Shiv, with Rajni Kothari. 1985. Bhopal: The imagination of disaster. *Lokayan Bulletin* 3 : 4 –5.

Ward, Ken, Jr. 1993. Workers blame layoffs on drive for union. *Charleston Gazette,* 19 April, 1A, 11A.

————. 1994a. Laid-off workers protest Union Carbide. *Charleston Gazette,* 7 May, 3A.

————. 1994b. NLRB finds union discrimination at Brown & Root. *Charleston Gazette,* 19 April, 3A.

Wargo, John. 1996. *Our children's toxic legacy: How science and law fail to protect us from pesticides.* New Haven, Conn.: Yale University Press.

Weakland, John. 1976. The double-bind theory by self-reflexive hindsight. In *Double-bind: The foundation of the communicational approach to the family,* edited by C. Sluzki and D. Ransom. New York: Grune & Stratton.

Weir, David. 1988. *The Bhopal syndrome: Pesticides, environment, and health.* San Francisco: Center for Investigation Reporting.

Weir, David, and Mark Schapiro. 1981. *Circle of poison: Pesticides and people in a hungry world.* San Francisco: Institute for Food and Development Policy.

White, Hayden. 1973. *Metahistory: The historical imagination in nineteenth-century Europe.* Baltimore: Johns Hopkins University Press.

White, James Boyd. 1984. *When words lose their meaning: Constitutions and reconstitutions of language, character and community.* Chicago: University of Chicago Press.

White-Stevens, Robert. 1964. Rachel Carson dies of cancer: Author of Silent Spring was 56. *New York Times,* 15 April, 1.

Whitehead, Alfred North, and Bertrand Russell. 1967 [1910]. *Principia mathematica.* New York: Sydics of the Cambridge University Press.

Whitten, Jamie. 1966. *That we may live.* Princeton, N.J.: Van Nostrand.

Wilden, Anthony, and Tim Wilson. 1976. The double-bind: Logic, magic and economics. In *Double-bind: The foundation of the communicational approach to the family,* edited by C. Sluzki and D. Ransom. New York: Grune & Stratton.

Wilkins, Lee. 1987. *Shared vulnerability: The media and American perceptions of the Bhopal disaster.* Westport, Conn.: Greenwood Press.

Woodrum, Robert. 1995. Union racer, trade foundation takes its message to the street. *Charleston Gazette,* 2 May, D1.

Woolard, E. 1992. An industry approach to sustainable development. *Issues in Science and Technology,* Spring, 29–33.

World Commission on Environment and Development. 1987. *Our common future.* New York: Oxford University Press.

Wynne, Lyman. 1976. On the anguish, and creative passions, of not escaping double-binds: A reformulation. In *Double-bind: The foundation of the communicational approach to the family,* edited by C. Sluzki and D. Ransom. New York: Grune & Stratton.

INDEX